Advances in Intelligent Systems and Computing

Volume 984

The series "Advances in Intelligent Systems and Computing" contains publications on theory, applications, and design methods of Intelligent Systems and Intelligent Computing. Virtually all disciplines such as engineering, natural sciences, computer and information science, ICT, economics, business, e-commerce, environment, healthcare, life science are covered. The list of topics spans all the areas of modern intelligent systems and computing such as: computational intelligence, soft computing including neural networks, fuzzy systems, evolutionary computing and the fusion of these paradigms, social intelligence, ambient intelligence, computational neuroscience, artificial life, virtual worlds and society, cognitive science and systems, Perception and Vision, DNA and immune based systems, self-organizing and adaptive systems, e-Learning and teaching, human-centered and human-centric computing, recommender systems, intelligent control, robotics and mechatronics including human-machine teaming, knowledge-based paradigms, learning paradigms, machine ethics, intelligent data analysis, knowledge management, intelligent agents, intelligent decision making and support, intelligent network security, trust management, interactive entertainment, Web intelligence and multimedia.

The publications within "Advances in Intelligent Systems and Computing" are primarily proceedings of important conferences, symposia and congresses. They cover significant recent developments in the field, both of a foundational and applicable character. An important characteristic feature of the series is the short publication time and world-wide distribution. This permits a rapid and broad dissemination of research results.

** Indexing: The books of this series are submitted to ISI Proceedings, EI-Compendex, DBLP, SCOPUS, Google Scholar and Springerlink **

More information about this series at http://www.springer.com/series/11156

Radek Silhavy
Editor

Software Engineering Methods in Intelligent Algorithms

Proceedings of 8th Computer Science
On-line Conference 2019, Vol. 1

 Springer

Editor
Radek Silhavy
Faculty of Applied Informatics
Tomas Bata University in Zlín
Zlín, Czech Republic

ISSN 2194-5357 ISSN 2194-5365 (electronic)
Advances in Intelligent Systems and Computing
ISBN 978-3-030-19806-0 ISBN 978-3-030-19807-7 (eBook)
https://doi.org/10.1007/978-3-030-19807-7

This Springer imprint is published by the registered company Springer Nature Switzerland AG
The registered company address is: Gewerbestrasse 11, 6330 Cham, Switzerland

Preface

This book constitutes the refereed proceedings of Software Engineering Methods in Intelligent Algorithms section of the 8th Computer Science On-line Conference 2019 (CSOC 2019), held in April 2019.

Particular emphasis is laid on modern trends, concepts and application of intelligent systems into a software engineering field. New algorithms, methods and application of software engineering techniques are presented.

CSOC 2019 has received (all sections) 198 submissions; 120 of them were accepted for publication. More than 59% of accepted submissions were received from Europe, 34% from Asia, 5% from America and 2% from Africa. Researches from more than 20 countries participated in CSOC 2019 conference.

CSOC 2019 conference intends to provide an international forum for the discussion of the latest high-quality research results in all areas related to computer science. The addressed topics are the theoretical aspects and applications of computer science, artificial intelligence, cybernetics, automation control theory and software engineering.

Computer Science On-line Conference is held on-line and modern communication technology, which are broadly used to improve the traditional concept of scientific conferences. It brings equal opportunity to all the researchers around the world to participate.

I believe that you will find the following proceedings interesting and useful for your own research work.

March 2019   Radek Silhavy

Organization

Program Committee

Program Committee Chairs

Petr Silhavy
Faculty of Applied Informatics, Tomas Bata University in Zlin

Radek Silhavy
Faculty of Applied Informatics, Tomas Bata University in Zlin

Zdenka Prokopova
Faculty of Applied Informatics, Tomas Bata University in Zlin

Roman Senkerik
Faculty of Applied Informatics, Tomas Bata University in Zlin

Roman Prokop
Faculty of Applied Informatics, Tomas Bata University in Zlin

Viacheslav Zelentsov
Doctor of Engineering Sciences, Chief Researcher of St. Petersburg Institute for Informatics and Automation of Russian Academy of Sciences (SPIIRAS)

Program Committee Members

Boguslaw Cyganek
Department of Computer Science, AGH University of Science and Technology, Krakow, Poland

Krzysztof Okarma
Faculty of Electrical Engineering, West Pomeranian University of Technology, Szczecin, Poland

Monika Bakosova
Institute of Information Engineering, Automation and Mathematics, Slovak University of Technology, Bratislava, Slovak Republic

Pavel Vaclavek	Faculty of Electrical Engineering and Communication, Brno University of Technology, Brno, Czech Republic
Miroslaw Ochodek	Faculty of Computing, Poznan University of Technology, Poznan, Poland
Olga Brovkina	Global Change Research Centre Academy of Science of the Czech Republic, Brno, Czech Republic; Mendel University, Brno, Czech Republic
Elarbi Badidi	College of Information Technology, United Arab Emirates University, Al Ain, United Arab Emirates
Luis Alberto Morales Rosales	Head of the Master Program in Computer Science, Superior Technological Institute of Misantla, Mexico
Mariana Lobato Baes	Superior Technological of Libres, Mexico
Abdessattar Chaâri	Laboratory of Sciences and Techniques of Automatic Control & Computer Engineering, University of Sfax, Tunisian Republic
Gopal Sakarkar	Shri. Ramdeobaba College of Engineering and Management, Republic of India
V. V. Krishna Maddinala	GD Rungta College of Engineering & Technology, Republic of India
Anand N. Khobragade	Maharashtra Remote Sensing Applications Centre, Republic of India
Abdallah Handoura	Computer and Communication Laboratory, Telecom Bretagne, France

Technical Program Committee Members

Ivo Bukovsky	Roman Senkerik
Maciej Majewski	Petr Silhavy
Miroslaw Ochodek	Radek Silhavy
Bronislav Chramcov	Jiri Vojtesek
Eric Afful Dazie	Eva Volna
Michal Bliznak	Janez Brest
Donald Davendra	Ales Zamuda
Radim Farana	Roman Prokop
Martin Kotyrba	Boguslaw Cyganek
Erik Kral	Krzysztof Okarma
David Malanik	Monika Bakosova
Michal Pluhacek	Pavel Vaclavek
Zdenka Prokopova	Olga Brovkina
Martin Sysel	Elarbi Badidi

Organizing Committee Chair

Radek Silhavy   Faculty of Applied Informatics, Tomas Bata
                                     University in Zlin

Conference Organizer (Production)

OpenPublish.eu s.r.o.
Web: http://www.openpublish.eu
Email: csoc@openpublish.eu

Conference Web site, Call for Papers

http://www.openpublish.eu

Contents

Literature Review on Database Design
Testing Techniques

Abdullahi Abubakar Imam[1,2(✉)] iD, Shuib Basri[1], Rohiza Ahmad[1],
and María T. González-Aparicio[3]

[1] SQ2E Research Cluster, Computer and Information Sciences Department,
Universiti Teknologi PETRONAS,
Bandar Seri Iskandar, 32610 Seri Iskandar, Perak, Malaysia
aiabubakar3@gmail.com,
{abdullahi_g03618,shuib_basri,
rohiza_ahmad}@utp.edu.my
[2] Ahmadu Bello University, Zaria, Nigeria
[3] Computing Department, University of Oviedo, Gijon, Spain
maytega@uniovi.es

Abstract. Database driven software applications are becoming more sophisticated and complex. The behavior of these systems solely depends on the data being used. Whereas this data has now become so massive, varyingly connected, distributed, and stored and retrieved with different velocity, era of big data. To make these systems operate in every anticipated environment with the required usability, durability and security, they are subjected to rigorous testing using the available Software Testing Techniques (STT). This test is described as a process of confirming correct behavior of a piece of software which consists of three parts, namely, interface (GUI), back-end (codes) and data-source (database). The purpose of this study is to identify and analyze existing STT in the context of databases design structures. Primary studies related to ST were identified using search terms with relevant keywords. These were classified under journal and conference articles, book chapters, workshops, and symposiums. Out of the search results, 23 Primary studies were selected. Database testing has been significantly discussed in the software testing domain. However, it was discovered that, existing software testing techniques suffer from several limitations which includes: incompatibility with new generation databases, lack of scalability and embedded SQL query detection issues. In addition, application of existing techniques into a full-fledged software system has not been reported yet.

Keywords: Software testing · SQL databases · NoSQL databases ·
Software quality

1 Introduction

As software applications get more complex and intertwined and with increasing usage of massive and varieties of data from different sources, cybercrimes and varieties of platforms, it is more important than ever to have a durable, robust, secured and consistent database system [1]. Similarly, methodologies that are adopted to make sure that

© Springer Nature Switzerland AG 2019
R. Silhavy (Ed.): CSOC 2019, AISC 984, pp. 1–13, 2019.
https://doi.org/10.1007/978-3-030-19807-7_1

database driven applications being developed are absolutely tested and meet their specified requirements need to be vigorous and flexible such that software applications can be screened to successfully operate in every anticipated environment with the required usability and security [2]. If software is deployed and database operations such as Create, Retrieve, Update and Delete (CRUD) are performed without testing the application's database performance, security and reliability, the company risks a crash of the entire system, internal logical structure deteriorations or broken insertions, deletions or updates [3].

According to Sree [4], Software Testing (ST) can be defined as a process of identifying defects from a software, isolating them, and subjecting (sending) them for rectification. On the words of [5], software testing is a process of providing information about the quality of system to stakeholders by investigating the system under test. The process ensures that there is no single defect associated to the product through testing any of the three main software components, namely; interface, codes and database. For these reasons among others, techniques such as black box, white box and gray box testing came into view to test software applications through the user interface or internal logical structure or both respectively. Also, approaches to particularly test the SQL database driven applications were proposed since analyzing codes only is insufficient for threats detection and mitigation [6, 7].

Although these approaches have proven useful in detecting problems associated with software design and implementation and SQL database operations as well as security problems such as SQL injection and cross-origin attacks, they should be adjusted to detect specific vulnerabilities database driven software applications [7].

In this article we analyzed and compartmentalized the existing studies with respect to software testing techniques in the context of database design structure. Primary studies related to ST were identified using search terms with relevant keywords. These were classified under journal and conference articles, book chapters, workshops, and symposiums. Out of the search results, 23 Primary studies were selected. Database testing has been significantly discussed in the software testing domain. However, it was discovered that, existing software testing techniques suffer from several limitations which includes: incompatibility with new generation databases, lack of scalability and embedded SQL query detection issues. In addition, application of existing techniques into a full-fledged software system has not been reported yet.

The remainder of this paper is structured as follows: Sect. 2 explains the method adopted. Section 3 presents the results. Section 4 discussed the findings. Section 5 highlights some recommendations. Section 6 concludes the paper.

2   Research Method

In this section, the method adopted to conduct this research is presented. It consists of research questions, literature sources and study selection process.

2.1 Research Questions

To commence the investigation about the state of the art in this area, three research questions were devised and put into words as follows:

RQ1 What are the existing software testing techniques, models or algorithms used to assess the storage (database) component of software?

RQ2 How do the identified solutions detect the menace associated with database design and what are their strengths and weaknesses?

RQ3 How to mitigate an existing technique to comprise the aspect of big data datasource as part of the testing components of a software application?

2.2 Literature Sources

In this study, a comprehensive and detailed search was conducted using the electronic libraries available such as Science Direct, Web of Science, IEEE Xplore, Springer, Google Scholar and ACM. These libraries were used to search for the relevant materials across the globe. The search yield several categories of materials starting from symposiums, conference proceedings, book chapters as well as journals papers.

2.3 Study Selection

Study selection is one of the key components of SLR and it's done during or after search process is completed, as such, a set of rules are engineered and applied to appropriately select the right studies. The rules are, manuscript can only be selected if it's:

- tackling any of the key words of this research.
- tackling any of the questions in this research or attempt to describe its nature.
- either published or submitted to a journal or conference. Book chapters as well as technical reports are also considered.
- written in English or fully translated to English language.
- related to topics such as SQL and NoSQL database evaluation and testing, software engineering and applications, software testing, CRUD operations testing, open source software development, teaching and education.

3 Results

The results of this study are presented in this section. These results are categorized into two different categories. We started by presenting the results related to software testing as a whole. In the second part, the results for SQL and NoSQL database testing are presented.

3.1 Software Testing Techniques

The following table (Table 1) presented the available software testing techniques in a summery form.

Table 1. Testing techniques

Components testing	Unit testing	Verification (Process Oriented)	**White box Testing** (Tests that are derived from knowledge of the program's structure and implementation)
	Module testing		
Integrated testing	Sub-system testing		
	System testing		
User testing	Acceptance testing	Validation (Product Oriented)	**Black Box Testing** (Tests are derived from the program specification)
Components testing + Integrated testing + User testing	Unit, module, system, sub-system and acceptance testing	Verification and validation	**Gray Box Testing** (Test are derived from both the knowledge of the program structure and specification)

In black-box testing, errors such as halting and a testing dependent functions independently, the functions may fail when the original sequence is changed at run time during the execution of other functions. As such, it is indispensable to consider a technique beyond black box testing.

However, white-box does not explicitly consider SQL statements which are embedded in application programs. It treats SQL statements as black boxes. Commonly, SQL semantics are not intentionally included in the test cases. Therefore, it is believed that, faults related to the internal database changes might be missed by traditional white box testing [8].

On the other hand, because gray-box testing technique is based on black-box and white-box testing techniques, it inherits all the problems associated with its parents (black-box and white-box) without exclusions. Although it combined several functionalities in one place, it is measured tedious, cumbersome and time consuming.

In view of the above, it can be concluded that the authors focus mainly on testing software applications through the user interface by adopting black box technique while others gave much emphasis to the coding side where white box technique is most suited. Alternatively, some authors believe that when the two techniques are combined, more reliable testing can be achieved. However, all these techniques and their associated works are only concentrating on the software itself while neglecting its datasource which contributes significantly to the quality and performance of any database driven software application. The following section presents the existing approaches for testing design structure of SQL and NoSQL databases driven applications.

3.2   SQL and NoSQL Database Testing

Table 2 below presents the approaches used to test the database driven software applications with respect to SQL and NoSQL databases.

In consideration of the related works on SQL database testing, it can be concluded that relational/traditional database have received considerable attention where several techniques with various approaches are reported. However, these techniques are either

Table 2. SQL & NoSQL database testing approaches

Author & Year	Techniques/Solution	Strength	Weaknesses
Tsumura et al. (2016)	Plain Pairwise Coverage Testing (PPCT) and Selected Pairwise Coverage Testing (SPCT)	No predicates Uses few elements is SQL query	–Limited test cases Only RDBMS
Hamlin and Herzog (2014)	SPAR Test Suite Generator (STSG)	Test suits are generated automatically Supports benchmarking and correctness-checking	–Relies on user specifications –Only RDBMS –No model checking
Setiadi and Lau (2014)	Data consistency framework based on system specifications	Data inconsistency detection	–Only RDBMS –Through GUI only –Starts after the execution of applications
Zou (2014)	Cloud storage area	Distribute data among the available computing devices	–Only RDBMS –Data privacy & security
Sarkar, Basu, and Wong (2014)	Novel framework called iConSMutate	Reusing existing DB states Generate test cases automatically	–Only RDBMS –No constraints solving –No model checking
Marin (2014)	Data-agnostic framework for test automation	Coverage model and fault model for OLAP cubes Uses record-and-replay	–OLAP –RDBMS
Sctiadi and Lau (2014a)	Structured model that enables the automatic generation of consistency rules	Derive the data consistency rules from business rules, system specifications and database schema	–Only RDBMS –Can't apply on existing system –Single repository
Grechanik, Hossain, and Buy (2013)	Novel approach for Systematic Testing in Presence of Database Deadlocks (STEPDAD)	Based database deadlocks	–Only RDBMS –No data privacy –No constraints solving
Pan, Wu, and Xie (2013)	The use of mutation scores to test the fault-detection capabilities	Ability to detect faults in real world database applications	–Only RDBMS
McCormick II et al. (2012)	MutGen to generate test for mutation testing for database applications	Ability to kill the mutant Detect DB constraints	–Only RDBMS

(continued)

Table 2. (*continued*)

Author & Year	Techniques/Solution	Strength	Weaknesses
Ron, Shulman-peleg, and Ibm (2016)	Comparative study of NoSQL DBs based NoSQL injections	Proved NoSQL security vulnerabilities	Awareness of NoSQL injections only. No measures
Gonzalez-Aparicio et al. (2016)	Context-aware model for CRUD operations	Trade-off analysis between availability & consistency of NoSQL DBs	Monitors the execution of CRUD operations for best selection of NoSQL DBs only
Bhogal and Choksi (2015)	Comparison study of NoSQL databases	NoSQL DBs evaluated based on variety of data	Volume, Velocity and veracity not considered
Truica et al. (2015)	Performance evaluation on CRUD operation for NoSQL databases	Ability to analyze features of document oriented databases	Compared 3 document oriented DBs (mongoDB, couchDB, & Couchbase) only
Klein et al. (2015)	Method to perform technology selection among NoSQL DBs	NoSQL databases are evaluated in a specific context	Not testing any BD variable Used to select NoSQL DB only
Abramova, Bernardino, and Furtado (2014)	Analysis of scalability properties of cloud serving NoSQL engine	Ability to test put, get and scan operations	Focus on scalability only
Naheman (2013)	Comparison between NoSQL and SQL DBs, and between NoSQL products	Performance testing on HBase	Comparison between SQL & NoSQL DBs only

embracing white-box technique (i.e. codes) or providing solutions that can work with SQL based database designs only. Also, the solutions are independent of software applications, so, operations like CRUD, security, database connection and data access schemas incorporated with application codes are not verified, thus incomplete testing of software.

Oppositely, it can be seen (from Table 3) that NoSQL databases are described as poor in the areas such as consistency, reliability, technical support and hardware/software compatibility which proves the use of BASE rather than ACID. Because of its complexity, insecurity, distributed nature of data stores and very high storage capacity, there is need to confirm its design, codes and implementation properly before system deployment [22, 23]. Though, techniques or approaches that focus on testing the big data (NoSQL) database driven software applications are yet to be reported.

Therefore it remains challenging to produce an algorithm, method, technique or model that can be used to test software applications with big data category of datasource, in particular, to be able to test big data (NoSQL) databases driven software applications with respect to NoSQL dependent variables such as CRUD loading time, security, data models, privacy, scalability and variability among others [23]. This is highly imperative as the databases are becoming larger and complex [20].

4   Results and Discussion

According to [24, 25], database testing refers to the analysis of database system's performance, consistency, availability, reliability and security. This test is usually independent of the application and is done to verify the ACID or BASE properties of a database management system. It consists of process that are layered in nature such as business layer, user interface layer, database itself and most importantly data access layer which deals directly with the database. Data access layer is the layer that is used for testing strategies like quality assurance and quality control. There are four stages to test database, namely: (1) Set fix, (2) Test run, (3) Outcome Verification and, (4) Tear down [24]. The solutions proposed for database testing are categorized based on SQL and NoSQL databases some of which are discussed below.

4.1   SQL Database Testing Techniques

This Define Relational databases are based on a set of principles to optimize performance which are derived from Consistency, Availability, Partition tolerance (CAP) theorem [26]. ACID which stands for Atomicity, Consistency, Isolation, and Durability, is a principle based on CAP theorem and used as set of rules for relational database transactions [26].

In line with Subsection 2.1.2 of this section where white-box testing is discussed, [8] believed that the same technique can be extended to cover database aspect of software applications. As such, WHODATE approach for database testing is proposed to semantically reflect SQL statements in the test case. As shown in the following figure, SQL queries are transformed into the traditional white box testing for test case generation. This is only applicable to SQL based databases as shown in Fig. 1.

At first, SQL queries are retrieved and passed to transformation unit where the queries are translated to best suit the format of white box technique. Thereafter, the white box generates its test cases as usual.

While [16] proposed the use of mutation scores to test the fault-detection capabilities of the real world database applications such as MySQL, SQL server and Oracle. Using this approach, sample schemas are selected from databases, and then queries from the schemas are run against the mutation operators. On the contrary, MutGen was introduced to generate test for mutation testing on database applications [15]. It incorporates weak-mutant-killing constraints in the code which queries mutant-killing constraints into the transformed code. This is done to kill the mutants by generating program inputs and sufficient database states. In contrast, [12] believed that the higher the mutation score the better because it indicates higher quality in identifying programming errors, thereby produced iConSMutate to generate test cases which will reuse the existing database state for database applications to improve quality in codes coverage and mutant detection. It should be noted that the mutant concept is only applicable to relational databases rather than the NoSQL databases.

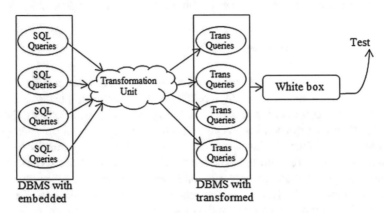

Fig. 1. WHODATE approach for SQL statement transformation

It is also discovered that After-State Database Testing (ASDT) can be used to detect data inconsistency in traditional databases only [10]. It is a black box based framework that uses design choices and system specifications to test the consistency state after the execution of the application. Using ASDT, a structured model that enables the generation of data consistency rules from business rules, database schema and system specifications. Its main aim is to minimize the use of resources and efforts in ASDT.

On the word of [6], multi-dimensional databases (OLAP cubes) can be tested when compared with record-and-replay technique using show-case. The solution is called data-agnostic which works together with coverage model and fault model for OLAP cubes. This is only applicable to multi-dimensional databases.

Contrariwise, since test suits are often produced and used for testing purposes, [9] produced SPAR Test Suite Generator (STSG) to automate the process of generating test suits used by SQL style database systems. Ground-truth answers are incorporated in the produced test suite. This approach supports benchmarking and correctness-checking on SQL based databases.

In [1] believed that avoiding the use of predicates while testing databases can yield more promising results. In substitute, two approaches were proposed: Plain Pairwise Coverage Testing (PPCT) and Selected Pairwise Coverage Testing (SPCT). These approaches use parameters selected from elements in the SQL SELECT and they are only applicable to SQL queries.

In consideration of the related works on database testing, it can be concluded that relational/traditional database have received considerable attention where several techniques with various approaches are reported. However, these techniques are either embracing white-box technique (i.e. codes) or providing solutions that can work with SQL based database designs only. Also, the solutions are independent of software applications, so, operations like CRUD, security, database connection and data access schemas incorporated with application codes are not verified, thus incomplete testing of software.

4.2 NoSQL Database Testing Techniques

NoSQL (not only SQL) is a trending term in modern data stores that accommodate big data; it refers to non-relational databases that rely on different storage mechanisms such as document store, key-value store, columner and graph [7].

Unlike the SQL databases that based on ACID, Not Only SQL (NoSQL) focuses on BASE principle in order to have robust and correct database with huge amount of data [7, 26]. BASE stand for Basically Available, Soft state, Eventually consistent. BASE follows CAP theorem and two of three guarantees must be chosen if the system is distributed (Brewer [27]).

[7] discovered that, the aforesaid NoSQL databases are highly vulnerable to cross-origin attacks. Since NoSQL databases use different query languages which make SQL injections irrelevant, attackers find a new opportunity to insert malicious code easily. Therefore, it's recommended to prevent all possible attacks via careful code and database examination before complete system deployment. Although the existing tools and techniques have proven useful in detecting SQL injection attacks, they should be adjusted to detect the specific NoSQL database vulnerabilities [7].

However, it is believed that, CRUD operations can be used to test data availability and consistency of each of the NoSQL databases. Hence [17] proposed a context-aware model for CRUD operations to monitors the execution of CRUD operations for best selection of NoSQL databases. Although this will give programmers the chance to make appropriate choice that corresponds to the client needs, it does not have the capability of detecting any erroneous practice in the logical structure of the application code that communicates with the designed database. Errors such as functions call order cannot be detected.

In the work of [18], a comparison was conducted between the NoSQL databases based on data variety as one of the 5, 6 or 7 Vs of big data. While in [19] the comparison was made between the document oriented NoSQL databases (MongoDB, CouchDB, and Couchbase) for asynchronous replications on CRUD operations. The outcome of the test was compared with famous relational databases such as Microsoft SQL Server, MySQL, and PostgreSQL to measure the scalability and flexibility of systems. Nevertheless, the two inventions were only making comparison between the existing SQL and NoSQL databases not the design principles of particular database driven software application.

In contrast, [20] considered all the NoSQL data models (key-value, column, document, graph) in testing a particular healthcare system requirements. One database was considered from each of the NoSQL data model category (Riak, Cassandra and MongoDB) with exception of graph as it does not support horizontal partitioning. A systematic method to perform this technology selection is proposed to evaluate the products in the specific context of use which do not focus on database design principles violations.

It is reported that existing NoSQL databases are not matured enough [22], which is why it is necessary to examine all the NoSQL database features based on system requirements before selecting one [23]. In [22], performance testing was conducted on the NoSQL databases, HBase in particular, for relationship mapping between number of column families and performance.

Table 3. Comparison between NoSQL and SQL Databases Naheman [22]

	NoSQL databases	SQL databases
Read & write concurrency	Fast	Slow
Data storage	Mass storage	General storage
Expansion mode	Horizontal expansion	Vertical expansion
Consistency	Poor	Good
Reliability	Poor	Good
Availability	Good	Good
Expansion costs	Low	High
Maturity	Low	High
Programmatically	Complex	Simple
Data mode	Free	Fixed
Human overhead	Relatively high	Medium
Technical support	Poor	Good
Upgrade costs	High	Low
Hard/software compatibility	Poor	Good
Flexibility	Good	Poor

It can be seen that NoSQL databases are described as poor in the areas such as consistency, reliability, technical support and hardware/software compatibility which proves the use of BASE rather than ACID. Because of its complexity, insecurity, distributed nature of data stores and very high storage capacity, there is need to confirm its design and implementation properly before system deployment [22, 23].

5   Findings and Recommendations

In general, this research has produced several findings as presented in Sects. 3 and 4. However, for quick reference, some findings are summarized as follows:

- Database Logical Structure: A technique that will detect incorrect logical implementation of CRUD operations and security measures based on the predefined functional and non-functional requirements is needed.
- Auto-problem solving technique: A technique that will suggest proper approach of restructuring CRUD operations and security measures based on the best practice.
- A technique that will propose structural amendments for better system performance and better data security approach.

- No indexing of files: the use of Btree selection concept to group and sample appropriate files out of the pool of distributed file will not only make the testing accurate but it will also make the testing process faster.
- Slow processing: there is need to establish a new approach for sorting files from smallest to the largest to enhance records searching capabilities during test.
- Security: existing techniques need to be revisited for security problems such as SQL injection and cross-origin attacks [7].

Erroneous database design: As data becomes huge, the loading time gets affected proportionately [6]. So, there is need to test software applications with varieties of data sizes to confirm the correct behavior of a system.

6 Conclusion and Future Focus

In this article we analyzed and compartmentalized the existing studies with respect to software testing techniques in the context of database design structure. Primary studies related to software testing were identified using search terms with relevant keywords. These were classified under journal and conference articles, book chapters, workshops, and symposiums. Out of the search results, 23 Primary studies were selected. Database testing has been significantly discussed in the software testing domain. However, it was discovered that, existing software testing techniques suffer from several limitations which includes: incompatibility with new generation databases, lack of scalability and embedded SQL query detection issues. In addition, application of existing techniques into a full-fledged software system has not been reported yet.

Acknowledgment. This paper/research was fully supported by Ministry of Higher Education Malaysia, under the Fundamental Research Grant Scheme (FRGS) with Ref. No. FRGS/1/2018/ICT04/UTP/02/04.

References

1. Tsumura, K., Washizaki, H., Fukazawa, Y., Oshima, K., Mibe, R.: Pairwise coverage-based testing with selected elements in a query for database applications. In: 2016 IEEE Ninth International Conference on Software Testing, Verification and Validation Workshops, pp. 92–101 (2016)
2. Inflectra, B.: Software Testing Methodologies (2016). https://www.inflectra.com/Ideas/Topic/Testing-Methodologies.aspx. Accessed 01 Jan 2016
3. Reza, H., Zarns, K.: Testing relational database using SQLLint. In: Proceedings - 2011 8th International Conference on Information Technology: New Generations, ITNG 2011, pp. 608–613 (2010)
4. Sree, U.: Software Testing Life Cycle: Defects and Bugs (2016). https://olaiainforarch.wordpress.com/. Accessed 11 July 2016
5. Berger, D., Fröhlich, P.: Software testing techniques. Power Point Lecture, 20 pages (2016)
6. Marin, M.: A data-agnostic approach to automatic testing of multi-dimensional databases. In: Proceedings of - IEEE 7th International Conference on Software Testing, Verification and Validation, ICST 2014, pp. 133–142 (2014)

7. Ron, A., Shulman-Peleg, A., Puzanov, A.: Analysis and mitigation of NoSQL injections. IEEE Secur. Priv. **14**(2), 30–39 (2016)
8. Chan, M.Y., Cheung, S.C.: Testing database applications with SQL semantics. In: Proceedings of 2nd International Symposium on Cooperative Database Systems for Advanced Applications, March, pp. 363–374 (1999)
9. Hamlin, A., Herzog, J.: A test-suite generator for database systems (2014)
10. Setiadi, R., Lau, M.F.: Identifying data inconsistencies using after-state database testing (ASDT) framework. In: Proceedings of the International Conference on Quality Software, pp. 105–110 (2014)
11. Zou, J.: Research and application of testing technology of the cloud computing database. In: Proceedings - 2014 IEEE Workshop on Electronics, Computer and Applications, IWECA 2014, pp. 699–702 (2014)
12. Sarkar, T., Basu, S., Wong, J.: IConSMutate: concolic testing of database applications using existing database states guided by SQL mutants. In: Proceedings of 11th International Conference on Information Technology: New Generations, ITNG 2014, pp. 479–484 (2014)
13. Setiadi, R., Lau, M.F.: A structured model of consistency rules in After-State Database Testing. In: 38th IEEE International Computer Software and Applications Conference Workshops, no. 2, pp. 650–655 (2014)
14. Grechanik, M., Hossain, B.M.M., Buy, U.: Testing database-centric applications for causes of database deadlocks. In: Proceedings - IEEE 6th International Conference on Software Testing, Verification and Validation, ICST 2013, vol. 191242, pp. 174–183 (2013)
15. Pan, K., Wu, X., Xie, T.: Automatic test generation for mutation testing on database applications. In: 8th International Workshop on Automation of Software Test (AST), pp. 111–117 (2013)
16. McCormick II, D.W., Frakes, W.B., Anguswamy, R., McCormick, D.W.: A comparison of database fault detection capabilities using mutation testing. In: 2012 ACM-IEEE International Symposium on Empirical Software Engineering and Measurement (ESEM), pp. 323–326 (2012)
17. Gonzalez-Aparicio, M.T., Younas, M., Tuya, J., Casado, R.: A new model for testing CRUD operations in a NoSQL database. In: 2016 IEEE 30th International Conference on Advanced Information Networking and Applications (AINA), vol. 6, pp. 79–86 (2016)
18. Bhogal, J., Choksi, I.: Handling big data using NoSQL. In: Proceedings - IEEE 29th International Conference on Advanced Information Networking and Applications Workshops, WAINA 2015, pp. 393–398 (2015)
19. Truica, C.O., Radulescu, F., Boicea, A., Bucur, I.: Performance evaluation for CRUD operations in asynchronously replicated document oriented database. In: Proceedings - 2015 20th International Conference on Control Systems and Computer Science, CSCS 2015, pp. 191–196 (2015)
20. Klein, J., Gorton, I., Ernst, N., Donohoe, P., Pham, K., Matser, C.: Performance evaluation of NoSQL databases: a case study. In: Proceedings of the 1st Workshop on Performance Analysis of Big Data Systems, pp. 5–10 (2015)
21. Abramova, V., Bernardino, J., Furtado, P.: Testing cloud benchmark scalability with cassandra. In: 2014 IEEE World Congress on Services, pp. 434–441 (2014)
22. Naheman, W.: Review of NoSQL databases and performance testing on HBase. In: 2013 International Conference on Mechatronic Sciences, Electric Engineering and Computer, pp. 2304–2309 (2013)

23. Cai, L., Huang, S., Chen, L., Zheng, Y.: Performance analysis and testing of HBase based on its architecture. In: IEEE 12th International Conference on Computer and Information Science (ICIS), 2013 IEEE/ACIS, pp. 353–358 (2013)
24. Henry, K.: Database System Concepts. Macgraw-Hill, New York (2010)
25. Silberschatz, S., Korth, Sudarshan: Database System Concept: Homogeneous Distributed Databases. Cent. Wiskd. Inform., pp. 19.3–19.125 (2007)
26. Abramova, V., Bernardino, J.: NoSQL databases: MongoDB vs cassandra. In: Proceedings of the International C* Conference on Computer Science and Software Engineering ACM 2013, pp. 14–22 (2013)
27. Brewer, E.: CAP twelve years later: how the 'rules' have changed. Comput. (Long. Beach. Calif) **45**(2), 23–29 (2012)

Implication of Artificial Intelligence to Enhance the Security Aspects of Cloud Enabled Internet of Things (IoT)

Ramesh Shahabadkar[1(✉)] and Krutika Ramesh Shahabadkar[2]

[1] Vardhaman College of Engineering, Shamshabad, Kacharam, Hyderabad,
Telangana, India
ramesh.shahabadkar@gmail.com
[2] Accenture Solutions Pvt Ltd, Global Village, Mallasandra, Bengaluru,
Karnataka, India

Abstract. The large-scale deployment of data driven internet of things (IoT) leads to employing a number of self-directed mobile or static sensor nodes in various areas of interest where the mobile nodes specifically operates on collecting data and cooperatively transmit that data back to the integrated cloud systems. The cloud enabled IoT has its wide range of applicability into various business, commercial and military applications. The wireless links can be easily disrupted by the adversaries due to their large attack surface. Adversaries in this regard ranges from hackers with a laptop to corporations and government officials. Due to the complex nature of localization of dynamic IoT nodes specifically the low power sensor devices, it becomes compelled to make themselves vulnerable for reprogramming and capture by an unauthorized user. On the other hand the deployment of low power sensor devices implicates operational constraints on conventional high level strong encryption approaches due to their limited processing power and computational capability. Owing to these issues the security challenges in current and futuristic IoTs are more. The study conceptualized an analytical system well-capable of protecting data and IoT driven cloud systems and also detect network intrusion considering artificial intelligence (AI) systems. The study also theoretically exhibits the extensive analysis of AI algorithms in cloud network intrusion detection. The performance analysis further conveyed the superiority of the proposed model.

Keywords: Internet of Things (IoT) · Intrusion Detection System (IDS) · Artificial Intelligence · Fuzzy based Decision Tree (FDT)

1 Introduction

In last two decades detection of network attacks has been exclusively performed by humans operating from remote locations. The current deployment of various types of electronic devices connected to both wired and wireless networks creates a large attack surface for the intruders [1, 2]. A notable question arises in this context that how progressive and promising the current security trends are in successfully defending the security threats from both known and unknown vulnerable resources. Hence, it can be

© Springer Nature Switzerland AG 2019
R. Silhavy (Ed.): CSOC 2019, AISC 984, pp. 14–24, 2019.
https://doi.org/10.1007/978-3-030-19807-7_2

said that the answer is not so straightforward considered the fact that the number of threats are rising every year [3].

The large scale deployment of cloud services enable providing convenient and demand based access to the shared pool of resources e.g. CPU, memory, storage along with network applications (i.e. applications which are easily scaled up or down by incorporating very less management efforts) in a collaborative manner [4, 5]. Owing to the distributed nature, the data driven cloud computing (CC) become an integral part of Internet of Things (IoT). IoT integrates the conventional network of physical objects such as sensors, actuators etc. with distributed CC to make an effective and smart approach towards embedded collaborative computing. It tracks down each and every object connected to the network with a unique identity. However, the distributed, open, and collaborative structure of IoT oriented services are highly suspicious and become an attractive target for potential intruders who can perform cyber-attacks irrespective of any operational constraints.

Therefore, integrity of valuable internet sources and their respective data should be maintained with much more authenticity. All the attempts meant more gaining unauthorized access, exposing, alter, disable or steal particular significant data should be permanently forbidden. Hence, it leads to a situation where the three prime entities regarding data security such as privacy, veracity and obtainability of data sources remain intact [6]. IDS thereby conceptualized exclusively to detect the malicious activities that could take place in any network for the purpose of extensive security infringements. Since many years IDSs played a vital role in wide area of network security applications by becoming a stronger defensive mechanism against Denial of Services (DoS) or Distributed Denial of Services (DDoS) Attacks. IDS generate control signals which notify the system administrators about the malicious activities. IDS generate true positive alerts while intrusion is taking place or false-positive alerts in the time of wrong intrusion detection. However, the conventional IDSs are mostly designed on the basis of signature based scheme and mostly single threaded and lacks the potential of detecting variant attacks in a large multi-threaded surface like IoT enabled cloud systems. Therefore, there is a need of developing multi-threaded defensive systems in IoT which could be a potential solution towards variant attacks in a large network surface. The current research trends intuitively stands out to adopt the machine learning (ML) algorithms which are basically a part of artificial intelligence (AI) systems for IoT enabled network intrusion detection [7, 8]. AI provides a computer, an ability of learning different patterns of attacks without intervention of explicit computer programming for a particular purpose. The study introduced a conceptual modeling subjected to perform intrusion detection in a distributed IoT. The design aspects of the proposed system mostly emphasized on different aspects such as scalability, adaptability for an ensemble-learning based IDS in IoT systems. The proposed system exhibits a proof of concept prototyping by introducing two different AI based learning paradigm such as (1) Decision Trees and (2) C-Fuzzy based predictive Algorithm to make the proposed IDS system more scalable, effective and multi-threaded in a distributed IoT. The experimental outcomes obtained further exhibited the effectiveness of the proposed system. The paper is organized as follows, Sect. 2 discusses about the background aspects of AL and ML algorithms. Section 3 talk about the evolution of AI algorithms in network intrusion detection in brief and also talk about the existing research trends of IDSs involved into cloud and IoT security. In Sect. 4 the analytical design aspects of the proposed ensemble-learning and decision tree based IDS has been presented, finally Sect. 5 illustrated the experimental outcomes followed by Sect. 6 which offers concluding the remarks.

2   Evolution of Artificial Intelligence Algorithms

This section extensively discusses about the evolution of AI and ML algorithms and their fundamental aspects. AI is basically a field where a system is trained to collaborate with various mental faculties and patterns through the use of different computational modeling. The prime attributes of AI research mostly incudes (1) reasoning, (2) knowledge extraction using pattern recognition, (3) automated planning and (4) scheduling. Apart from all these, in the research evaluation of AI brought a new trend on ML, natural language processing (NLP), computer vision etc. The study mostly emphasized on MI as it is one of the most promising sub-domain and application area of AI and also further adopted a MI based principal to conceptualize the proposed IDS [9–11].

ML algorithms basically developed on the purpose of learning and generalizing from a limited set of data. It deals with different aspects where a limited set of data are used to train the system model. The system mainly takes a set of input data and using those data it further perform predictions and decisions. However, it doesn't intend to follow any explicit programmed instructions to operate on. Having these features make an ML algorithm suitable for ID tasks [12]. ID systems are implicitly designed, to make a successful attempt towards solving different problems including classification, clustering and regression problems. The Classification operations primarily defined to explore and recognize the membership exist between group elements while regression analysis deals with predicting a response. Clustering perform grouping of entities or objects based on their certain characteristics. If the output pipeline considers class labels then it comes to the classification, however, it output becomes continuous numerical values then it is referred to regression and it is called as clusters if and only if the output values are subsets. Therefore, regression analysis subjected to perform prediction based estimation which is quite opposed to the operational and design aspects associated with the classification and clustering problems [13, 14].

ML algorithms are classified based on their different way of learning paradigms. It is splitted into three different categories such as (a) supervised, (b) unsupervised and (c) reinforcement learning.

- Supervised Learning (SL): It refers to the learning mechanism which is commonly used in various classification problems. The prime objective of this kind of learning paradigm is to train a computing device to learn the classification system.
- Unsupervised Learning (UL): It seems much harder as compared to the other learning paradigms. The prime objective of this type of learning is to train a computing system that how to do something without explicitly defining it. It has become a powerful tool to detect the pattern structures identifiable in unlabelled data structure. However, it also reflects the statistical properties associated with input patterns of data [15, 16].

- Regression Modeling (ML): It considers establishing a relationship between a particular environment and the learning agent. The earning agent in this regards interact with the environment and perform a self-learning based on various consequences of its actions in terms trials and errors rather than from being explicitly taught.

The extensive analysis of the conventional algorithms conveyed a matter of fact that no single AI system can achieve best accuracy in all the situations. The prime use of these AI algorithms could be usage of combined multiple algorithms to obtain better quality of reasoning instead of any single one. There exist two different types of learning approaches which are thereby (1) ensemble learning and (2) hybrid learning approaches respectively. In case of ensemble learning for different classifiers various homogeneous weak models are combined according to their different degree of individual output (e.g. Majority Voting). On the other hand hybrid approaches combines different heterogeneous AI approaches such as cascading models [17].

3   AI Algorithms in Network Intrusion Detection

The conceptual background of ID usually built on the top of a belief that it is defined to detect the attempt made by some intruders towards violating the security policies such as compromising confidentiality, integrity and availability of a resource. IDS are a software system that can run on an individual computer or an overall network. It basically gathers different types of information from various sources and performs analytics in order to identify the possible security loopholes [18]. It also intends to identify both intrusion (i.e. attack performed by an intruder from outside of an organization) and misuse (i.e. attacks within an organization). The design approaches of conventional IDS a system varies according to the goal of detecting suspicious or malicious network activities. This section also discusses about few of the significant conventional ML algorithms, exclusively built for the purpose of intrusion detection. It also aims to review few of the significant IDS techniques for IoT cloud environment. IT also exhibits various ML algorithms involved into IDS for grid and cloud environment to some extent. There exist various significant literatures talk about neural network based distributed IDS, clustered ensemble methods for IDS in distributed IoTs also Big-distributed IDS (B-DIDS). Apart from these literatures there exists various review manuscripts available related to IDS techniques in CC environment [19], ensemble based distributed IDS [20], various pattern recognition techniques using ML [21], crowd analysis in computer vision as well [22].

The following section exhibits Different methods developed till date for Intrusion Detection:

The study of Kleber et al. [23] evaluated various IDS for grid and CC paradigm by incorporating Grid-M, a middleware prototype developed for the efficient intrusion detection. The study utilized a set of audit packages to create a scenario of policy violation. It is defined with a set of rules and the analyzer defined here operates on two different types of attributes (1) Retrieval of improper contents and (2) comparing numerical intervals. The outcomes obtained from extensive experimental simulation shows that real time analysis is possible up to certain extent and also for a limited number of rules defined. The validation of their prototype model exhibits its performance efficiency as it accesses very lower cost of computation while providing satisfactory outcomes. Study in similar direction has been conducted by Li et al. [24] where a CC adaptive architecture adopted to enable efficient resource provisioning without affecting the system performance. The study performed detection of malicious attacks with higher accuracy. It further thoroughly studied the better applicability of machine learning tools and further integrated Neural Network (NN) to achieve better scalability, flexibility and performance of the system. The following figure exhibits an NN based IDS architectural blueprint for IoT networks (Fig. 1).

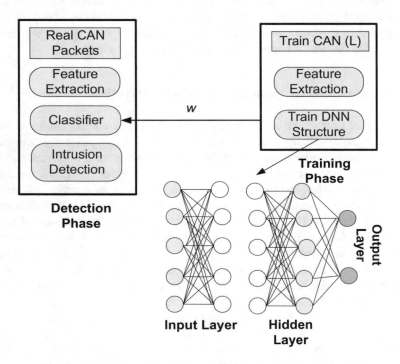

Fig. 1. Orientation of conventional IDS designed using Deep NN

The performance analysis of the above stated study exhibited its superiority in detecting various attacks by consuming very less memory and processing power. The study found lacks detection stability and couldn't perform well while a wide variety of attacks concerned. A framework using cluster enabled IDS detection introduced in the study of Wu et al. [25]. The proposed design for security systems combined multiple-partitions and embedded that into one optimal process. The study claimed to excel better performance in terms of knowledge reuse and privacy protection. Masarat et al. [26] designed and conceptualized a map reduce engine namely HAMR which has been built on the top of a clustered labelled IDS alarm mechanism. It also enables dynamic and incorporates different algorithms for intrusion detection.

4   System Model

The aim of the proposed study is to formulate a conceptual methodology for efficient IDS for IoT enabled CC. The performance analysis of the proposed system performed on the basis of different performance parameters including execution time and accuracy. The study adopted an ensemble learning paradigm which enables combining individual learner's prediction. Various evidences are proof of the fact that combinations of individual predictions give more accuracy in prediction rather than individual classifiers. The design principles of a classifier exclusively built on the basis of a policy where more emphasize laid on estimation of error rate rather than random prediction. The proposed conceptual model incorporated an ensemble model which integrates a decision tree approach for predicting potential attacks on IoT based cloud environments. The following Fig. 2 exhibits a tentative architecture of the proposed methodology in terms of various operational components.

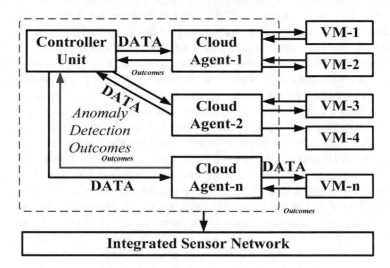

Fig. 2. Proposed research methodology

The proposed system utilizes decision tree based weak classifiers integrated with cloud based virtual machines VMs to detect any potential attempt made by any attacker to violate the confidentiality, integrity and availability of the resource (e.g. VM) in any scenario. The proposed system is composed of different attributes that are having individual and active role of participation in the detection of malicious or suspicious network anomalies. The different attributes are thereby (1) Controller Unit, (2) Cloud Agent, Several slave nodes as (3) VMs as workers. The controller unit plays a significant role and act as an integral part of the system it becomes an interface between the agent scheduler and the database and keeps a track of agent scheduler or manager. Controller is the prime module where aggregation of data, generated from various resources along with the adaption of learned models from various ensemble learning models take place.

As our proposed methodology emphasized on IoT network and CC is a prime attribute of IoT. Hence, the study in this regards mostly concerns detecting intrusion in the VM workflows from different aspects. VMs are deployed as workers, which are continuously supervised by the cloud agent. All the data considered in this study are mostly sensor generated which are aggregated in a base station or sink node and further proceeded to the cloud network and envisioned for better analytics. In the above stated model another prime attribute is agent scheduler or tracker which continuously keeps a track of agent working under it, whether an agent within a particular time stamp is available or not. It that agent notifies its manager that it ready to take up job then that implies that cloud agent's (CA) corresponding VM worker is active and poses its availability. Therefore the system further assigns tasks to the VMs based on a round robin fashion. However, checking of cloud model's resource availability before assigning any task also implies that the system has got a positive impact on the load-balancing scenario. The system has been formulated by means of computing the system availability with respect to its weighted average of CPU utilization and memory consumption factors in percentage. The system implements the learning models in the VMs which act as workers. The study incorporated fuzzy based decision tree algorithms for the learning procedures, in this procedure the learning classifiers are trained with various patterns of workflows with respect to memory and CPU constraints. The deployed agents further collect various workflows related prediction results from the learning models and notify the Controller unit, further controller unit compute intrusion detection results by incorporating a trapping algorithm. Agents in this system act as interface element between corresponding VMs and the agent scheduler. The system basically uses three different tasks in order to distinguish different pattern of attacks within a normal connection. Therefore, the categorized tasks are thereby (1) Training tasks which are used to train the learning models with different labeled data, (2) Validation tasks, which are used to validate the adaptability of the system to the change of patterns of inputs as we use ensemble learning which combines different learning models. The system also prompted a built-in capacity which initiates fault-tolerance, in case, the work performance of any VM goes down at that point of time the agent scheduler will deallocate the respective VM and distribute that particular scheduled task to other active VMs. The agent scheduler in that scenario will also distribute the task among other active nodes and imitates a transition so that the system immediately performs some decision making for system failure handling.

The process flow of the system implies that initially the system will establish a connection between the controller unit and other VMs. Once this is done further agent manages the communication between the controller and VMs. Each agent reports their respective agent scheduler about the prediction results and also it notifies about the suspicious anomaly detection results. This operation takes place in the controller unit in the presence of agent scheduler. A VM continuously update its corresponding agent about the status info which includes number of CPU, execution power of each CPU, memory utilization graph etc. Considering the availability of a VM, if the VM is available then the system push tasks to the VM for execution. Further the VM runs the learning model based on the algorithm defined by controller. The computed results further forwards back to agent and it conveys the agent scheduler. The agent scheduler finally collects results from all the VMs. The analytical modeling also integrated with supervisor unit and its underlying deceiving algorithm to determine the existence of suspicious anomaly, further the study also analyses the accuracy. The following section demonstrates the experimental outcomes obtained after simulating the proposed model with respect to a standard intrusion detection data set.

5   Results Discussion

The performance evaluation of the proposed system has been done using a validation model namely decision tree models. The comparative analysis carried out by comparing the outcomes with the random tree based decision models. The experimental prototyping utilized a η-fold cross validation model which uses KDD99 data set. In our case $\eta = 5$. The key parameters of the decision tree were (1) minimum instances per leaf, (2) confidence factor and (3) binary split. The prediction model used by decision tree classifier also incorporated a fuzzy classifier and abbreviated as Fuzzy Decision Tree (FDT). The prediction made by a DT classifier will have the following possible outcomes such as True Positive (T_P), False Negative (F_N), False Positive (F_P) and True Negative (T_N). The performance metrics introduced in the experimental analysis for both proposed FDT based ensemble learning approach and decision tree is computed with the following mathematical expression.

$$\text{Accuracy} = \frac{T_p + T_N}{T_p + F_p + T_N + F_N} \tag{1}$$

The experimental analysis used a system which having a controller of 8 GB RAM with an Intel CPU with 4 cores. The VMs are initialized with 2 GB RAM and Intel CPU with dual cores. The study considered different test cases to evaluate the performance efficiency of the proposed system. The comparative analysis in the following Tables 1 and 2 clearly depict that Decision Tree (DT) being evaluated in proposed system exhibits better outcomes in comparison with DT running on a single system. The Tables 1 and 2 depicts the fact as follows:

Table 1. Outcomes obtained by assessing DT on a single system

Tests	Test 1	Test 2	Test 3	Test 4
Execution time (sec)	5100	5478	5876	**6547**
Accuracy		96.12%	97.48%	97.45%	98.6%

The proposed ensemble based learning model produced the following results under a condition of implementing the scenario with 3 deployed slave VMs. A decision tree based paradigm played a significant role in this. The results are summarized in the following Table 2. The system evaluated controller for running a trapping algorithm which can calculate the results of anomaly detection and further convey about the prediction accuracy.

Table 2. Outcomes obtained by assessing DT on proposed system

Tests	Test 1	Test 2	Test 3	Test 4
Execution time (sec)	3567	4578	4676	**3489**
Accuracy	98,56%	98.38%	99.45%	**99.47%**

The proposed system utilized a fuzzy based decision tree and obtained very significant outcomes in terms of accuracy of prediction for intrusion detection in an IoT enabled cloud system. The following figures exhibits the comparative analysis which is carried out to exhibit the robustness of the system in terms of accuracy and processing speed. The comparison has been performed in between the proposed model and the single system-decision tree model. The comparative analysis is exhibited as follows (Fig. 3):

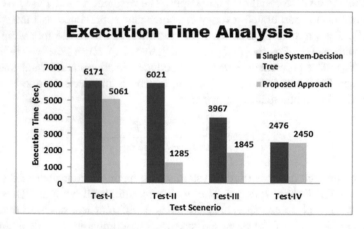

Fig. 3. Execution time analysis

A closer look into the above figure clearly depicts the fact that the proposed system is very much computationally cost effective and exhibits very negligible complexity. The next Fig. 4 will show the average accuracy found in the case of proposed system. The comparative analysis in this regards is exhibited below:

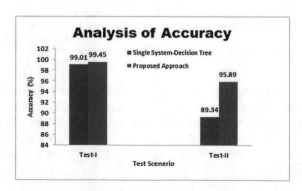

Fig. 4. Evaluation of accuracy

The above Fig. 4 also shows that the numerical outcomes obtained by simulating the proposed methodology exhibited the conventional baseline (Single-tree) in terms of both accuracy and execution time. Therefore, it can be said that the proposed ensemble based learning exhibited negligible computational complexity and also accomplishes better accuracy during the detection of any suspicious activity. The fault-tolerant system integrated also exhibited better outcomes than the single system.

6 Conclusion

In this manuscript a conceptual model for intrusion detection in a cloud enabled IoT is introduced. The conceptual IDS incorporated a fuzzy based decision tree classifier model to ensure better accuracy of network anomaly detection. The performance analysis of the proposed system further conveyed that the ensemble learning based FDT achieves better outcomes. The aforementioned potential aspects of the proposed model clearly depict its extensibility into on demand futuristic IoT applications. It also ensured a scalable operation on on-demand IDS services.

References

1. Ensemble Learning: Wiki Data Mining Ensemble Learning. https://sites.google.com/a/kingofat.com/wiki/datamining/classification
2. Li, K., Han, Y.: Study of selective ensemble learning method and its diversity based on decision tree and neural network. In: Control and Decision Conference (CCDC) (2010)
3. www.cs.waikato.ac.nz/ml/weka
4. KDD99: KDDCup1999data (1999). http://kdd.ics.uci.edu/databases/kddcup99/10percent.gz
5. Mell, P., Grance, T.: The NIST Definition of Cloud Computing. Special Publication 800-145, September 2011
6. Yassin, W., Udzir, N.I., Muda, Z., Abdullah, A., Abdullah, M.T.: Cloud based intrusion detection service framework. In: IEEE International Conference Cyber Security, Cyber Warfare and Digital Forensic (CyberSec) (2012)
7. Zhang, J., Zulkernine, M.: Network intrusion detection using Random Forests. In: Proceedings of the Third Annual Conference on Privacy, Security and Trust, pp. 53–61 (2005)
8. Kai, L., Ping, Z.Z.: Using an ensemble classifier on learning evaluation for E-Learning system. In: International Conference on Computer Science and Service System (2012)
9. Janeja, V.P., Azari, A., Heilig, B.: B-dids: mining anomalies in a Bigdistributed intrusion detection system. In: IEEE International Conference on Big Data, pp. 32–34 (2014)
10. Tang, Y., Wang, Y., Cooper, K.M.L., Li, L.: Towards big data bayesian network learning - an ensemble learning based approach. In: 2014 IEEE International Congress Big Data (BigData Congress), pp. 355–357 (2014)
11. Jiang, Y.: Selective ensemble learning algorithm. In: International Conference on Electrical and Control Engineering (2012)
12. Su, L., Liao, H., Yu, Z., Zhao, Q.: Ensemble learning for question classification. In: IEEE International Conference on Intelligent Computing and Intelligent Systems, ICIS 2009, vol. 3 (2009)

13. Zhang, J., Zulkernine, M.: A hybrid network intrusion detection technique using random forests. In: The First International Conference on Availability, Reliability and Security (2006)
14. Abdulsalam, H., Skillicorn, D.B., Martin, P.: Classification using steaming random forests. IEEE Trans. Knowl. Data Eng. **23**(1), 22–36 (2011)
15. Tavallaee, M., Bagheri, E., Lu, W., Ghorbani, A.A.: A detailed analysis of the KDD CUP 99 data set. In: Proceedings of IEEE Symposium on Computational Intelligence for Security and Defense Applications (CISDA 2009), pp. 1–6. IEEE Press, July 2009
16. Boughaci, D., Herkat, M.L., Lazzazi, M.A.: A specific fuzzy genetic algorithm for intrusion detection. In: Proceedings of ICCIT (2012)
17. Debie, E.: Reduct Based Ensemble Learning System for real-valued classification problem. In: IEEE Symposium on Computational Intelligence and Ensemble Learning (CIEL) (2013)
18. Tesfahun, A., Bhaskari, D.L.: Intrusion detection using random forests classifier with SMOTE and feature reduction. In: International Conference on Cloud & Ubiquitous Computing & Emerging Technologies (2013)
19. MeeraGandhi, G., Appavoo, K., Srivatsa, S.K.: Effective network intrusion detection using classifiers decision trees and decision rules. Adv. Netw. Appl. **2**(3), 686–692 (2010)
20. Pedrycz, W., Sosnowski, Z.A.: C-fuzzy decision tree. IEEE Trans. Syst. Man Cybernet. Part C **35**(4), 498–511 (2005)
21. Pedrycz, W., Sosnowski, Z.A.: Designing decision tree with the use of fuzzy granulation. IEEE Trans. Syst. Man. Cybern. A Syst. Hum. **30**(2), 151–159 (2000)
22. Modi, C., Patel, D., Patel, H., Borisaniya, B., Patel, A., Rajarajan, M.: A survey of intrusion detection techniques in Cloud. J. Netw. Comput. Appl. **36**(1), 42–57 (2013)
23. Kleber, V., Schulter, A., Westphall, C.B., Westphall, C.K.: Intrusion detection for grid and cloud computing. IT Prof. **12**(4), 38–43 (2010)
24. Li, Z., Sun, W., Wang, L.: A neural network based distributed intrusion detection system on cloud platform. In: IEEE 2nd International Conference on Cloud Computing and Intelligent Systems (CCIS), pp. 75–79 (2012)
25. Wu, S., Er, M.J., Gao, Y.: A fast approach for automatic generation of fuzzy rules by generalized dynamic fuzzy neural networks. IEEE Computational Intelligence Society, pp. 578–594, August 2001
26. Masarat, S., Taheri, H., Sharifian, S.: A novel framework, based on fuzzy ensemble of classifiers for intrusion detection systems. In: International Conference on Computer and Knowledge Engineering (ICCKE), pp. 165–170 (2014)

S-DWF: An Integrated Schema for Secure Digital Image Watermarking

R. Radha Kumari[1(✉)], V. Vijaya Kumar[2], and K. Rama Naidu[3]

[1] JNT University, Anantapur, India
phd13ravadaradha@gmail.com
[2] Department of CSE & IT and CACR, Anurag Group of Institutions,
Hyderabad, India
[3] Department of ECE, Jawaharlal Nehru Technological University,
Anantapur, India

Abstract. The recent advancement in the field of information and communication technologies requires potential security solutions to provide efficient
information protection. The proposed study conceptualizes a secure digital
watermarking framework also abbreviated as *S-DWF* which performs an '*image
dithering*' form of decomposition followed by a key-based watermarking procedure. The proposed S-DWF implemented in a numerical computing platform
with respect to the addition of different noisy level and also checked for performance assessment. The outcome of the study shows that the proposed S-
DWF attains better information protection in terms of Peak *Signal to Noise
Ratio (PSNR)* and *Bit Error Rate (BER)* even in the presence of any noisy attack
as compared to the existing baseline. It also balances the trade-off between
computational effort required and the visual and perceptual quality of the
retrieved watermark image signal information.

1 Introduction

The present age of technology has opened very flexible ways for exploring and
exchanging of the resources through global connected networks which is popularly
known as 'Internet channel'. The exchanges of information and resources are mostly
process in the form of digital media formats such as Image, video, Audio and text
[1, 2]. The advancement of technologies enables an individual to access the information
that takes place over the internet and which may result in terms of performing unauthorized action such as copyright and Intellectual Property Rights issues, modification
of content, faking of data content, privacy leakage and etc. Therefore, there is need of
powerful security mechanism to protect the privacy and important content of digital
data's [3]. Most important, the prime focus of this paper is on Image security because
the images are widely used as means of digital communication for the information
exchanges between the government, business, medical and private sectors. One of the
main advantages of the image communication is that it takes less transmission time
with containing lots of media contents [4]. Thus, many research efforts and researchers
have arrived with some security techniques which are as cryptography, steganography,
digital signature and watermarking in order to preserve the data confidentiality,

© Springer Nature Switzerland AG 2019
R. Silhavy (Ed.): CSOC 2019, AISC 984, pp. 25–34, 2019.
https://doi.org/10.1007/978-3-030-19807-7_3

authenticity and integrity [5, 6]. Although the steganography is practice of hiding original information into cover image, cryptography is technique of protecting data content by encryption and also digital signature is an art of confirming the authenticity of digital image. However, these all techniques have their own scope and limitations. Hence, a watermarking is introduced to be more reliable and efficient image security mechanism for preserving the image originality and content form duplication. Watermarking is kind of data hiding technique which is embedded into the electromagnetic signals of digital image. Furthermore, it has been observed that many existing researches have carried toward in the domain of image watermarking with their advantages and limitations [7–10]. Also, there are various and different watermarking schemes are discussed in review of literature section for achieving good image security.

The proposed study introduces a novel digital watermarking process namely a secure digital watermarking framework (*S-DWF*), which optimizes the computation by incorporating a '*image dithering*' form of decomposition. The result analysis portion further subjected to perform a comparative analysis which shows that the proposed method obtain potentially superior results as compared to the existing system when the watermark image signal is retrieved at the receiver end. The overall paper structure is organized as follows: Sect. 2 exhibits the conventional state of art studies carried out in the domain of digital watermarking, followed by Sect. 3- problem description, Sect. 4- formulation of proposed methodology, Sect. 5 design analysis of *S-DWF*, finally it exhibits Sect. 6- result analysis and Sect. 7- conclusion.

2 Problem Description

In the recent time security has become one of the prime considerations to enhance reliable communication performance in digital image processing as well as in networking. Digital image watermarking significantly gained much attention from the researchers since last one decade but lacks computational efficiency during information embedding process. Although there exist different image watermarking solutions in practice but most of them do not impose enough security solutions when it comes to the matter of information embedding using decomposition mechanism. As most of the time it is found that information embedding process leads to numerical instability which further results desired watermark image information in the receiver end which is visually not perceptible. This fact is realized in this proposed work where dedicated effort inclines towards developing an efficient cost effective digital watermarking technique well capable of offering significant amount of resistance by not getting affected much by types of noisy attack in the transmission channel.

3 Related Work

The presented study discusses about various research attempts towards research-based techniques on Image watermarking. The work adopted by Laouamer and Tayan [11] have presented a new watermarking scheme based on mathematical interpolation tool for interfere detection and efficient recovery from an image. The proposed method is compared with other existing technique and is found to be effective and resistant to various attacks. Ernawan and Kabir [12] have used DCT quantization process in their work for achieving strong and efficient watermarking technique for content copyright security. Loan et al. [13] have utilized chaotic encryption technique with DCT in order to provide strong watermarking security for both color and grayscale image processing operations. Guo et al. [14] have introduces a new image watermarking techniques based on the content protection and error diffusion approach. Shehab et al. [15] have uses matrix factorization and transformation based watermarking scheme for providing authentication and self-recovery operation for the medical image application. Ahmaderaghi et al. [16] have designed a secure framework for digital image based on the discrete shearlet transformation and decision theory in order to provide great payload factor and higher imperceptibility. The study of Liu et al. [17], have presented a blind watermarking scheme which established on DCT and color space approach. The objective of presented approach is providing full copyright protection and image authentication. The experimental effects of proposed study reveal that it achieves great potential to identify tampered area and resist various attacks. Wang et al. [18] have used histogram approach and post processing technique for depth image protection. In [19] the Kim et al. have applied decomposition method with curvelet transformation approach in order to design quality aware watermarking technique. Guo et al. [20] have developed a different watermarking scheme that adopts concept of nature inspired algorithm with discrete wavelet transformation. The novelty of proposed technique is that it provides good visual quality with higher PSNR values and can be efficiently extracted from watermarked images under various attacks. Liu et al. [21] have introduced a novel framework that composed of fractional transformation and eigenvalue approach for developing robust and imperceptible watermarking algorithm. In [22] Su has brought Hessenberg decomposition method with blind watermarking approach to integrate watermarked color image into host image. Bhowmik et al. [23] have offered a novel framework based on the visual attention to construct noise free and qualitative watermarking method. Similarly, Guo et al. [24] have presented another watermarking technique that utilizes a class of wavelet transformation and QR factorization matrix approach. Zong et al. [25] have also worked to construct effective watermarking algorithm against various attacks that based on DCT method. In [26] Roldan et al. have made his effort in the way of achieving image authentication, tamper localization and recover process. The work of Makbol et al. [27] have, constructed a novel watermarking scheme based on the factorization method and human visual characteristics with DWT in order to provide efficient digital protection and improved security by applying AES algorithm. In study of Dragoi et al. [28] have constructed reversible watermarking approach based on pair-wise pixel to achieve robust watermarked algorithm. Chaudhari et al. [29] have tried to provide full copyright protection for multispectral images by employing both encryption and watermarking technique. In the same way the study of Cheng and Shen [30] have, presented an improved watermarking model that based on the feature extraction and association rules techniques which exploit the feature of micro images from both host images and watermarked. The author has also suggested that the proposed model can also be applicable to other watermark techniques without affecting quality of images content.

4   Formulation of Proposed Methodology

The proposed system basically offers an efficient computational modelling namely a secure digital watermarking framework (*S-DWF*) well-capable of performing digital image watermarking by means of different intermediate computational steps involved. However, the prime purpose is to secure an input image data ownership without affecting its visual integrity and originality. Thereby, it is further processed through an embedded watermarking process. In this phase of the study the watermark image is made imperceptible and robust using two different methods such as (1) a continuous tone imagery process and (2) A key based encoding process to make the watermark image visually unreadable. The proposed conceptual modelling basically hides significant information of the message $m = \{0, 1\}^n$ in a watermark image, where n represents length of data which is basically transmitted through a communication channel after performing embedding watermarking process. The following Fig. 1 shows a schematic design adopted for the assessment of the proposed conceptual modelling of *S-DWF*.

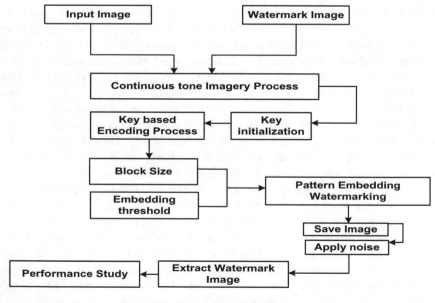

Fig. 1. Design of *S-DWF*

The above Fig. 1 clearly depicts that the proposed watermarking process considers block size ranges between (3–8) and an embedding threshold value = 0.1 prior performing an optimization process namely '*image dithering*' form of decomposition. This procedure basically subjected to divide the entire image into blocks (4 × 4) by computing block indexes which are required to embed bit stream. The process of optimization in '*image dithering*' decomposition significantly reduces the computational effort while embedding the bit-stream into the blocks with respect to R, G, and B matrix attributes. The processed watermark image further get retrieved from the local hard disk drive and undergone though different noise levels/attacks e.g. Gaussian noise, image resize, salt and pepper noise etc. and checked for whether this affects visual or perceptual quality of the watermark image or not. Finally the proposed system extract the watermark image followed by an inverse key-based encoding and observe the outcome. The study also performed a comparative analysis with respect to different performance metrics computed such as peak signal to noise ratio (*PSNR*), bit error rate (*BER*) and also correlation coefficient estimation and justify the outcome obtained. The simulation outcome further exhibited that the proposed watermarking technique exhibits higher *PSNR* and also significant reduction in *BER* which shows that the obtained watermark image doesn't get affected much by the channel coefficients and other network parameters and also by any redundant information unlike conventional baseline which the work carried out by Amiri et al. [31]. The segment of the study further presents design analysis of the proposed *S-DWF*.

5   Design Analysis of *S-DWF*

This section highlights different design attributes associated with the proposed *S-DWF* to assess an efficient watermarking for the purpose of protecting data ownership. The design phase of the proposed system is multi-fold and consists of different component execution phases. In the initial phase of the implementation the proposed system considers an input image and a watermark image and further subjected to perform the following consecutive operations as follows:

- *A continuous tone imagery process*: In this process *S-DWF* performs image toning by means of separating three different such as R, G, and B color spaces. All the computed individual color space data are further processed through an error computation and thresholding process prior performing concatenation. The following Fig. 2 shows the steps involved in this entire process.

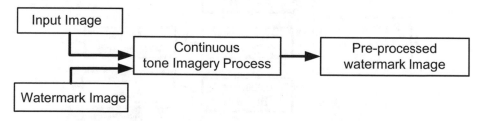

Fig. 2. A continuous tone imagery process

- *A key based encoding process*: This computational process takes the watermark image *w* (64 × 64) and extracts number of pixel coefficients. Then it also considers a non-negative integer seed which is an input key to produce control random number generation. This is a composed structured data consisting of three different fields such as (a) type, seed and state [625 × 1 unit 32]. Further, the computational flow involves random permutation, re-shaping of matrix size to encrypt *w* (64 × 64) (Fig. 3).

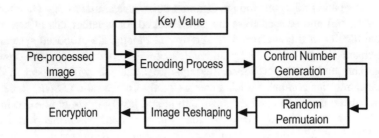

Fig. 3. Proposed key based Encoding Mechanism

- **Proposed Pattern Embedding Watermarking Process**: In this process the study has performed an optimization paradigm namely *'image dithering'*, the process basically computes the total number of blocks from the input image and generate a randomized seed. It further extract the image blocks corresponding to the R, G and B color space and also performs a decomposition procedure in order to produce a *dithering* matrix and an unitary matrix. The formation of tridiagonal matrix reduces the computational complexity and at the same time the tridiagonal matrix consists has the same eigenvalues as the original input image. On the other hand the proposed **S-DWF** also extracts respective bits from the watermark image and further the input processed/decomposed image and the watermark image information bits are get embedded into respective blocks to hide the original watermark information to significantly maintain data ownership. The system also performs a complementary operations and store back the respective blocks into a watermarked image. The following block representation exhibits the processes involved in the proposed embedding watermarking (Fig. 4).

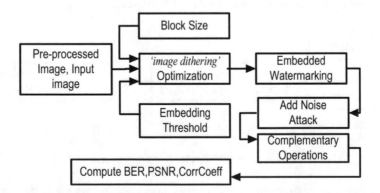

Fig. 4. Proposed Optimized embedded watermarking mechanism

Finally, the processed watermarked image is passed though different noisy environment followed by complementary steps of computations to get back the original watermark image with its hidden information. The numerical outcomes further exhibited in the next segment clearly show that the proposed system produces better signal with superior visual representation when optimized watermarking considered. It is also shown that it achieves significantly less BER and more PSNR as compared to the existing system.

6 Results Discussion

This section discusses about the simulation outcomes obtained after implementing the proposed **S-DWF** in a numerical computing environment. The input image and the watermark image combinely processed through a set of computational environment to perform cost effective and improved information embedding for the protection of significant information. The study considered a similar digital watermarking approach proposed by Amiri et al. [31] as a baseline and implemented its core embedding process to address the numerical instability and complexity problems during data protection and also when noise is added in the presence of different channel coefficients. The minimum system requirements to assess the performance of the proposed system are 64-bit operating system (windows) with x64-based process supported with i5-8250U CPU and 4.00 GB RAM.

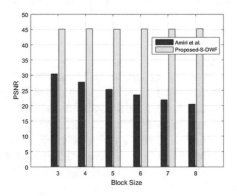

Fig. 5. Performance assessments with respect to PSNR

The above highlighted Fig. 5 shows a comparative performance analysis which is carried out to assess the PSNR obtained after simulating the proposed **S-DWF** in a numerical computing environment. A closer look into the above figure clearly exhibits that proposed system attains better PSNR value in the obtained watermark image which is approximately 45 with respect to different size of block operations, on the other hand the watermarking approach introduced by Amiri et al. [31] produces an image signal where the visual quality of the watermark gradually decreases when the size of the blocks up-scaled. It clearly shows that the proposed optimization procedure obtain an well-balanced trade-off between computational effort and the visual outcome of the retrieved watermark image perceptual quality even if any nosey attack imposed in the channel.

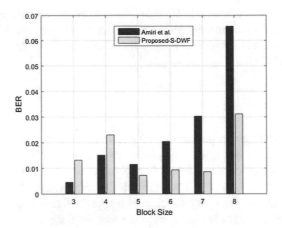

Fig. 6. Performance assessment with respect to BER

The above Fig. 6 also exhibits that the proposed *S-DWF* offers very less information loss in the presence of any noisy environment. Om the other hand BER is quite higher found in the watermarking approach of Amiri et al. [31] as it doesn't impose any optimization procedure in the input image blocks associated with color space.

7 Conclusion

The proposed study addresses the computational complexity associated with conventional watermarking algorithms and imposed an optimized decomposition mechanism namely '*image dithering*' to obtain a well-balanced trade-off between computational effort required in the phase of embedding watermarking and the communication requirements where retrieving watermark image with efficient visual quality plays a crucial role, The performance enhancement of the proposed *S-DWF* has been carried out by formulating an conceptual design modelling which is further assessed in a numerical computing platform. The comparative performance analysis clearly shows that the proposed *S-DWF* attains superior visual outcome in the retrieved watermark image with very less computational effort unlike the work of Amiri et al.

References

1. Patsakis, C., Zigomitros, A., Papageorgiou, A., Solanas, A.: Privacy and security for multimedia content shared on OSNs: issues and countermeasures. Comput. J. **58**(4), 518–535 (2015)
2. Marqués, I., Graña, M.: Image security and biometrics: a review. In: International Conference on Hybrid Artificial Intelligence Systems, 28 March 2012, pp. 436–447. Springer, Heidelberg (2012)
3. Supriya, A.V., Shetty, S.: A survey on multimedia content protection (2017)

4. Korus, P.: Digital image integrity–a survey of protection and verification techniques. Digit. Signal Proc. **71**, 1–26 (2017)
5. Taleby Ahvanooey, M., et al.: A comparative analysis of information hiding techniques for copyright protection of text documents. Secur. Commun. Netw. **2018**, 22 pages (2018)
6. Nyeem, H., Boles, W., Boyd, C.: Digital image watermarking: its formal model, fundamental properties and possible attacks. EURASIP J. Adv. Signal Process. **2014**(1), 135 (2014)
7. Singh, A.K., Kumar, B., Singh, G., Mohan, A.: Medical image watermarking techniques: a technical survey and potential challenges. In: Medical Image Watermarking, pp. 13–41. Springer, Cham (2017)
8. Rani, B.U., Praveena, B., Ramanjaneyulu, K.: Literature review on digital image Watermarking. In: Proceedings of the 2015 International Conference on Advanced Research in Computer Science Engineering & Technology (ICARCSET 2015), 6 March 2015, p. 43. ACM (2015)
9. Tew, Y., Wong, K.: An overview of information hiding in H.264/AVC compressed video. IEEE Trans. Circ. Syst. Video Technol. **24**(2), 305–319 (2014)
10. Bianchi, T., Piva, A.: Secure watermarking for multimedia content protection: a review of its benefits and open issues. IEEE Signal Process. Mag. **30**(2), 87–96 (2013)
11. Laouamer, L., Tayan, O.: Performance evaluation of a document image watermarking approach with enhanced tamper localization and recovery. IEEE Access **6**, 26144–26166 (2018)
12. Ernawan, F., Kabir, M.N.: A robust image watermarking technique with an optimal DCT-psychovisual threshold. IEEE Access **6**, 20464–20480 (2018)
13. Loan, N.A., Hurrah, N.N., Parah, S.A., Lee, J.W., Sheikh, J.A., Bhat, G.M.: Secure and robust digital image watermarking using coefficient differencing and chaotic encryption. IEEE Access **6**, 19876–19897 (2018)
14. Guo, Y., Au, O.C., Wang, R., Fang, L., Cao, X.: Halftone image watermarking by content aware double-sided embedding error diffusion. IEEE Trans. Image Process. **27**(7), 3387–3402 (2018)
15. Shehab, A., et al.: Secure and robust fragile watermarking scheme for medical images. IEEE Access **6**, 10269–10278 (2018)
16. Ahmaderaghi, B., Kurugollu, F., Rincon, J.M.D., Bouridane, A.: Blind image watermark detection algorithm based on discrete shearlet transform using statistical decision theory. IEEE Trans. Comput. Imaging **4**(1), 46–59 (2018)
17. Liu, X.L., Lin, C.C., Yuan, S.M.: Blind dual watermarking for color images' authentication and copyright protection. IEEE Trans. Circ. Syst. Video Technol. **28**(5), 1047–1055 (2018)
18. Wang, S., Cui, C., Niu, X.: A novel DIBR 3D image watermarking algorithm resist to geometrical attacks. Chin. J. Electron. **26**(6), 1184–1193 (2017)
19. Kim, W.H., Nam, S.H., Lee, H.K.: Blind curvelet watermarking method for high-quality images. Electron. Lett. **53**(19), 1302–1304 (2017)
20. Guo, Y., Li, B.Z., Goel, N.: Optimised blind image watermarking method based on firefly algorithm in DWT-QR transform domain. IET Image Process. **11**(6), 406–415 (2017)
21. Liu, X., Han, G., Wu, J., Shao, Z., Coatrieux, G., Shu, H.: Fractional Krawtchouk transform with an application to image watermarking. IEEE Trans. Signal Process. **65**(7), 1894–1908 (2017)
22. Su, Q.: Novel blind colour image watermarking technique using Hessenberg decomposition. IET Image Process. **10**(11), 817–829 (2016)
23. Bhowmik, D., Oakes, M., Abhayaratne, C.: Visual attention-based image watermarking. IEEE Access **4**, 8002–8018 (2016)

24. Guo, Y., Li, B.Z.: Blind image watermarking method based on linear canonical wavelet transform and QR decomposition. IET Image Process. **10**(10), 773–786 (2016)
25. Zong, T., Xiang, Y., Guo, S., Rong, Y.: Rank-based image watermarking method with high embedding capacity and robustness. IEEE Access **4**, 1689–1699 (2016)
26. Rosales Roldan, L., Cedillo Hernandez, M., Chao, J., Nakano Miyatake, M., Perez Meana, H.: Watermarking-based color image authentication with detection and recovery capability. IEEE Latin Am. Trans. **14**(2), 1050–1057 (2016)
27. Makbol, N.M., Khoo, B.E., Rassem, T.H.: Block-based discrete wavelet transform-singular value decomposition image watermarking scheme using human visual system characteristics. IET Image Process. **10**(1), 34–52 (2016)
28. Dragoi, I.C., Coltuc, D.: Adaptive pairing reversible watermarking. IEEE Trans. Image Process. **25**(5), 2420–2422 (2016)
29. Zope-Chaudhari, S., Venkatachalam, P., Buddhiraju, K.M.: Secure dissemination and protection of multispectral images using crypto-watermarking. IEEE J. Sel. Top. Appl. Earth Obs. Remote Sens. **8**(11), 5388–5394 (2015)
30. Chang, C.S., Shen, J.J.: Features classification forest: a novel development that is adaptable to robust blind watermarking techniques. IEEE Trans. Image Process. **26**(8), 3921–3935 (2017)
31. Amiri, S.H., Jamzad, M.: Robust watermarking against print and scan attack through efficient modeling algorithm. Signal Process. Image Commun. **29**(10), 1181–1196 (2014)

Semantic Richness of Tag Sets: Analysis of Machine Generated and Folk Tag Set

Faiza Shafique[✉], Marhaba Khan, Fouzia Jabeen, and Sanila

Department of Computer Science, Shaheed Benazir Bhutto Women University,
Peshawar, Pakistan
faizashafiq46@gmail.com, marhabakhan907@gmail.com,
fouzia.jabeen@sbbwu.edu.pk, sanilariaz905@gmail.com

Abstract. Social networking sites like Flickr, YouTube and Del.icio.us have been gaining more popularity on the internet. Users can create, evaluate, and distribute the information over the internet with the change of web to a medium. A social tagging system allows users to share sources and enlighten them with different descriptive tags. Folksonomy is a contribution of social tagging. It allows users to tag online content/resource for common accessibility and resource searching. Users can freely type any form of text or keywords when tagging a resource. The reason behind popularity of folksonomy applications is all users create tags without having any technological skills and experience. With the social tagging simplicity, it can gather huge amount of user contributed tags. Tag sets can also be generated with the help of different online websites. Different relationships like similarity, co-occurrence etc. appears among tags of folk tag set. In this work, we have tested that whether the relationships that exists in folk tag sets are also present in the tag sets generated by automatic tools. For our testing, we took five automatic tag set generating websites, which includes To cloud, Word It Out, Tag crowd, Word sift and Word Art. The result of this work helps to conclude the semantic richness of tag set generated by automatic tools.

Keywords: Collaborative tagging · Folksonomy · Automatic tagging ·
Tag clouds · Tags · Search · Retrieval

1 Introduction

Social tagging gains more popularity with the invention of different sites like Del.icio. us and Flickr. Since, different social systems have been designed that supports tagging to variety of resources. Tagging is the process where each user has the ability to give a tag to an object or resource. The three-place terminology of tagging includes resource, tag, and tagger. Resource is any digital Information/objects such as web pages, photos or videos clip etc. Tag is a string generally single keyword that is created by users to every resource in the work of tagging. Tagger is the user who hand over tag to each target resource [1]. In Del.icio.us, a user give tag to particular uniform resource locators (URLs), each user have allowed to have their own set of tags for each URL. Flickr allows users to give tags to photos that are uploaded by them or also by other users.

© Springer Nature Switzerland AG 2019
R. Silhavy (Ed.): CSOC 2019, AISC 984, pp. 35–47, 2019.
https://doi.org/10.1007/978-3-030-19807-7_4

Blog sites like Blogger, Wordpress and Livejournal, allows blogs authors to assign tag to their posts. On social networking sites, like Facebook users can also give tagging information by marking something as "Like" [1].

In the last few years, many kinds of collaborative tagging systems have experienced which, is the system that allows internet users to manage online resources such as photo, videos etc. It allows sharing of these resources and users are free to add their personalized tags to these resources [2]. Taggers are actively involved in the process of pointing out and cataloguing resources of interest assigning them one or more descriptive keywords or tags. They exploit the growing amount of information collected to improve their searches and content discovery process.

When every user can assign a freely defined set of tags to a resource, the tag collection reflects the social attitudes of the community of users and a shared social organization. The result of this process has recently been defined as folksonomy [3]. The term folksonomy comes from the "Folk" and "Taxonomy", was adopted by Thomas Vander Wal in 2004 [4]. The reason behind popularity of folksonomy applications is all users without any technical skills and experience create tags. Folksonomy is useful in many contexts. A user can find his/her social bookmarks about hardware through the tag of "are". Tag provides a way to find out similar and useful resources. Users are connected by collaborative tagging and are capable to allocate resources. By this way, they can find users having similar interest via shared tags.

Folk tag set contains tags given by different users in response to collaborative tagging activity. There are many types of relationships that exists between folk tags (tags given to a resource by collaborative tagging activity) [5]. These are subsumption, similarity, co-occurrence, equivalence etc.

A subsumption relation between tags i.e., tag_a and tag_b can be defined as: tag_a subsumes tag_b, whenever tag_b is used, tag_a can also be used without ambiguity [6]. Subsumption relation is directional, that is, $t_a \rightarrow s\ t_b$ does not imply $t_b \rightarrow s\ t_a$. For example, literature$\rightarrow$s Chinese literature, since for any document annotated with Chinese literature, we can also annotate it with literature. However, if we swap the two tags, the statement would not hold [6]. Semantic similarity refers to the similarity among two senses of a tag. Two tags are said to be as similar tags whenever they share a certain level of common characteristics [6]. Co-occurring tags are tags which frequently appear together. For example, it may show super-class/sub-class relationship like 'mammal' can be taken as a super-class of both 'dog' and 'cat' [5]. The equivalence relation can be defined as if the meanings of two tags are completely same, or they referred to the same target. In social tagging, there survive four types of equivalence relation among tags. The most common type of equivalence is singulars and plurals. For example, book-books, store-stores and man-men are the singular-plural sets, in which each referred to completely the same target. The second type of equivalence relation is of verbs or adjectives. For example, sit-sitting, ring-ringing that sounds different parts of speech but have exactly the same meaning. Nouns and their abbreviations is the third type of equivalence relation. For example, ATM is the full form of Automated Teller Machine and Mister is equal to Mr. The last and fourth type is the equivalence of underscore or hyphen (_). For example play-group-playgroup and chat-room-chatroom [7].

Tag clouds are visual presentations of a set of keywords, usually a set of "tags" preferred on basis of some validation, in which, attributes of the text such as size, weight, or color are used to characterize features of the related terms. There are many automatic tools which, include To cloud, Word It Out, Word sift, Tag-crowd, Tag-crowd and Tag-generator etc., that generate tag sets and represent it in the form of tag cloud. Figure 1 represents tag set in the form of tag cloud generated by automatic tool Tag Cloud.

2 Related Work

Lee et al. [7] derived in their paper subsumption relationships between tags based on folksonomy visualization using Wikipedia texts. They targeted Del.icio.us tag data. They suggest a technique named FolksoViz for deriving relationships and of folk-sonomy. For mapping every tag in a Wikipedia text they have used tag sense disam-biguation (TSD) technique. They show in their results that FolksoViz managed to display the correct subsumption pairs and find relationships to complete folksonomy visualization.

Si et al. [6] suggested three different methods for discovering subsumption relations named tag-tag, tag-word, tag-reason co-occurrences. They also used Layered Directed Acyclic Graph (Layered DAG) constructing algorithm for the elimination of redundant tags in relations. They perform their work on two data sets, first named BLOG and the second named BOOK. In result, they showed that the tag-reason method has excellent performance for discovering subsumption relation than tag-tag and tag-word methods. Their result also show that Layered DAG construction algorithm works successfully as expected. Rêgo et al. [8] detected subsumption relationships between tags in folkson-omy. For this purpose, they identified several similarity measures (support and confi-dence, mutual overlapping, cosine similarity). In this computation this problem brings basic complexities such as class imbalance and class overlapping, for solving these complexities they have used two basic methods oversampling (SMOTE) and under-sampling (CNN). They concluded in their experiments that mutual overlapping is best performing baseline. They also prove that both methods give appropriate correctness.

Simpson et al. [9] recommended tag similarity method which uses the Jaccard measure to standardize tag co-occurrences. The tags are then structured in a co-occurrence graph, which is fed to an iterative discordant clustering algorithm to dis-cover clusters of associated tags. By looking at the obtained clusters it can be observed that clusters are too big to be utilized for human navigation, however, by removing unpopular tags this problem can be reduced. Between's used to select edges to be removed, but it performs poorly on densely inter-related tags in Del.icio.us dataset.

Fig. 1. Tag set generated by automatic tool that is represented in the form of Tag Cloud

Carolina et al. [10] described different types of context (co-occurrence and spatial) to categorize objects using its existence, location and appearance. They have used a new method of object categorization that include both types of contexts named COLA (Co-occurrence Location and Appearance) which uses CRF (Conditional random field) to make better the performance of simple recognition. They evaluate their results on two demanding datasets: PASCAL 2007 and MSRC. The results showed that combine co-occurrence and spatial context improves accuracy [13].

Thomee et al. [11] proposed in their paper an algorithmic structure to automatically detect equivalence relationships in geo-spatial patterns in folksonomy. They used Expectation-Maximization (EM) algorithm and a Flicker dataset. Their result demonstrate that method detect correct and useful equivalence relationships.

3 Methodology

The proposed approach is shown in Fig. 2. In the first step we took tag sets generated by To cloud, Word It Out, Word sift, Tag-crowd, Word art tools. In subsequent steps, we have checked the presence and absence of four core relationships, which includes subsumption, similarity, and co-occurrence. With mathematical formulas, we have measured existence of these relationships. The detail is discussed in Sects. 3.1–3.4.

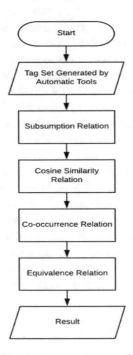

Fig. 2. Proposed approach

3.1 Subsumption Relation

Tag x_i is said to subsume Tag x_k if Tag x_i refers to a more general concept than Tag x_k. Then the two tags form a subsumption relation.

We measured the subsumption relation between two tags i.e. Tagx_i and Tag x_k using Eq. 1.

$$\mathbf{TF(x_k|Wiki(x_i)) < TF(x_i|Wiki(x_k)), \mu < TF(x_i|Wiki(x_k))} \tag{1}$$

Tag x_i Subsume Tag x_k, if Eq. 1 gives True value. Where $Wiki(x_i)$ is the Wikipedia page of Tag x_i, $TF((x_k|Wiki(x_i))$ is the term frequency of Tag x_k in the Wiki (x_i), $Wiki(x_k)$ is the Wikipedia page of Tag x_k, $TF(x_i|Wiki(x_k))$ is the term frequency of Tag x_i on $Wiki(x_k)$. μ is the threshold value [11].

3.2 Similarity Relation

Two tags are considered similar to each other if they share a certain degree of common characteristics. Then the two tags form a similarity relation.

In order to find similarity among two tags we used the cosine similarity that can be computed using Eq. 2.

$$\text{Cosine } (tag_a, tag_b) = \frac{\overrightarrow{tag_a}.\overrightarrow{tag_b}}{\|tag_a\|.\|tag_b\|} = \frac{\sum_{k=1}^{n} tag_a * tag_b}{\sqrt{\sum_{k=1}^{n}(tag_a)^2}\sqrt{\sum_{k=1}^{n}(tag_b)^2}} \tag{2}$$

The term tag_a and tag_a are the $\overrightarrow{tag_a}$ and $\overrightarrow{tag_b}$ vectors. The two vectors $\overrightarrow{tag_a}.\overrightarrow{tagx_b}$ represent the dot product of tag_a and tag_b as $dp_{tag_a tag_b}$. The $(tag_a)^2$ is the dot product of tag_a with itself represented as dp_{tag_a}. The term $(tag_b)^2$ is the dot product of tag_b with itself represented as dp_{tag_b}. The $\|tag_a\|.\|tag_b\|$ shows the vector length as l_{tag_a} and l_{tag_b} respectively.

In this evaluation, for the sake of clarity we adopt the following conversions:

- tag_a and tag_b are the $\overrightarrow{tag_a}$ and $\overrightarrow{tag_b}$ vectors.
- $dp_{tag_a tag_b}$ is the $\overrightarrow{tag_a} \cdot \overrightarrow{tag_b}$ dot product between tag_a and tag_b.
- $dp_{tag_a tag_a}$ is the dot product of tag_a itself.
- $dp_{tag_b tag_b}$ is the dot product of tag_b itself.
- l_{tag_a} and l_{tag_b} are the $\|\overrightarrow{tag_a}\|$ and $\|\overrightarrow{tag_b}\|$ vector length.

3.3 Co-occurrence Relation

Co-occurrence between two tags is computed by using the Eq. 3. Here we are finding that how many times two tags occur within a document or blog.

$$\text{Co-occur}(tag_a/tag_b) = \frac{|\mathbf{tag_a} \cap \mathbf{tag_b}|}{|\mathbf{tag_b}|} \tag{3}$$

In Eq. 3, the $|tag_a \cap tag_b|$ is the number of occurrences of tag_a with another tag_b among all resources, $|tag_b|$ dividing by the total number of occurrences of tag_b [12].

3.4 Equivalence Relation

Two tags are called to be equivalent when the meanings of the tags are identical, or also they direct to the identical target. In this case, we can say that two tags form an equivalence relation.

For the presences of equivalence relation between tags, we survive that there are four types of equivalence relation among tags. The first and most commonly type of equivalence is in the form of singulars and plurals. For example Book-Books, Store-Stores and Man-Man, this directs to exactly the same target. The second type is the equivalence of verbs or adjectives. For example, Sit-Sitting, and Ring-Ringing. The third type is the nouns and their abbreviations. For example ATM-Automated Teller Machine and Mister is equal to MR. The fourth and last type of equivalence is underscore or hyphen (_). For example, play-group-playgroup and chat-room-chatroom.

In order to detect this type of equivalence relation, we have to use some rules and also heuristics approach according to their types. In this paper, we evaluate the first two types of equivalence relation (i.e. singulars, plurals and verb, adjective). The remaining two types of equivalence relation exist between tags of tag set in very rare cases. Therefore, we only focused on the first two types of equivalence. The first equivalence type of singulars and plurals can be dig out by checking the existence of –s, -es or –ies at the end of tags. The second equivalence type of nouns and their abbreviations can be obtained by checking the existence of –ing, -ation, or –y at the end of a tag. The third equivalence type of abbreviation can be dig out by checking whether each letter of one tag is placed in the same order in the other tag. For example, ATM consists of A, T and M, and they are placed in Automated Teller Machine in a same order. Unluckily, this rule is not ideal. The last equivalence type of – or _ embed can be easily detect by checking whether by removing – or _ from, two tags are same or not [7].

4 Experimental Results and Discussions

This Section explains the experimental results. The Sect. 4.1 briefly explains and analyzes the experimental datasets, we have chosen for experiments. In Sect. 4.2, were have presented the results. Then in Sect. 4.2.1 we have analyzed the semantic depth.

4.1 Experimental Datasets

The experiments are performed on five tag sets that are made by five different automatic tools. The tag sets are generated for five different blogs, which contains at least thirty tags.

The Table 1 show automatic tools that we have chosen, along with the URL of page and topic description. Tag set generated is presented in the form of Tag Cloud. Figures 3, 4, 5, 6, and 7 represents the generated tag sets.

Table 1. URLs used by automatic tools

Tool	Url	Topic
To Cloud	https://en.wikipedia.org/wiki/Ccomputer_network	Computer network
Word It Out	https://en.wikipedia.org/wiki/The_Three_Little_Pigs	The three little pigs
Tag Crowd	https://en.wikipedia.org/wiki/Communication	Communication
Word Sift	http://en.wikipedia.org/wiki/folksonomy	Folksonomy
Word Art	https://en.wikipedia.org/wiki/Communication	Communication

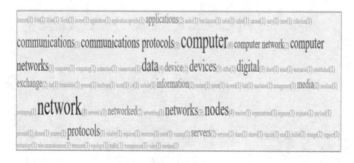

Fig. 3. Tag set generated by automatic tool To Cloud

Fig. 4. Tag set generated by automatic tool Word It Out

Fig. 5. Tag set generated by Word Sift

Fig. 6. Tag set generated by automatic tool Tag-Crowd

Fig. 7. Tag set generated by automatic tool Word Art

4.2   Results

This Section demonstrates the experimental results, In Table 2, 3, 4, 5 and 6, we shown in (✓) and (✗) the presence and absence of relationship there is no relationships between two tags. The computation results are between 0 and 1. Therefore, to decide presence or absence we need to choose threshold value. We opt for value greater than 0.6 to mark presence of certain relationship.

Table 2. Comparison of Relationship exists in Tag Set Generated by To Cloud

Tool	Relationships		
To Cloud	Subsumption	Similarity	Co-occurrence
Application, computers	✓	✗	✓
Signals, storage	✗	✗	✓
Telecommunications, communication	✗	✓	✗
Connection, collection	✓	✓	✓
Web, WiFi	✗	✗	✓

Table 3. Comparison of Relationship exists in Tag Set Generated by Word It Out

Tool	Relationships		
Word It Out	Subsumption	Similarity	Co-occurrence
Pigs, phrases	✗	✗	✗
Made, date	✗	✓	✓
Sticks, bricks	✓	✓	✓
House, houses	✗	✓	✓
System, story	✗	✓	✓

Table 4. Comparison of Relationship exists in Tag Set Generated by Word Sift

Tool	Relationships		
Word Sift	Subsumption	Similarity	Co-occurrence
Tagging, tag	✗	✓	✗
Collaborative, classification	✗	✓	✗
Folksonomy, bookmarking	✗	✓	✓
Visualize, Vander	✓	✗	✓
Folksonomy, taxonomy	✗	✓	✗

Table 5. Comparison of Relationship exists in Tag Set Generated by Tag-Crowd

Tool	Relationships		
Tag-Crowd	Subsumption	Similarity	Co-occurrence
Development	✗	✓	✗
Meaning, message	✗	✓	✗
Received, reception	✗	✗	✓
Signs, sources	✗	✓	✓
Sequence, share	✗	✗	✓

Table 6. Comparison of Relationship exists in Tag Set Generated by Word Art

Tool	Relationships		
Word Art	Subsumption	Similarity	Co-occurrence
Digital, data	✓	✗	✗
Communications, transmission	✓	✓	✓
Network, exchange	✗	✗	✗
Allows, more	✗	✗	✓
Computer, computers	✓	✓	✗

In Table 7, the existence of equivalence relation between tags of tag set generated by automatic tool To Cloud is presented. The two types of equivalence relations are examined i.e. singular/plural equivalence and verb/adjectives equivalence.

Table 7. Equivalence Relationship exists in Tag Set Generated by To Cloud

Tool	Relationships			
To Cloud	Equivalence			
	Singular, Plural		Verb or Adjective	
Application applications, computers computer	1, 1	✓, ✓	Application, computers computing	0, 1
Signals signal, storage	1, 0	✓, ✗	Signals, storage	0, 0
Telecommunications telecommunication, communications communication	1, 1	✓, ✓	Telecommunications, communications	0, 0
Connection connections, collection	1, 0	✓, ✗	Connection, collection	0, 0
Web, WiFi	0, 0	✗, ✗	Web, WiFi	0, 0

4.2.1 Evaluation of Semantic Depth

To represent the semantic depth of each considered relationship, we shown it in graph. Figures 8, 9, 10, 11 and 12 shows in percentage the depth of existence of each relationship (i.e. Subsumption, Similarity, Co-occurrence and Equivalence).

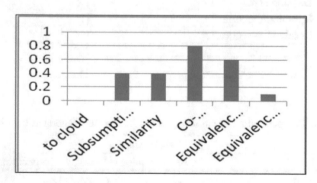

Fig. 8. Depth of relationship between tags in automatic tool To Cloud

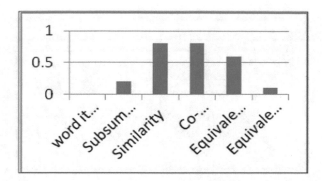

Fig. 9. Depth of relationship between tags in automatic tool Word It Out

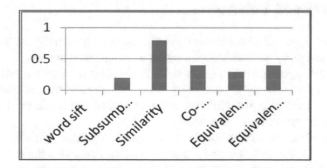

Fig. 10. Depth of relationship between tags in automatic tool Word Sift

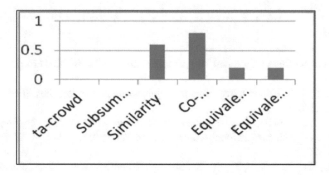

Fig. 11. Depth of relationship between tags in automatic tool Tag-Crowd

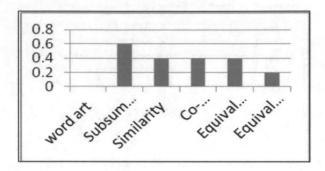

Fig. 12. Depth of relationship between tags in automatic tool Word Art

5   Summary and Conclusion

There are many types of relationships that exist among tags of a tag set. Folks can give tags or there are many automatic tools, which also generates tag sets. In this paper we have performed experiments to answer the question that whether the relationship that exist among tags of folk tag set also exist in tag set generated by automatic tools? The answer of this question in yes will raise another question that if relationship exists than what is the semantic depth of that relation.

We can conclude after experiments that that similarity relationship is more common in the tags of tag sets than the co-occurrence relationships, also the existences of co-occurrence relationship is more frequent than the subsumption relationship. The equivalence relationship also exists between the tags of tag sets but its existence is rare as compared to other relations.

References

1. Gupta, M., et al.: Survey on social tagging techniques. ACM SIGKDD Explor. Newsl. **12**(1), 58–72 (2010)
2. Li, Q., Lu, S.C.Y.: Collaborative tagging applications and approaches. IEEE Multimed. **15** (3), 14–21 (2008)
3. Marchetti, A., Rosella, M.: SemKey: a semantic collaborative tagging system. In: Proceedings of 16th International Conference on World Wide Web, WWW 2007, vol. 7, pp. 8–12 (2007)
4. Nazim, M., Hai, T., Thanh, N., Qi, X., Sik, G.: Expert Systems with Applications Semantic similarity measures for enhancing information retrieval in folksonomies. Expert Syst. Appl. **40**(5), 1645–1653 (2013)
5. Jabeen, F., Khusro, S., Majid, A., Rauf, A.: Semantics discovery in social tagging systems: a review. Multimed. Tools Appl. **75**(1), 573–605 (2016)
6. Si, X., Liu, Z., Sun, M.: Explore the structure of social tags by subsumption relations, August, pp. 1011–1019 (2010)

7. Lee, K., Kim, H., Kim, H.: FolksoViz : a semantic relation-based folksonomy visualization using the wikipedia corpus. In: 10th ACIS International Conference on Software Engineering, Artificial Intelligences, Networking and Parallel/Distributed Computing, SNPD 2009, pp. 24–29. IEEE (2009)
8. Rêgo, A.S.C., Marinho, L.B., Pires, C.E.S.: A supervised learning approach to detect subsumption relations between tags in folksonomies. In: Proceedings of the 30th Annual ACM Symposium on Applied Computing, SAC 2015, April, pp. 409–415 (2015)
9. Mousselly-Sergieh, H., Egyed-Zsigmond, E., Gianini, G., Döller, M., Pinon, J.-M., Kosch, H.: Tag relatedness in image folksonomies. Document numérique **17**(2), 33–54 (2014)
10. Galleguillos, C., Rabinovich, A., Belongie, S.: Object categorization using co-occurence, location and appearance. In: Proceedings of CVPR (2008)
11. Thomee, B., De Francisci Morales, G.: Automatic discovery of global and local equivalence relationships in labeled geo-spatial data. In: Proceedings of 25th ACM Conference on Hypertext Social Media, HT 2014, pp. 158–168 (2014)
12. Eisterlehner, F., Hotho, A., Jäschke, R.: ECML PKDD Discovery Challenge (2009) (DC09), CEUR-WS.org (2008), (2009)

Analysis of Behaviour Patterns of the Residents of a Smart House

Antonín Kleka[1], Radim Farana[2](✉) (iD), and Ivo Formánek[3]

[1] University of Ostrava, 30. dubna 22, 701 03 Ostrava, Czech Republic
ankleka@gmail.com
[2] Mendel University in Brno, Zemědělská 1, 613 00 Brno, Czech Republic
radim.farana@mendelu.cz
[3] University of Entrepreneurship and Law, Michálkovická 1810/181,
710 00 Ostrava, Czech Republic
ivo.formanek@vspp.cz

Abstract. The market of smart homes, offices, generally smart buildings has been growing very much in recent years. At the same time, the market offers many opportunities for both technical applications and business cases. The paper shows how the smart homes technology data and its analysis can be used to determine the behaviour patterns of the residents; how the important data can be saved; how to compare the behaviour patterns with actual situation; how to evaluate the difference between the patterns and reality; how to calculate the intervention of the expert systems; how to inform about a change in behaviour of the residents; how to report an alarm when house space seems being disrupted and how to react when serious malfunctions occurs. A fuzzy-expert system was used with benefit for these tasks.

Keywords: Smart home · Behaviour · Analysis · Pattern · Fuzzy-logic · Expert system

1 Introduction to the Problem

The market of smart homes, offices and smart buildings generally has been growing in recent years. We can see a real boom of this market. Although the number of really smart homes and households still counts only in percentages in the Czech Republic (only approximately 2% of all households meet the criteria of so-called intelligent housing), the number of houses and flats equipped with intelligent control elements for local heating and local security systems is growing.

One of the features of a smart house is, besides other things, the ability of them to record different technological data of the house variables (e.g. room temperature, energy consumption, etc.). These functionalities make use of many sensors placed in the house. Smart house behaviour data is predominantly used to generate statistics and charts, which are then presented to the user in a comprehensible form by means of the house control panel graphical interface or by means of another smart device [1, 2].

The main objective of this paper is to present the possibilities of further utilization of the acquired technological data. Specifically, to analysis and generation of typical

© Springer Nature Switzerland AG 2019
R. Silhavy (Ed.): CSOC 2019, AISC 984, pp. 48–54, 2019.
https://doi.org/10.1007/978-3-030-19807-7_5

behaviour patterns that would describe a certain real state in the house. The typical behaviour patterns created in such a way should allow us to assign different technological data (the data acquired on daily base, on hourly base, or in real time) to typical behaviour patterns. It is logical that the data we want to assign to the typical behaviour patterns can be more or less different from the typical data of the typical behaviour patterns. And that is a task for an expert system to be able to make a decision whether the difference of acquired technological data is caused by standard conditions (e.g. by inherent data variability), or is caused by a malfunction, security breach, or other unusual situation.

Smart housing can be categorized into different groups, see e.g. [3–5]. These categorizations are gradually shifting with the increasing possibilities of control systems and application of bus systems, see e.g. [6–8]. In the contribution, we focus on houses equipped with a central management system with the possibility of archiving and subsequent processing of technological data. Thanks to cooperation with our industrial partner, we could use the technological data of a particular real house:

The house is equipped with its own weather station, which measures the outside temperature, humidity, wind speed and direction, precipitation and intensity of sunshine.

Inside, the temperature, humidity and CO_2 (carbon dioxide) levels in ppm (parts per million) and volatile organic compound (VOC) in percent are recorded for each room. (VOC means the degree of so-called volatile organic matter in the air.) In the living room, due to the location of the fireplace, there is also a CO (carbon monoxide) sensor that arises from the incomplete burning of fossil fuels or biomass. Many logical values are also stored in the database to indicate whether the circuits (e.g. heating) or equipment (e.g. pumps) are in operation.

It also records whether the active bulk remote control (BRC) is active. The power provider during the day sends the pulses that receive the BRC receiver. If BRC is equal to 1, it means that BRC is active and electricity is taken at a discounted rate.

The database also contains data on the course of the regulation and how to heat a smart home.

When analyzing the data, it turned out that the processing of data on electricity, and in particular temperature control are not easily usable, because they require a correlation with the current external conditions etc.

2 Behaviour Patterns Analysis

As mentioned above, a smart house allows us to track technological data from various sensors located in the house and then react to specific thresholds by a specific pre-set operation. However, the achievement of the limit value itself is not sufficient for the proper operation of the system. The technological data should be viewed more comprehensively. This means that a given pre-programmed operation or a state cannot be trigged by the only one certain value, but must be trigged by the whole trend, which shows signs of certain regularity.

This trend, which we are able to identify and classify in the data, is called a behaviour pattern. As a behaviour pattern, it is not necessary to represent just only one course of one quantity, but also a combination of the development of more quantities.

One of the problems in defining patterns is that a given pattern, or state, can change over time by varying effects.

The behaviour patterns are therefore only appropriate for a certain period of time, such as seasonal behaviour patterns due to the influence of outdoor temperature and the amount of rain or snowfall on house processes such as ventilation, heating, etc.

Another way to respond to changing conditions is to create patterns dynamically each day only from records that precede the day immediately (days, weeks).

The basic premise of these patterns is to find out whether the house is occupied at a particular time or not, and only by analyzing the development of the quantities over time. The measured quantities are CO2, VOC, humidity and temperature in the room, or throughout the house. Identifying the presence of individuals of these variables should be possible because the house residents naturally affects their surroundings [9, 10]:

- Thermal losses of the organism.
- Evaporation of water.
- Exhaled CO_2.
- VOC, sweat and breathing. In smoking areas, cigarette smoke is the largest source of VOC. However, volatile substances are also part of various detergents, etc.

Detection of people should therefore be possible by monitoring trends and, above all, by increasing the quantities described above. The presence of one person in the room should be clear in these respects. A possible problem with this method may be manual window ventilation/opening, which affects all of the above-mentioned variables if the house is not equipped with an automatic ventilation system.

To identify the behaviour patterns of individual rooms, the course with no activity of the residents were compared, see Fig. 1, with the course corresponding to the use of the room, Fig. 2.

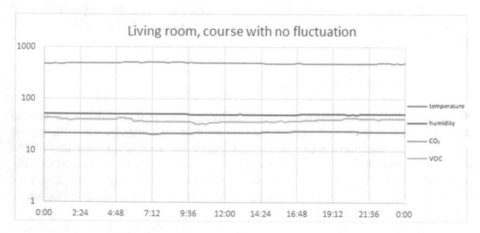

Fig. 1. Living room, occupied with residents, course of the quantities with no fluctuation.

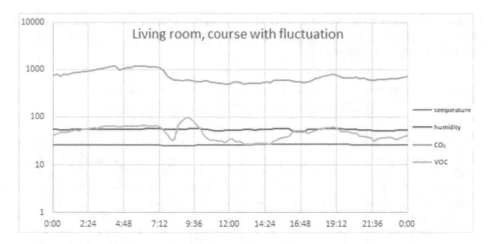

Fig. 2. Living room, occupied with residents, course of the quantities with fluctuation.

Behaviour patterns were then classified according to different variables, such as season, day of the week, holiday day, etc. Several statistical methods were used to identify typical patterns, resp. for identification of significant variables for their determination [11] including the use of multidimensional data analysis methods [12]. In particular, the following methods were used:

- Cluster analysis.
- Hierarchical methods based on a matrix of object similarity.
- K-means method, from non-hierarchical methods.

As an example, a cluster of seven courses of one month of the winter period on the working days for a bedroom is presented, see Fig. 3. From the clusters created, individual patterns were created as the mean or median of the individual values at

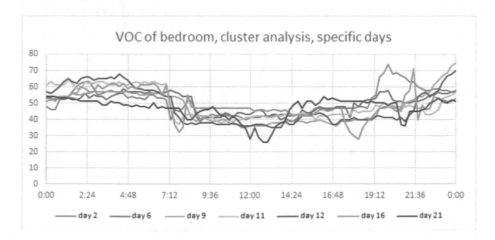

Fig. 3. VOC of a bedroom, specific courses creates a cluster, specific days.

sampling times, see Fig. 4. To determine matching the current pattern with the typical behaviour pattern was then also applied the envelope method, when the maximum and minimum values were stored.

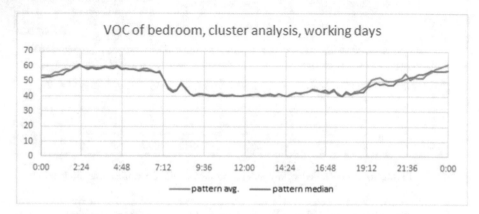

Fig. 4. VOC of a bedroom, cluster analysis, behaviour patterns, winter season, working days.

3   Creation of Behaviour Patterns Database

The most challenging task is to create a pattern base, manage it and, above all, compare current patterns with saved patterns. In particular, the interval method was applied where either an interval was created by constant deviation from the average pattern, or the envelope method of all curves was used, see Fig. 5.

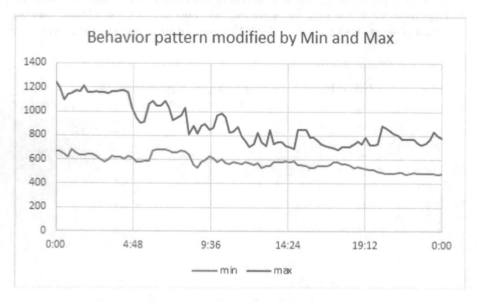

Fig. 5. Cluster envelope of courses in Fig. 4.

A course that does not go out of the value range is considered as the same. It turned out that this method is quite complicated, especially for the necessity to keep all historical patterns available for updating patterns, and especially the time-consuming comparison of good match with behaviour patterns.

Therefore, the fuzzification of values based on the expected number of persons in the room by fuzzy sets was applied:

```
{Z-zero, A-one, B-two, C-three, D-four, E-five, F-six,
H-more}
```

The behaviour pattern is then described by a sequence of values at each sampling point, e.g.:

```
ZZZZZZZACCCDDDDBBBBAAAAAAAAZZZZZZZZZZZZZ ...
```

This greatly simplifies storing the patterns into behaviour patterns database. At the same time, the ability to react easily (e.g. to a change in the properties of the sensors, causing a change in the range of values) has arisen.

Or, we can fuzzificate difference between the values of the pattern and the current course of the fuzzy sets:

```
{Z-zero, S-small, M-medium, H-huge}
```

and the actual behaviour pattern:

```
ZZZZZZZABCCCDDCBBBBBAAAAAAAZZZZZZZZZZZZZ, ...
```

can be easily evaluated as a vector of difference values:

```
ZZZZZZZSZZSZZSZZZSZZZZZZZZZZZZZZZZZZZZZ.
```

We can see that the courses differ only slightly in three values. Subsequently, using the fuzzy-expert system, we will evaluate the degree of disagreement and the resulting conclusion. For a simple analysis, we can evaluate three fuzzy variables: *NumberOfDifferences* – describing the number of deviations, *LongestDifference* – describing the length (number of deviations) of the longest continuous sequence of deviations and the *HighestDifference* – describing the magnitude of the greatest deviation.

Example of saved rule:

```
IF NumberOfDifferences IS Small AND LongestDifference IS
Short AND HighestDifference IS Small
THEN ActivateUser IS No, ChangePattern IS No;
```

In this particular case, the difference was evaluated as a small shifting of the pattern over time and was not pronounced as a suspect, nor was the need for a pattern change evaluated.

4 Conclusions

The paper presents the possibilities of data analysis of current smart buildings. On a concrete example of data of a real smart home, the behaviour patterns definition has been shown. The contribution also demonstrated the application of behaviour patterns for changes identification - i.e. either due to suspicious unusual behaviour or due to system failure or due to an object violation. The methods used to determine behaviour patterns have proven themselves and the proposed fuzzy-expert system has been able to quickly compare the pattern variations and evaluate the corresponding conclusions.

Information about the movement of people in the analysed house was not available due to data anonymization. However, laboratory tests were performed to verify the relationship between the measured values and the number of persons in the room. As a consequence, which was not originally expected, it turned out that based on the analysis of the technological data from the behaviour of the house, it is possible to determine (with great precision) the number of persons currently in the individual rooms. Unfortunately, this is a loss of privacy resulting from the use of increasingly sophisticated housing support systems.

Acknowledgment. This work was supported during the completion of a Student Grant with student participation, supported by the Czech Ministry of Education, Youth and Sports. Presented work used anonymized data provided by Brand-Tech s.r.o.

References

1. Kleka, A.: Searching for Intelligent House Behaviour Models. University of Ostrava, 94 p. Diploma thesis, supervisor: Farana, R (2017). (in Czech)
2. Filip, E.: Intelligent House Management. CTU in Prague, 78 p. Diploma thesis, supervisor: Hlinovský, M (2010). (in Czech)
3. Průcha, J.: Smart Housing. Intelligent House. Insight Home, a.s. (2012)
4. Harper, R. (ed.): Inside the Smart Home. Springer, London (2003). ISBN 1-85233-688-9
5. Valeš, M.: Intelligent House. Brno: ERA, 136 p. (2006). ISBN 80-7366-062-8. (in Czech)
6. Loxone Electronics GmbH: Loxone Smart Home. Loxone Electronics GmbH. https://www.loxone.com/cscz/. Accessed 31 Aug 2018
7. LazNet s.r.o.: Online Technology: SDS – Network Monitoring System Intelligent Domestic Systems and Energy Metering. http://www.onlinetechnology.cz/clanky/inteligentni. Accessed 31 Aug 2018
8. Haluza, M.: Classic Versus Smart Wiring. (in Czech). http://elektro.tzb-info.cz/domovni-elektroinstalace/7842-klasicka-versus-inteligentni-elektroinstalace. Accessed 31 Aug 2018
9. Jue, T. (ed.): Biomedical Applications of Biophysics, 1st edn., 237 p. Humana Press, New York (2010). ISBN 978-1-60327-232-2
10. Yao, Y., Yu, Y.: Modeling and Control in Air-Conditioning Systems, 1st edn., 479 p. Springer, Heidelberg (2017). ISBN 978-3-662-53311-6
11. Polikar, R.: Pattern recognition. In: Wiley Encyclopedia of Biomedical Engineering, Wiley (2006). http://users.rowan.edu/ ~ polikar/RESEARCH/PUBLICATIONS/wiley06.pdf. Accessed 31 Aug 2018
12. Borg, I., Lingoes, J.: Multidimensional Similarity Structure Analysis, 390 p. Springer, New York (1987). ISBN 978-1-4612-9147-3

Combined Method of Monitoring and Predicting of Hazardous Phenomena

M. V. Orda-Zhigulina[1](✉), E. V. Melnik[1], D. Ya. Ivanov[2],
A. A. Rodina[1], and D. V. Orda-Zhigulina[1]

[1] Southern Scientific Center of the Russian Academy of Sciences,
St. Chehova, 41, Rostov-on-Don 344006, Russia
jigulina@mail.ru
[2] SRI MVS SFU, Rostov Region,
St. Chekhov, 2, GSP-284, 347928 Taganrog, Russia

Abstract. The article suggested the combined method of monitoring of hazardous phenomena and predicting of hazardous processes and security of coastal infrastructure and providing of safety of human lives based on "fog computing", blockchain and Internet of Things. These modern information technologies use to connect the large number of various sensors using the new proprietary software operating system with mobile phones, processors and technology of analyzing messages of the local population in social net-works and media. These technologies could be used for decentralization of information processing and scheduling, providing system openness, reducing reaction time of the information systems and increasing reliability and scalability of the information systems. So, the suggested combined method could be used for new monitoring systems of hazardous phenomena, predicting of hazardous processes, security of coastal infrastructure and providing of safety of human lives. The main advantage of the method is using of existing information infrastructures as mobile networks and etc.

Keywords: Processing systems · Industrial internet of things ·
Distributed registry · Fog computing · Blockchain ·
Monitoring and predicting of hazardous phenomena · Hydrological ·
Meteorological · Biological monitoring

1 Introduction

Developing and research of methods of monitoring of hazardous phenomenon, predicting of hazardous processes, security of coastal infrastructure and providing of safety of human lives is actual task. It is important to increase reliability and to decrease reaction time to events of monitoring systems of hazardous phenomenon and predicting of hazardous processes in coastal zones. Obviously, current climate changings are reasons of increasing number of hazardous phenomena such hurricanes, droughts, floods, fires. Frequency of hazardous phenomena has tripled last 20 years. More than 3 billion people have been died, and more 800 billion people have been suffering [1]. Usually, hazardous phenomena lead to adverse consequences of climate change for

© Springer Nature Switzerland AG 2019
R. Silhavy (Ed.): CSOC 2019, AISC 984, pp. 55–61, 2019.
https://doi.org/10.1007/978-3-030-19807-7_6

population and infrastructure of coastal zone. Level of living is getting worse as a lack of fresh water and sanitation even in coastal zones with traditionally good infrastructure. For example, the maximal water salting and shallowing are new effects for delta of Don river [2].

2 Methods and Facilities for Monitoring and Predicting of Hazardous Phenomena

According published articles hydrological, meteorological and biological monitoring are the most popular methods of monitoring of hazardous nature phenomena. Hydrological monitoring concludes study of coastal erosion and landslide processes, geomorphological and geophysical studies, monitoring of the geological environment of the coastal shelf zone, coastal routes, leveling over the network of reference profiles (including on-line after severe storms), granulometric analysis of beach sediments and underwater coastal slope; sonar profiling on coastal shallows, echo sounding, continuous seismic acoustic profiling and etc. The main advantage of hydrological monitoring is full information about large areas of water surface. It makes sense to analyze real time data from meteorological sensors, remotely-piloted aerial vehicle, LANDSAT satellite photographs. So, the Southern Scientific Center of Russian Academy of Sciences has systems for obtaining meteorological data in real time (salinity and water level, temperature and wind speed at reference points on the coast of the Taganrog Gulf and in delta of the Don River) [4].

There are biological monitoring systems allow to analyze nature phenomena associated with bottom sampling and underwater video filming to confirm the geophysical data and to obtain material and technical data for laboratory studies.

Benthic organisms are the most conservative and suitable for biomonitoring. They can be used to assess the state of coastal ecosystems in monitoring. According to the state of the bottom communities, it is possible to estimate changing of insolubility of water. So, number of freshwater benthic invertebrates has been decreased in the Taganrog Gulf of the Sea of Azov since 2012 [5]. On the one hand, biological monitoring is inertial, it is impossible to identify the dangerous phenomenon in a real time. On the other hand, biological monitoring predicts the possibility of large hazardous events in long-term period.

The new combined approach to design of monitoring systems of hazardous phenomena and predicting of hazardous processes and security of coastal infrastructure and providing of safety of human lives connects analyzing of messages from the local population in social networks, biological and hydrological monitoring, meteorological data and information about human's health. Fog computing, Internet of Things and distributed registry are the base methods of initial data processing in the suggested combined approach. The main advantage of the combined approach is possibility to use existing information infrastructure. It is possible to connect existing regional systems of monitoring of hazardous phenomenon and predicting of hazardous processes, which are realized at the base of hydrological, meteorological and biological monitoring separately.

For instance, there is a meteorological system which contain a large number of meteorological sensors at the base of wireless technologies [6]. There are a lot of methods which are based on of combining wireless sensors for controlling water supply and transport. In this case the transport vehicles combined to the clusters for a convenience of data processing [7]. Different elements, sub-systems and methods of fresh water monitoring can be used [8–12]. It is interesting to use methods for building a system for monitoring blue-green algae [13], etc.

Figure 1 demonstrates local parameters of hydrological, meteorological and biological processes for the delta of the Don River and the Taganrog Gulf [5].

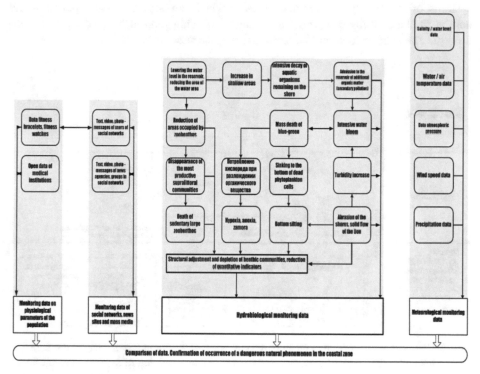

Fig. 1. The local parameters of hydrological, meteorological and biological processes for monitoring of hazardous phenomena.

The combination of Internet of Things, fog computing and blockchain to verify of initial data sensors could be used in the suggested combined approach for monitoring of hazardous phenomenon, predicting of hazardous processes, security of coastal infrastructure and providing of safety of human lives. Technology of distributed registry is popular multifunctional information technology recent days. It provides reliable recording of various data and it provides a distributed storage of records of the all system's operations, such a transactions and so on [15]. The distributed registry technology could be used to verify initial hydrological, meteorological and biological data from all members of the monitoring system of hazardous phenomenon and predicting of hazardous processes.

Some types of monitoring systems contain different sensors and controllers which are installed at the nodes of the system in different parts of industrial objects. For instance, the controllers provide transmitting and visualization of the collected data. The systems have powerful analytical tools for interpreting the information received and many other intellectual components [16, 17]. This class of systems are realized at the base of Internet of things technology. So, elements of such systems could be used for collection of initial data about real-time meteorological information and information about the social networks people's activity for the monitoring system of hazardous phenomenon and predicting of hazardous processes.

According literature and patent review it is need to combine different existing systems of monitoring one complex system which contains different types of initial data at the base of fog computing and Internet of Thigs and distributed registry technology. The suggested combined approach is presented at Fig. 2.

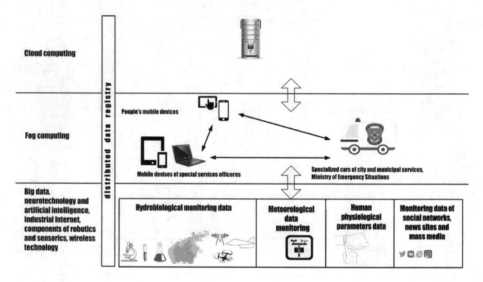

Fig. 2. The suggested combined method of monitoring of hazardous phenomenon, predicting of hazardous processes, security of coastal infrastructure and providing of safety of human lives.

The main advantage of suggested approach is possibility to learn new data from existing monitoring systems. The initial data is collecting from various sources. The sources of the initial data are satellite video and photography, weather forecasts and so on. The initial data analysis allows to determine the place and time of possible hazardous phenomena accurately. There is no verification of the proposed combined approach, which should have given more weight to the paper, so it the system is still in developing and it will be in future to improve the quality and significance of research.

Currently information and control systems are built on the network technologies of information exchange such as Ethernet, CAN, ZigBee, Wi-Fi, technologies of big data processing, such as cloud and fog computing, Internet of Thigs and so on. These information technologies are called digital economy technologies and could be used for the suggested combined method of monitoring of hazardous phenomena. There is a level of sensors for collecting information, a level of controllers for preprocessing, a level of application servers in such systems. However, the processing of the main volume of the initial data is centralized. The dispatching of the system is also centralized. On the one hand, the centralized system is simple. On the other hand, latency in data transmission is a disadvantage of the centralized system. This is low reliability of centralized system. Since the failure of the central element of the system leads to the failure of the whole system. It is impossibility of scaling centralized system. As there are restrictions of bandwidth of the systems' communication channels. It is possible to computing initial data from a limited number of sensors only. And there is a problem of connecting of different types of centralized information systems. Therefore, it is better to use decentralized systems for new monitoring systems of hazardous phenomena [18]. Fog computing could be used to reduce the load on the network and reduce the response time to events in the decentralized systems [19–28].

3 Conclusions

The authors suggested the combined method of monitoring of hazardous phenomenon and predicting of hazardous processes and security of coastal infrastructure and providing of safety of human lives. The suggested method is based on fog computing, blockchain and Internet of Things. These modern information technologies connect the various sensors using the new soft-ware systems, mobile phones, processors and processing of messages from local population in social networks. Such information technologies could be used for decentralization of information processing and scheduling, providing system openness, reducing reaction time and increasing reliability and scalability of information system. The combined method could be used for new monitoring systems of hazardous phenomenon and predicting of hazardous processes and security of coastal infrastructure and providing of safety of human lives as the base of existing information infrastructures. The publication was prepared as part of the RFBR project № 18-05-80092 and GZ SSC RAS N GR project 00-19-04 0256-2019-0031 № AAAA-A19-119011190173-6.

References

1. Moskin, K.D.: Problemy dal'neyshego razvitiya sistemy monitoringa, laborator-nogo monitoringa i prognozirovaniya chrezvychaynykh situatsiy prirodnogo i tekhnogennogo kharaktera. Problemy prognozirovaniya chrezvychaynykh situ-atsiy. XV Vserossiyskaya nauchno-prakticheskaya konferentsiya. 13-14 oktyabrya 2016 goda. Doklady i vystupleniya. - M.: FKU Tsentr "Antistikhiya" MCHS Rossii, s.13 (2016)
2. Matishov, G.G.: Nuzhny li tikhomu Donu novyye plotiny? Priroda, № 1, 2018, pp. 25–34 (2018)
3. Ryabchuk, D.V., Sergeyev, A.Y., Kovaleva, O.A., Leontyev, I.O., Zhamoyda, V.A., Kolesov, A.M.: Problemy abrazii beregov vostochnoy chasti finskogo zaliva: sostoyaniye, prognoz, rekomendatsii po beregozashchite. Uchenyye zapiski Rossiyskogo gosu-darstvennogo gidrometeorologicheskogo universiteta. № 44, pp. 187–203 (2016)

4. Data of meteorological stations SSC RAS in real time. http://archive.ssc-ras.ru/meteo/?st=kag
5. Biryukova, S.V., Bulysheva, N.I., Savinkin, A.I., Semin, V.L.: Donnyye soobshchestva taganrogskogo zaliva letom 2017 g. Materialy nauchnykh meropriyatiy, priurochennykh k 15-letiyu Yuzhnogo nauchnogo tsentra Rossiyskoy akademii nauk: Mezhdunarodnogo nauchnogo foruma "Dostizheniya akademicheskoy nauki na yuge Rossii"; mezhdunarodnoy molodezhnoy nauchnoy konferentsii "Okeanologiya v 21 veke: sovremennyye fakty, modeli, metody i sredstva" pamyati chlena-korrespondenta ran D.G. Matishova; Vserossiyskoy nauchnoy konferentsii "Akvakul'tura: mirovoy opyt i Rossiyskiye razrabotki". Rostov-na-Donu: Yuzhnyy nauchnyy tsentr RAN, pp. 299–301 (2017)
6. Finogeyev, A.A., Finogeyev, A.G., Nefedova, I.S.: Raspredelennaya obrabotka dannykh v besprovodnykh sensornykh setyakh na osnove mul'tiagentnogo pod-khoda i tumannykh vychisleniy. Trudy mezhdunarodnogo simpoziuma nadezh-nost' i kachestvo, Penzenskiy gosudarstvennyy universitet (Penza), vol. 1, pp. 258–260 (2016)
7. Aleisa, E.: Wireless sensor networks framework for water resource management that supports QoS in the Kingdom of Saudi Arabia. Procedia Comput. Sci. **19**, 232–239 (2013)
8. Pat. CN203502404 (U), China, Ocean water quality monitoring data acquisi-tion system. Inventors Zhou Zhonghai; Yuan Jian et al., applicants OCEANOGRAPHIC INSTR RES INST et al., publ. 26.03.2014. https://worldwide.espacenet.com/publicationDetails/biblio?DB=EPODOC&II=0&ND=3&adjacent=true&locale=en_EP&FT=D&date=20140326&CC=CN&NR=203502404U&KC=U
9. Pat. CN105974863 (A), China, Ocean pasture platform-based microwave ob-servation system. Inventors Li Peiliang; Chen Dong et al., applicants OCEAN UNIV CHINA et al., publ. 28.09.2016. https://worldwide.espacenet.com/publicationDetails/biblio?DB=EPODOC&II=0&ND=3&adjacent=true&locale=en_EP&FT=D&date=20160928&CC=CN&NR=105974863A&KC=A
10. Pat. CN105739345 (A), China, Marine ranching shore-based monitoring system. Inventors Li Peiliang; Chen Dong et al., applicants OCEAN UNIV CHINA et al., publ. 06.07.2016. https://worldwide.espacenet.com/publicationDetails/biblio?DB=EPODOC&II=0&ND=3&adjacent=true&locale=en_EP&FT=D&date=20160706&CC=CN&NR=105739345A&KC=A
11. Pat. 9014983, USA, Platform, systems, and methods for obtaining shore and near shore environmental data via crowdsourced sensor network / Inventor Michael G. (Solana Beach, CA), applicant Blue Tribe, Inc (Solana Beach, CA, US), publ. 21.04.2015. http://patftuspto.gov/netacgi/nph-Parser?Sect1=PTO2&Sect2=HITOFF&u=%2Fnetahtml%2FPTO%2Fsearch-adv.htm&r=5&p=1&f=G&l=50&d=PTXT&S1=9014983&OS=9014983&RS=9014983
12. Pat. 9223058, USA, Platform, systems, and methods for utilizing crowdsourced sensor networks to generate environmental data reports / Inventor Michael G. (Solana Beach, CA), applicant Blue Tribe, Inc. (Solana Beach, CA, US), publ. 29.12.2015 http://patft.uspto.gov/netacgi/nph-Parser?Sect1=PTO2&Sect2=HITOFF&u=%2Fnetahtml%2FPTO%2Fsearch-adv.htm&r=1&p=1&f=G&l=50&d=PTXT&S1=9223058.PN.&OS=PN/9223058&RS=PN/9223058
13. Pat. 9983115, USA System and method for monitoring particles in a fluid using ratiometric cytometry. Inventors Sieracki Ch. K. (Edgecomb, ME), Wolfe P. (Falmouth, ME) et al., applicant Fluid Imaging Technologies, Inc. (Scarborough, ME, US), publ. 29.05.2018. http://patft.uspto.gov/netacgi/nph-Parser?Sect1=PTO2&Sect2=HITOFF&u=%2Fnetahtml%2FPTO%2Fsearch-adv.htm&r=1&p=1&f=G&l=50&d=PTXT&S1=9983115.PN.&OS=PN/9983115&RS=PN/9983115
14. Programma Pravitel'stva RF ot 28.07.2017, № 1632 (2017)

15. Pryanikov, M.M., Chugunov, A.V.: Blokcheyn kak kommunikatsionnaya osnova formirovaniya tsifrovoy ekonomiki: preimushchestva i problemy. Int. J. open Inf. Technol. **5**(6) (2017)
16. Sadiku, M.N.O., et al.: Industrial internet of things. IJASRE **3** (2017)
17. Jeschke, S., et al.: Industrial Internet of Things, pp. 3–19. Springer, Cham (2017)
18. Kaliaev, I.A., Melnik, E.V.: Detsentralizovannyye sistemy komp'yuternogo upravleniya: monografiya, E.V. Mel'nik. – Rostov n/D: Izd-vo YUNTS RAN, p. 196 (2011)
19. Kalyayev, A.I., Kalyayev, I.A.: Metod detsentralizovannogo upravleniya raspredelennoy sistemoy pri vypolnenii potoka zadaniy. Mekhatronika, avtomatizatsiya, upravleniye **16**(9), 585–598 (2015)
20. Yi, S., et al.: Fog computing: platform and applications. In: Proceedings - 3rd Workshop on Hot Topics in Web Systems and Technologies, HotWeb 2015, pp. 73–78 (2016)
21. Stojmenovic, I., Wen, S.: The fog computing paradigm: scenarios and security issues. In: Proceedings of 2014 Federated Conference on Computer Science and Information Systems, vol. 2, pp. 1–8 (2014)
22. Hosseinpour, F., Westerlund, T., Meng, Y.: A review on fog computing systems. Int. J. Adv. Comput. Technol. **8**(5), 48–61 (2016). HannuTenhunen
23. Tekhnologii raspredelennogo reyestra dannykh. Walport, M. (2015)
24. Kuo, T.T., Kim, H.E.: Distributed ledger technology: beyond block chain. Government Office for Science, Ohno-Machado, L. pp. 1–88 (2017)
25. Wu, H., Li, Z., King, B., Miled, Z., Wassick, J.: Blockchain distributed ledger technologies for biomedical and health care applications. J. Am. Med. Inf. Assoc. Tazelaar, J. (2017)
26. Bonomi, F., et al.: Fog computing and its role in the Internet of Things. In: Proceedings of the First Edition of the MCC Workshop on Mobile Cloud Computing, pp. 13–16 (2012)
27. Atzori, L., Iera, A., Morabito, G.: The internet of things: a survey. Comput. Networks. **54** (15), 2787–2805 (2010)
28. Kaliaev, I.A., Gayduk, A.R., Kapustyan, S.G.: Samoorganizatsiya v mul'tiagentnykh sistemakh. Izvestiya Yuzhnogo federal'nogo universiteta. Tekhnicheskiye nauki. Tom № 3 (104), pp. 14–20 (2010)

Analysis of Existing Concepts of Optimization of ETL-Processes

Sarah Myriam Lydia Hahn[✉]

Computers and Information Technology, Technical University Cluj Napoca,
Cluj-Napoca, Romania
sarah-hahn@gmx.net

Abstract. Extract-Transform-Load (ETL) describes the process of loading data from a source to a destination. The source and the destination can be separated physically and transformations may take place in between. Data preparation happens regularly. To minimize interference with other business processes and to guarantee a high data availability these processes are often run during night times. Therefore the demand for shorter processing times of ETL-processes is increasing steadily. Besides data availability and actuality another reason is the transition to real- or near-time analysis of data and the growing data volume. There are several approaches for the optimization of ETL-processes which will be highlighted in detail in this article. A closer look will be taken on the advantages and disadvantages of the presented approaches. Concluding each approach will be set into competition and a recommendation depending on the use case is given.

Keywords: ETL process · Optimization · Business intelligence · State-space · Fault tolerance · Sorters · Shared cache · Parallelization · Cost based

1 Introduction

Day after day more and more data is generated. The data volume which is generated each day is increasing exponentially. According to Gantz and Reinsel [2], it is estimated that by 2020 the total data volume will climb to 35 zettabytes. Data are characters which can be interpreted by a machine. For the transmission of information, the information itself is represent as data. Information is created at any point in time, for example, by surfing the internet, sensors detecting status notices on machines, purchasing products in a store or during a check-up in a hospital stay. All of this information is collected and saved as data on disk storages. If the data is not deleted it can be accessed and information can be generated out of it. Data should only be collected if it will be used later on. Otherwise disk storage has to be paid without any further benefits. In general, the cost of disk storage is decreasing gradually and is also depended on the type of disk used. The decreasing cost of storage entails that in reality more data is

© Springer Nature Switzerland AG 2019
R. Silhavy (Ed.): CSOC 2019, AISC 984, pp. 62–76, 2019.
https://doi.org/10.1007/978-3-030-19807-7_7

saved for any use case and not deleted. Besides existing use cases more and more are emerging. Examples for such use cases are the following:

- The analysis of clinical data, saved on electronic health care records, radiation oncology information systems, treatment planning systems and spreadsheets [9]
- Automatically displaying customized content on a website based on the current context of the user [10]
- Optimization of the supply chain trough the analysis of the different chain parts [18]

The biggest challenge is to bring data from different sources together and to create a holistic, consolidated view. For building such a holistic view ETL-processes are used. ETL stands for extract, transform and load. According to Mehra, et al. [6], ETL describes a process for loading data with the following steps along the process.

- **Extract:** First the necessary data is extracted from different data sources into a working area. A data source could be a database, a flat file, an interface or any other source. The data itself is not adapted in any way. The data is only made accessible.
- **Transform:** In the working area the data is transformed to make it usable in the target database. On the one hand the data is adapted formally. This means that the syntax of the data corresponds to the syntax of the target database and that it is equal across different data sources. This step is necessary to combine the data from different sources. For example, the customer number across several source systems could be totally different. In the web shop it is saved as a text value, in the customer relationship management (CRM) system it is saved as a number. To unify the data it is much easier to use the same data type across all sources. The transformation would be a conversion of the customer number of each system to align them. On the other hand, data is modified regarding the content. Examples are translations of shortcuts, calculations of key performance indicators or the removal of duplicate entries.
- **Load:** In the last step the transformed data is loaded into the target database. This target database is the datawarehouse (DWH).

There is not only a trend in collecting more and more data but also to get more impacts out of it in real-time. To get closer to near- or even real-time analysis the time which is necessary to process the incoming data has to be optimized. According to Karagiannis et al. [4], to satisfy the requester of the data the processing times should be continuously reduced. Another reason for reducing the time of data processing is that usually there is only a defined time window available for the completion of an ETL-process. Outside of this time slot the data must be accessible for the applications. With more and more data the process itself will take longer and longer [14].

There are many different approaches to optimize an ETL-process – from using metadata to the development of a heuristic algorithm.

In the following Sect. 2 a definition of an ETL-process will be given including all necessary wordings to understand the presented concepts. In Sect. 3 the different existing approaches will be analyzed in detail regarding the method as well as the advantages and disadvantages. It will also be examined whether the approach can be used for all ETL-processes or not. Limitations can be the number of data sources or targets. In Sect. 4 the possibility of a combination of different approaches is considered. Last but not least, a recommendation will be presented.

2 Definition ETL-Process

In the following an ETL-process will be introduced as a graph and the necessary formulas will be provided. Furthermore the details of such a process will be shown and the activities defined. Definitions which are only necessary for specific concepts are introduced with the concepts itself.

ETL-processes can be displayed as directed acyclic graphs [15]. Such a graph is a sequence of several activities as loading and writing data and several other transformations. A graph consists of vertices and edges. An example of a directed acyclic graph can be seen in Fig. 1. A vertex (V) in the context of ETL-processes could be a data source, a transformation or the data target. In the example V_4 could be a union whereas V_7 is a data target. It consists of activities (A) and recordsets (RS). An activity has at least one input and one output schema, whereas a recordset has only one schema. Recordsets can be divided into source and target recordsets. Examples for recordsets are a relational table or a flat file. An edge (E) describes the relationship between two vertices. The formula of a graph (G) is the following:

$$G(V,E) \tag{1}$$

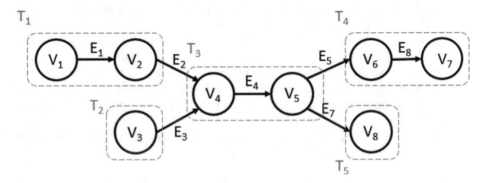

Fig. 1. Directed acyclic graph

A graph G consists of vertexes and edges. Such a graph can be divided into several execution trees (T). An execution tree is also a directed acyclic graph with the following definition:

$$T(V',E') \tag{2}$$

An execution tree consists as a whole graph G of vertexes and edges, not the whole graph but only a subset of it. A vertex with no indegree is the root of an execution tree. A vertex with no outdegree is the leaf. A graph or an execution tree is defined as a linear flow when every activity has exact one input and one output [15]. In the example in Fig. 1 T_1 is a linear flow whereas the whole graph is a complex flow.

Activities can be defined on a logical or on a physical-level. The logical level describes an activity with its function as well as the in- and output schema. The physical level is more detailed and defines how the function is implemented. The definition of the implementation contains the used algorithm and the code, the costs either by time or resources, input preconditions as well as extra constraints [17]. A logical defined activity consists of at least one physical defined activity. There can be different physical activities defined for the implementation of a logical activity. This causes a variation in the costs depending on the implementation. In most ETL-tools the design of an ETL-process is on the logical level. Further activities can be classified into different types of components [7]:

- **Row-synchronized** components are activities which can be processed for each row, like a filter. A comparison to other rows is not necessary.
- **Block** components are activities with only one input. They can be only executes when all the data rows are loaded because a comparison is necessary. An example is an aggregation.
- **Semi-block** components are activities with more than one input. A comparison is only possible when all data rows from all input sources are loaded into the cache. A union is such an activity.

3 Existing Concepts of Optimization of ETL-Processes

The process of designing and developing an ETL process is defined in seven steps. First the business requirements have to be specified. These requirements have to be put down in a concept. In the following the concept model is transferred into a logical model. Last but not least the code for the logical model has to be developed.

There are several possibilities to optimize an ETL process – in the conception, the logical model or the code. To measure whether an optimization has an impact or not, quality objectives have to be defined beforehand [1]. It is common that an ETL workflow is first designed and developed without any additional optimization. The optimization is made afterwards if the quality objectives are not fulfilled. For the measurement there are several criteria which should be mentioned – freshness, data consistency, resilience to failures, maintainability, speed of the overall process, partitioning, pipelining and overheads. For each criteria different indicators can be used [11]. Regarding to Smistsis et al. the criteria performance, recoverability, reliability, freshness, maintainability, scalability, availability, flexibility, robustness, affordability, consistency, traceability, and auditability should be mentioned. They also included them into their QoX metric suite [13].

Hereinafter existing concepts for the optimization of ETL-processes will be presented. Most of the following optimizations are optimizations on the logical model. The concepts are not dependent on a special code or programming language. To get a closer look into the concepts please refer to the original papers listed in the references. Advantages and disadvantages of the concepts will be highlighted during the analysis of each concept. For a comparison of the different methods the measuring results of each concept is used.

3.1 State-Space Optimization

Simitsis et al. and Tziovara et al. [14, 17] describe an ETL-process as a state space problem which has to be solved. Therefore, each workflow is a state S. In this case a workflow is comparable to a graph as described in Sect. 2. With a transition T a new state which is equivalent to the original one by having the same in- and output can be created. Such a new state S' can be defined as

$$S' = T(S) \tag{3}$$

With this formulation it is possible to depict the whole process with all steps. In an ETL tool there are unknown functions possible as self-written functions. It is not possible for such a tool to optimize the process holistically because of black box functionalities [16].

A transition can be one of the following three:

- **Swap:** Activities having only one input schema can be changed in their execution order.
- **Factorize and Distribute:** This transition is only possible with two activities having the same functionality in the first step followed by an activity with two input schemas. In this case the execution order of the two steps can be changed. As a result both activities from the first step are consolidated in the new activity which is now positioned at the second step. It is a so called factorization. The other way around is also possible and called distribution. Such a transition connects linear flows.
- **Merge and Split:** Merge describes the grouping of two activities into one. The functionality is still the same but now it is not separated into two activities. The other way around is called Split.

The objective is to find the optimal state. The optimum has to be measurable with a discrimination criteria. Common criteria are costs as time used or used hardware resources. The sum of the costs of all activities should be as minimal as possible. Therefore the costs of all possibilities have to be calculated. Simitsis et al. developed an exhaustive search as well as a heuristic search [14, 15]. The exhaustive search generates all possible states and calculates the costs for each of them. The heuristic search tries to generate only states which could be considered to be in the area of the optimum of the search space.

For both algorithms the pseudo code can be found in their concept. They also tested the algorithm by implementing them in C++ and tried it out with several ETL-processes. As a result the exhaustive search has to be stopped for medium and large workflows because it needed too much time. Also for small workflows the algorithm needed too much time regarding the heuristic search. The result of both algorithms were identical for small workflows. The algorithms are applicable for both – linear and complex workflows. Because the exhaustive search has no advantage compared to the heuristic search it is not recommended. In average the improvement of the heuristic search is over 70% compared to the initial ETL-processes and the time needed is less than one hour in the tests made.

For using the algorithm some manual work has to be done during the design of an ETL-process. It must be possible to know which activities can be reordered without changing the result of the whole process. To know the content of each column and if it is identical a mapping of all column names has to be made. This means that with the help of a mapping the column names of columns with the same content are identical. For example if there are different columns with the column name "profit" and "profit_EUR" they should be both mapped to "profit_EUR" if the content of both is the same. Only with an identical column name in the mapping it is clear, that they are can be swapped. Furthermore the input and output schema of all activities has to be defined. The collection of this metadata is very laborious and even more complex if it has to be made additionally. The evaluate whether it is worth it or not the possible optimization potential has the be set in comparison to the efforts. Which means that it has to be evaluated from case to case.

The concept of Simitsis et al. was used as basis from Kumar and Kumar [5] to develop their own efficient heuristic for optimizing an ETL-process. Instead of defining it as state-space problem their algorithm is predicated as a dependency graph.

Figure 2 shows a dependency graph of the execution trees T_3, T_4 and T_5 from Fig. 1. In a dependency graph the dependencies between the activities are visible. The graphic representation is a tree. It can be seen which activities are swappable. In the example V_6, V_7 and V_8 could be switched provided these are row-synchronized components. The dependency graph has to be created in advance. The developed algorithm consists of two parts. In the first part activities with high costs are moved to the end so that there is less amount of data in the input. Return activities, which filter a lot of data, are swapped to the root. This change in the execution order is only possible in linear flows. The second part takes a closer look on changing the execution order of activities between linear flows. It has to be proofed if an activity is transferable or not. An example would be a filter after a union which could be pushed backwards. The filter then has to be in both linear flows which are the input for the union. First this is made for all activities with no dependencies. It can be divided into forward and backward passes. A forward pass means an activity is changed into a following workflow, a backward pass is the opposite. When the best execution order regarding the costs for these activities is found the inter-dependent transferable activities follow. This step is separated because the violation of the dependencies is also checked.

For the tests the algorithm was implemented in Java but it is not implemented into an ETL-tool, yet. By testing the algorithm the optimization degree is around 70%. In a direct comparison with the exhaustive and the heuristic search the algorithm wins by a few percentage points. The time needed for the optimization is within seconds. This is much faster than the heuristic search. Because of these results the algorithm from Nitin Kumar and P. Sreenivasa Kumar is recommended in comparison to Simitsis et al.

3.2   Fault Tolerance

Simitsis, et al. [12] expanded the concept of the optimization of an existing ETL-process as described in Sect. 3.1 by fault-tolerance. In this case faults are defined failures in a network grid, power failures, human errors and resource limitations. Failures caused by wrong data or ETL design are not mentioned. The most common

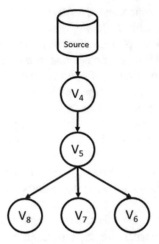

Fig. 2. Dependency graph

way in case of an error is to restart the process. A restart means that the execution time until the failure occurred is added on top to the normal process time. In relation to have the data faster available recovery points in the workflow can be set. A recovery point can be set between activities in the workflow at any position. In a workflow with n activities there are $n − 1$ possible positions for placing a recovery point. In total there are:

$$n * log\, n \tag{4}$$

possible states for a workflow. This means by doubling n the result will be a little bit more than the double of the result before. To find the best workflow an algorithm was developed.

The overall objective by adding these recovery points is functional correctness, performance, freshness and reliability. Functional correctness means that the result after the optimization has to be identical to the result before. The ETL-process has to be designed and implemented correctly before. Performance can be measured by the two indicators time and used resources. Freshness implies how fast the data from the sources are available. The higher the freshness should be the higher has to be the execution interval of the workflow. Higher intervals mean smaller batches and so the time of a workflow could be less. Reliability means that the process is finished successfully even when a failure occurs in the execution. Therefore repetition, recovery points or redundancy are helpful. Repetition is just a new start of the workflow. A recovery point can be synchronous or asynchronous. A synchronous recovery point makes a consistent snapshot at a given time. The workflow cannot be executed further by making the snapshot. Asynchronous recovery points only log the state of the data object. The workflow is not blocked, but a restart from the recovery point is more expensive. A replication has at least two identical instances. In the case of a failure the primary one can be switched with a back-up.

There are several types of costs which come along with the different actions of the optimization. The first one is processing costs. Each activity has operational costs depending on the data volume and fixed costs. By partitioning a process there are costs for splitting the flow into several partitions and merge it again at the end. Adding a recovery point is also an expense because it is another activity. A replication has an expense because the several flows have to communicate. The higher the freshness should be the more often the process has to be executed. With each execution the total costs of the whole workflow can be calculated. To find the perfect solution the optimization problem is set to a state-space optimization. For this use case the transition from Sect. 3.1 are supplemented with the following ones:

- **Partition:** By adding a partition this transition adds a split before and afterwards a merge in the graph.
- **Add recovery point:** The transition adds a new end point of the flow with the record set which should be used for the restart.
- **Replicate:** Before the replication the recordsets have to be copied and the flows have to be merged afterwards into one with a replicator.

The optimization problem has been solved with heuristic search. The algorithm contains the following steps:

- **Operator push-down:** Activities which are selective like a filter are pushed backwards to the root element. The following activities have less data as an input.
- **Operator grouping:** Row-synchronized activities are placed together. This is also important for parallelism.
- **Pipeline parallelism:** Execution trees are executed in parallel.
- **Partition parallelism:** The data is split into several partitions, which are executed in parallel.
- **Split-point placement:** If a blocking activity is able to work in parallel then a split activity is inserted before.
- **Merge-point placement:** A merge activity follows after the end of a parallel execution.
- **Blocking recovery point:** Recovery points after block or long time activities are inserted.
- **Phase recovery point:** Recovery points after the end of each phase in the ETL-process are inserted.
- **Recovery feasibility:** Recovery points can be inserted at other positions than the mentioned ones.
- **Reliable freshness:** If the objective is a high freshness replication it is better than recovery points to shorten the execution time. The insertion of recovery points takes additional time in the execution.

All in all in this algorithm not only the workflow itself is optimized but also a set of activities in case of a failure is inserted into the flow. If recovery points or a replication is useful for the ETL-process then this algorithm can be used. If only the existing workflow should be optimized it is suggested to use the methods from Sect. 3.1. The algorithm is implemented for Kettle and coded in C++. For other ETL-tools it has to be adapted. In their concept the pseudo code is available. If Kettle is already in use there is no development effort necessary. The suggested solution from the algorithm is almost as good as the best solution generated by an exhaustive search.

3.3   Addition of Sorters

Tziovara et al. [17] also uses the definition of state-space problem for adding sorts in an existing ETL-process to get an optimization. A sorter rearranges the order of recordsets. It does not have an impact on the functionality and the result of an ETL-process. A sort activity goes along with costs and depends on the amount of tuples n which have to be sorted. The costs are exponential regarding to the tuples and can be represented with the same equation than the possible states of a workflow in (4). On the other hand it has to be considered that the costs of other activities can decrease when the data is already sorted. For example a join between two record sets has lower costs if both record sets are sorted by the join column.

The objective is to place a sort activity at the correct position to decrease the costs of the whole ETL-process. A sorter can be inserted at source or staging tables or on edges which have activities before and afterwards. Depending on the following steps the attributes which should be sorted can be defined. As mentioned below the join columns of a join activity or the sort order of a view could be used as sort attributes. All attributes used by the different elements should be integrated into the sort activity. This is only possible if there are no concurrent requests. If, for example, two views are based on a product table - one ordered by price and another one ordered by article name – a sort column only makes sense on both column or maybe is not useful at all. Another definition is the sort order – ascending or descending. Vasiliki Tziovara et al. developed an exhaustive ordering algorithm which first gets all possible states from G based on a logical definition. This includes all possible positions for adding a sorter. Furthermore it calculates the costs for each state and returns the optimal state.

In a test cost optimization the benefit of the sorters was about 70%. It is not mentioned how the algorithm was implemented. Because the algorithm is an exhaustive one the limitation for the execution of it is the possible positions of sort activities.

3.4   Shared Caching and Parallelization

In a traditional ETL-process each activity has its own cache for the in- and output. The output of an activity in a cache has to be copied to the cache of the next activity. Otherwise the following activity does not have the necessary input available. Copying needs time and resources which can be saved. Liu and Iftikhar [7, 8] did an optimization by implementing a framework for shared cache and parallelization during execution. Their concept consists of the following three steps.

First execution trees are identified which utilize one cache. Row-synchronized components following each other can be summarized to one execution tree. Therefore Xiufeng Liu and Nadeem Iftikhar developed a recursive algorithm which starts from the root of the graph and always takes a look at the following activity. If it is a row-synchronized one it is added to the execution tree. If not, the execution tree is finished with the activity before and a new one starts. By using only one cache per execution tree there is no copying of data necessary in between an execution tree. The execution trees found are linear flows.

The next step is to parallelize the execution in between an execution tree. In an execution tree all activities can be executed per row. it is possible to parallelize the execution of the following activities for data rows. The maximal parallelization is the number of rows in the input recordset. The minimal parallelization is one. In the framework there is also an algorithm which detects the optimum degree of the parallelization. The optimum is the number of parallelization pipelines where the time for executing the whole execution tree is the fastest.

Last but not least the execution within one activity can also be parallelized. The execution is divided into threads. By finishing all threads the output is merged together by a row order synchronizer. With the row order synchronizer it is guaranteed that the order of the data rows of the input and the output are the same. This is necessary if for example a merge activity follows.

It is estimated that with this framework an ETL-process can be up to 4.7 times faster than before. In this approach the logic of the process is not adapted, only the resources are used more efficiently for saving time. The framework has been implemented in the open-source ETL-tools Talend and Kettle. It has not been implemented for other ETL-tools yet. The possibility of an implementation for other tools has to be evaluated before.

3.5 Cost Based Optimization

Another approach is a cost based optimization of an ETL-process as mentioned from Halasipuram et al. [3]. They developed a framework with seven steps:

1. **Identifying optimizable blocks:** The ETL-process is divided into several blocks. The further steps of the framework will have a closer look on the several blocks. There are different conditions to identify such blocks. When results are materialized afterwards a new block begins. This is necessary because otherwise it is not secured, that the results are the same. Another block boundary are transformation operators. These operators transform data and often custom code is used. Custom code which is not known to the optimizer cannot be handled. The optimizer can also not handle user defined functions as well as other custom operators.
2. **Identifying all execution trees:** In between a block all possible execution trees are identified. An execution tree is a valid one if the result is the same than before. To reduce the search space not all possible execution trees are identified by the optimizer but only the realistic ones.
3. **Identifying candidate statistic sets for each execution tree:** For each possible execution tree candidate statistic sets (CSS) is generated. These CSSs are generated by the optimizer. The pseudo code for the generation is available in their concept.
4. **Determining minimal set of statistics:** In this step the CSS for each execution tree with the minimal costs is chosen. The optimal CSS is depending on the cost metric used. For example this could be CPU costs or memory availability. Therefore the pseudo code is available, too.

5. **Instrument plan to gather statistics:** For getting the statistics they have to be implemented in the ETL-process. For the cardinality a counter can be set in each data flow which counts the number of passing tuples. For the distributions a histogram can be created from each tuple passing by.
6. **Run instrumented plan and gather statistics:** After the implementation of the logging of the statistics in step 5 the ETL-process is executed and the statistics from the execution is saved.
7. **Optimization based on statistics:** With the statistics from the execution the costs can be calculated exactly. With any cost based optimization algorithm the optimal workflow can be identified. This step is not part of the concept and has to be developed.

The algorithm is not implemented in an ETL-tool. It needs the XML-files from IBM InfoSphere DataStage. Because XML is a common file format for ETL-processes it can be easily adapted for XML files from other ETL-tools. It is depending on the workflow how many iterations are necessary for the optimal solution. In their tests most of them needed less than 6 iterations. It is not said how high the degree of optimization was.

4   Conclusion

Table 1 opposes the concepts to each other. Therefore few criteria have been chosen - for which kind of workflow each concept can be used, the development effort which has to be made until the method is usable and the optimization potential. A closer look on the compatibility between the concepts and a suggestion will also be taken.

All concepts can be disposed on all kind of workflows – linear and complex ones. There are parts of the methods used which work only on linear execution trees. If this is the case the workflows are divided beforehand into execution trees which are manageable for the algorithm. This is also applied for other restrictions. Because there is no difference between the different concepts this criteria has no impact on the suggestion.

The next criteria is the effort which has to be made to implement algorithm. If the effort is very high it is wise to choose another one with less effort to start with the optimization. This method will save resources. In all concepts there was the pseudo code of the algorithm available. With this pseudo code the algorithm can be implemented in any programming language like C++ or Java. Nevertheless the implementation will take time because development efforts are necessary and the pseudo code cannot be translated directly into a programming language. Some of the concepts are already developed in a certain programming language. It can be tried to get the code from the authors but it is nothing said about whether it is available or not. If it is, then the code maybe has to be translated into another language if necessary. The algorithms are also specified for certain tools. It also depends on the use case if an existing developed algorithm is an advantage. The development effort has no impact on the suggestion.

Table 1. Comparison

	State-space optimization – exhaustive search (1) [14, 15]	State-space optimization – heuristic search (2) [14, 15]	Efficient heuristic (3) [5]	Fault tolerance (4) [12]
Kind of workflows	No restrictions	applicable	applicable	applicable
Development effort	Pseudo code available, implemented in C++	Pseudo code available, implemented in C++	Pseudo code available, implemented in Java	Pseudo code available, implemented in Kettle (Java)
Compatibility	Not combinable with other algorithms for solving a optimization problem [(2), (3), (4), (5), (7)]	Not combinable with other algorithms for solving a optimization problem [(1), (3), (4), (5), (7)]	Not combinable with other algorithms for solving a optimization problem [(1), (2), (4), (5), (7)]	Not combinable with other algorithms for solving a optimization problem [(1), (2), (3), (5), (7)]
Optimization potential	>70%	>70%	>70%	>70%
Advantage	Optimum is found	Same results as (1)	Faster than (2)	Implementation of fault tolerance, implemented in Kettle
Disadvantage	Long duration time, only applicable for small workflows, manual work necessary, not implemented in an ETL-tool	Manual work necessary, not implemented in an ETL-tool	Dependency graph necessary, not implemented in an ETL-tool	Fault tolerance may not be necessary
Suggestion	Using (2) instead	Cost-value ratio has to be done before (regarding on the workflow), metadata should be maintained by designing the workflow	Should be used instead of (2)	Should only be used if fault tolerance is necessary or Kettle as ETL-tool is in usage

(*continued*)

Table 1. (*continued*)

	Addition of sorters (5) [17]	Shared caching and parallelization (6) [7, 8]	Cost based optimization (7) [3]
Kind of workflows	applicable	applicable	applicable
Development effort	Pseudo code available	Pseudo code available, implemented in Kettle and Talend	Pseudo code available, cost based optimization has to be developed
Compatibility	Not combinable with other algorithms for solving a optimization problem [(1), (2), (3), (4), (7)]	Executable after (1), (2), (3), (4), (5) or (7)	Not combinable with other algorithms for solving a optimization problem [(1), (2), (3), (4), (5)]
Optimization potential	>70%	4.7 times faster	
Advantage	Useful for ETL-processes and applications	Optimization on implementation level, not on logical order, implemented in Kettle and Talend	Existing cost based optimization can be used
Disadvantage	Long duration time, only applicable for small workflows, not implemented in an ETL-tool		Optimization potential is not known, not implemented in an ETL-tool
Suggestion	Should only be used if there are other applications too	Should be used after an optimization on logical order	Optimization potential has to be tested regarding to the other concepts. Not usable with less time left

With the criteria compatibility it is highlighted whether the different concepts can be combined to get a more optimized result. All concepts except the shared caching and parallelization [7, 8] are solutions for an optimization problem. For getting the solution of an optimization problem only one algorithm can be used. Such an algorithm can be an existing one or a newly created one. It is also possible to combine existing ones or adapt them. But it is not wise to execute different ones successively. If this has a better output then the two algorithms should be merged to a new one. This is the reason the algorithms for solving an optimization problem can only be combined with shared caching and parallelization. Shared caching and parallelization optimizes an ETL-process on the hardware level. It can be used for all ETL-processes. Therefore it is combinable with all other presented concepts. This leads to the suggestion, that the concept of shared caching and parallelization should always be applied after an optimization in logical order.

The optimization potential is for all algorithms over 70% and almost identical. Only for the cost based optimization there is no optimization indication. It has to be kept in mind that for the different concepts different test cases were used. The optimization potential is not directly comparable. If we presume that it is comparable there is no advantage for one of the optimization algorithms so that the optimization level has no impact on the suggestion.

The advantage of an exhaustive search is that the optimum is found because all possibilities are tested. The testing of all possibilities implicates a long duration. The duration time needed is not to be underestimated - therefore the algorithm is not suggested. If only a solution for an optimization problem should be found the efficient heuristic should be used because it is, in comparison to (1) and (2) the fastest one and no preparatory work is necessary. The cost based optimization is not suggested because it is not implemented yet and the optimization potential is not known. Methods (4) and (5) come along with additional functions and can only be recommended if these functions are needed.

The execution of such an optimization as well as the implementation is very time-consuming it should only be done if there is a high optimization potential and need. In case of the creation of a new ETL-process it is useful if the designer has the performance of the load in mind. This is less expensive than an adaption afterwards. Furthermore with an exhaustive search the optimum of the execution order of an ETL-process is found but the execution of the algorithm is very time-consuming. It is the objective to find the optimum in less time with other algorithms. There are also other methods for solving an optimization problem than the presented ones. An example for such an algorithm is a genetic algorithm. Because they were not analyzed yet these are possible topics for forthcoming research.

References

1. Castellanos, M.G., et al.: Quality-driven ETL design optimization, U.S. Patent No 8 (2014)
2. Gantz, J., Reinsel, D.: The 2011 Digital Universe Study: Extracting Value from Chaos, IDC IView (2011)
3. Halasipuram, R., Deshpande, P.M., Padmanabhan, S.: Determining essential statistics for cost based optimization of an ETL workflow. In: EDBT, pp. 307–318 (2014)
4. Karagiannis, A., Vassiliadis, P., Simitsis, A.: Scheduling strategies for efficient ETL execution. Inf. Syst. **38**(6), 927–945 (2013)
5. Kumar, N., Kumar, P.S.: An efficient heuristic for logical optimization of ETL workflows. In: International Workshop on Business Intelligence for the Real-Time Enterprise, pp. 68–83. Springer, Heidelberg (2010)
6. Mehra, K.K., et al.: Extract, transform and load (ETL) system and method, U.S. Patent No. 9 (2017)
7. Liu, X., Iftikhar, N.: An ETL optimization framework using partitioning and parallelization. In: Proceedings of the 30th Annual ACM Symposium on Applied Computing, pp. 1015–1022 (2015)
8. Liu, X., Iftikhar, N.: Optimizing ETL dataflow using shared caching and parallelization methods, arXiv preprint arXiv:1409.1639 (2014)

9. Mayo, C., et al.: Taming big data: implementation of a clinical use-case driven architecture. Int. J. Radiat. Oncol. Biol. Phys. **96**, E417–E418 (2016)
10. Orenga-Roglá, S., Chalmeta, R.: Social customer relationship management: taking advantage of Web 2.0 and Big Data technologies, SpringerPlus (2016)
11. Simitsis, A., et al.: Benchmarking ETL workflows. In: Technology Conference on Performance Evaluation and Benchmarking, pp. 199–220, Springer, Heidelberg (2009)
12. Simitsis, A., et al.: Optimizing ETL workflows for fault-tolerance. In: IEEE 26th International Conference on Data Engineering, pp. 385–396 (2010)
13. Simitsis, A., et al.: QoX-driven ETL design: reducing the cost of ETL consulting engagements. In: Proceedings of the 2009 ACM SIGMOD International Conference on Management of data, pp. 953–960 (2009)
14. Simitsis, A., Vassiliadis, P., Sellis, T.: Optimizing ETL processes in data warehouses. In: Data Engineering, pp. 564–575 (2005)
15. Simitsis, A., Vassiliadis, P., Sellis, T.: State-space optimization of ETL workflows. IEEE Trans. Knowl. Data Eng. **17**(10), 1404–1419 (2005)
16. Tziovara, V., Simitsis, A.: ETL workflows: from formal specification to optimization. In: East European Conference on Advances in Databases and Information Systems, pp. 1–11. Springer, Heidelberg (2007)
17. Tziovara, V., Vassiliadis, P., Simitsis, A.: Deciding the physical implementation of ETL workflows. In: Proceedings of the ACM tenth international workshop on Data warehousing and OLAP, pp. 49–56 (2007)
18. Wang, G., et al.: Big data analytics in logistics and supply chain management: Certain investigations for research and applications. Int. J. Prod. Econ. **176**, 98–110 (2016)

Multiagent Distribution of Roles in Communities with Limited Communications

Donat Ivanov[1]($\boxtimes$) ⓘ and Eduard Melnik[2] ⓘ

[1] Southern Federal University, 2 Chehova St., 3479328 Taganrog, Russia
`donat.ivanov@gmail.com`
[2] Southern Scientific Center of the Russian Academy of Sciences,
Rostov-on-Don, Russia

Abstract. The paper is aimed at the problem of the distribution of roles in communities of agents with limited communications and based on the principles of the swarm intelligence. The iteration approach to the distribution of roles in a group of agents is proposed. There are results of the study and computer simulation of the proposed approach.

Keywords: Swarm intelligence · Distribution of tasks · Distribution of roles · Multi-agent technologies

1 Introduction

The rapid development of computer science, communication and information technology led to the emergence of new design concepts, such as agent-based design [1–3], decentralized decision making [4–6], new concepts for distributed computing [7–9]. It allows to create the necessary tools, including for modeling complex systems. The concept of agent-based design has found wide application in systems related to modeling.

The study of the behavior of communities of various living organisms and artificial systems has a particular interest. The study of the interaction of ants, bees, bacteria led to the development of methods and algorithms for optimization [10, 11], which have found their application in various technical systems [12–15].

Drawing a parallel between living organisms and decentralized technical systems, we can consider several multi-agent systems. An agent is a software module that represents the interests of a certain actor: models of a living organism, vehicle, robot, etc.

It is necessary to distribute some duties or roles in the group in such a way as to provide a specified percentage of agents with each type of roles in the group for solving certain tasks by groups of agents, often with limited communications between the agents within the group. Thus, some practical applications needs a method that allows to get distribution (set in percentage) of a set of roles in a group of agents with limited communications.

© Springer Nature Switzerland AG 2019
R. Silhavy (Ed.): CSOC 2019, AISC 984, pp. 77–82, 2019.
https://doi.org/10.1007/978-3-030-19807-7_8

2 The Problem Statement

Consider the group R consisting of N agents r_i, where i – unique agent's identification number, $i = \overline{1, N}$. Each agent r_i in time t is described by a set of parameters: agent's coordinates $(x_i(t), y_i(t), z_i(t))$, velocity vector $\overline{v}_i(t)$, status (selected role) s_i from the set $s = \langle s_0, s_1, s_2, \ldots, s_k \rangle$ of available roles.

At the initial time t0 all agents are in a random state. Each agent can independently switch to any other state.

All agents of the group R move within a certain work area Z with width Z_w, length Z_l. Agents avoid collisions between each other. That is, there is a limit on the minimum allowable distance d_{min} between any pair of agents of the group:

$$\sqrt{\left(x_j(t) - x_i(t)\right)^2 + \left(y_j(t) - y_i(t)\right)^2 + \left(z_j(t) - z_i(t)\right)^2} \geq d_{min}, \ i = \overline{1, N}; j = \overline{1, N}; i \neq j \tag{1}$$

Onboard communication facilities allow each agent to obtain data on the status of other agents of the group falling within its scope, limited by the radius li of direct communication.

Distribution of $c_m\%$ of agents (with fault no more than Δc) to the status $s_m(m = \overline{1, k})$ is required. Wherein

$$\sum_{m=1}^{k} c_m = 100\% \tag{2}$$

3 The Proposed Method

It is proposed to use iterative process for solve the task.

At each iteration step, each agent r_i using its on-board communication facilities determines the number N_i of a subgroup R_i which contains agents from the group R falling within its scope, limited by the radius l_i of direct communication (agents-neighbors).

The agent r_i then calculates the quantity $n_m^i (m = \overline{1, k})$ of agents (from R_i) which have a status $s_m(m = \overline{1, k})$.

Then the agent calculates the current distribution of roles in the subgroup R_i:

$$c_m^i = \frac{n_m^i}{N^i} \cdot 100\%, \quad (m = \overline{1, k}) \tag{3}$$

After that, the agent r_i calculates the deficit Δc_m^i (surplus) for number of agents for each roles:

$$\Delta c_m^i = c_m - c_m^i, \quad (m = \overline{1, k}) \qquad (4)$$

Information on the status of the subgroup R_i available to the agent r_i is given in Table 1.

Table 1. Information on the status of the subgroup Ri available to the agent r_i.

The role	s_0	s_1	s_2	...	s_m	...	s_k
Required distribution (for all group), %	c_0	c_1	c_2	...	c_m	...	c_k
Current distribution (for sub-group) R_i, pieces	n_0^i	n_1^i	n_2^i	...	n_m^i	...	n_k^i
Current distribution, %	c_0^i	c_1^i	c_2^i	...	c_m^i	...	c_k^i
Deficit (or surplus), %	Δc_0^i	Δc_1^i	Δc_2^i	...	Δc_m^i	...	Δc_k^i

Then the agent finds the maximum $\max \Delta c_m^i$ and chooses the role s_m corresponding to this deficit.

In the event that all roles are distributed in a group with an error less than Δc the task is complete. Otherwise, the next step in the iterative process is repeated.

The algorithm of actions of the agent ri in the distribution of roles is shown in the Fig. 1.

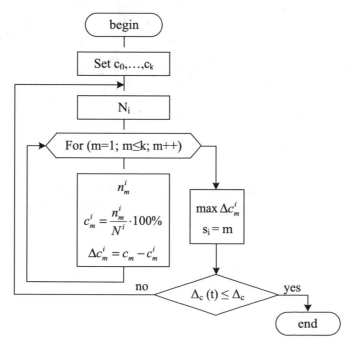

Fig. 1. A Algorithm of agent's r_i actions due to distribution of roles.

4 Simulation

To study the proposed method, computer simulation was carried out. The software that was used in modeling allows to specify the number of agents, the required distribution of roles, the radius of "visibility" of agents.

A series of program experiments was conducted, which confirmed the efficiency of the proposed approach.

In the article, we present some experimental results.

The group R of $N = 100$ agents at the initial moment of time is randomly distributed on the zone Z with a width $Z_w = 500$ m and a length $Z_l = 500$ m. Agents move in a random direction at a constant speed $|\overline{v}_i(t)| = 1$ m/c. The minimum allowable distance is $d_{min} = 5$ m. Radius of direct communication l_i was chosen in the interval $l_i \in [50; 300]$ with step 50 m.

It was necessary to distribute the three roles in such a way that $c_1 = 50\%$, $c_2 = 30\%$, $c_3 = 20\%$. At the initial moment, each agent was assigned one of these three roles at random.

For each selected set of parameters, a series of model experiments was carried out. This took into account the minimum, maximum and average (for the whole group) of the number of agents falling into one subgroup, the number of repetitions of the iterative process necessary to perform the distribution of roles, the minimum and maximum error for individual roles. Table 2 shows the results of first ten simulations with $l_i = 300$ m.

Table 2. Results of first ten simulations ($l_i = 300$ m)

Simulation's number	1	2	3	4	5	6	7	8	9	10
min(N_i)	34	31	29	27	31	27	32	36	27	22
max(N_i)	96	96	92	93	94	90	95	97	89	92
Average value N_i	65,2	63,9	59,5	60,1	60,8	55,3	60,5	65,9	58,1	62,1
Number of iteration steps	7	13	10	9	8	5	6	6	11	10
Minimum error of distribution, %	0	1	0	2	0	0	0	1	0	0
Maximum error of distribution, %	3	2	2	5	2	2	2	3	2	2

Similar tables were constructed for experiments with values $l_i = 250$, $l_i = 200$, $l_i = 150$, $l_i = 100$, $l_i = 50$ м. The average results of computer simulation are presented in Table 3.

Table 3. The average results of computer simulation

Radius l_i of direct communication	50	100	150	200	250	300
Average value $min(N_i)$	0	2,5	7,4	11,7	20,2	29,6
Average value $max(N_i)$	8,1	19,7	35,6	55	78,9	93,4
Average value N_i	2,96	10,50	21,44	34,30	48,15	61,13
Average value number of iteration steps	2,4	4,4	4,8	6,7	5,9	8,5
Minimum number of iteration steps	2	3	4	4	4	5
Maximum number of iteration steps	3	6	6	11	10	13
Average value minimum error of distribution, %	1,1	1,3	1,1	0,6	0,6	0,4
Average value maximum error of distribution, %	4,6	4,1	4,2	3,2	2,7	2,5

5 Conclusions and Future Work

An algorithmically simple method for distributing roles in agent groups based on the principles [16] of swarm intelligence [17] is proposed. The results of computer modeling confirmed the operability of this method. A few iterations provide an acceptable error for solving practical problems.

Given that the proposed method basically contains a probabilistic approach, it can be seen that the error in the distribution and the necessary number of iterative steps is less in those cases when the numbers of local subgroups are higher. Increase the number of local subgroups can be either by increasing the line-of-sight radius, or by forming a more compact system in a group of agents. Also in the future, it will be able to consider the possibility of routing when sending messages about the status of those agents that do not belong to one local subgroup.

Further research will be devoted to the search for methods and algorithms that will reduce the error in the distribution of tasks. Studies will be carried out with groups of different numbers, as well as with agents whose movement speed is high enough, which will result in many agents leaving one of the local subgroups and joining other local subgroups.

Acknowledgement. The reported study was funded by RFBR according to the research project № 18-29-22046.

References

1. Fisher, K.: Agent-based design of holonic manufacturing systems. Robot. Auton. Syst. **27**, 3–13 (1999)
2. Fujita, S., Hara, H., Sugawara, K., Kinoshita, T., Shiratori, N.: Agent-based design model of adaptive distributed systems. Appl. Intell. **9**, 57–70 (1998)
3. Gilbert, N., Troitzsch, K.: Simulation for the Social Scientist. McGraw-Hill Education, New York (2005)
4. Tsitsiklis, J., Athans, M.: On the complexity of decentralized decision making and detection problems. IEEE Trans. Autom. Control **30**, 440–446 (1985)

5. During, M., Pascheka, P.: Cooperative decentralized decision making for conflict resolution among autonomous agents. In: 2014 IEEE International Symposium on Innovations in Intelligent Systems and Applications (INISTA) Proceedings, pp. 154–161 (2014)
6. Lu, J., Han, J., Hu, Y., Zhang, G.: Multilevel decision-making: a survey. Inf. Sci. (Ny) **346**, 463–487 (2016)
7. Coulouris, G., Dollimore, J., Kindberg, T., Blair, G.: System models. In: Distributed Systems: Concepts and Design, vol. 5, pp. 37–80 (2011)
8. Peleg, D.: Distributed computing. In: SIAM Monographs on discrete mathematics and applications, vol. 5 (2000)
9. Jonas, E., Pu, Q., Venkataraman, S., Stoica, I., Recht, B.: Occupy the cloud: distributed computing for the 99%. In: Proceedings of the 2017 Symposium on Cloud Computing, pp. 445–451 (2017)
10. Colorni, A., Dorigo, M., Maniezzo, V.: Distributed optimization by ant colonies. In: Proceedings of the European Conference on Artificial Life, pp. 134–142 (1991)
11. Karaboga, D.: An idea based on honey bee swarm for numerical optimization. Technical Reports. TR06, Erciyes University 10 (2005)
12. Di Caro, G., Dorigo, M.: AntNet: distributed stigmergetic control for communications networks. J. Artif. Intell. Res. **9**, 317–365 (1998)
13. Di Caro, G., Ducatelle, F., Gambardella, L.M.: AntHocNet: an adaptive nature-inspired algorithm for routing in mobile ad hoc networks. Eur. Trans. Telecommun. **16**, 443–455 (2005)
14. Alba, E.: Parallel Metaheuristics: A New Class of Algorithms (2005)
15. Civicioglu, P., Besdok, E.: A conceptual comparison of the Cuckoo-search, particle swarm optimization, differential evolution and artificial bee colony algorithms. Artif. Intell. Rev. **39**, 315–346 (2013)
16. Kaliaev I., Kapustjan S., Ivanov D.: Decentralized Control Strategy within a Large Group of Objects Based on Swarm Intelligence. Presented at the (2011)
17. Dorigo, M., Birattari, M.: Swarm intelligence. Scholarpedia, vol. 2, p. 1462 (2007)

Mac-Cormack's Scheme for Shock Filtering Equation in Image Enhancement

P. H. Gunawan[✉] and Agung F. Gumilar

School of Computing, Telkom University,
Jl. Telekomunikasi No 1 Terusan Buah Batu, 40257 Bandung, Indonesia
phgunawan@telkomuniversity, agungfg@student.telkomuniversity.ac.id

Abstract. Mac-Cormack's scheme is elaborated to approximate the solution of shock filtering equation in image enhancement. This scheme is in second order approximation of spatial and time variables. Here, the comparison results of upwind and Mac-Cormack's scheme are given. The results show that Mac-Cormack's scheme is able to preserve the edge discontinuity. For evaluating the performance of numerical results, the discrete L^2 norm error for both numerical schemes is given. From several experiments, along the increasing of image sizes, the error of Mac-Cormack's scheme is observed getting smaller. For instance, using image sizes (64,64) the error is obtained 0.13762, meanwhile using (512,512) the error is observed 0.06640.

Keywords: Mac-Cormack's · Shock filtering · Enhancement · Image processing

1 Introduction

Image enhancement can be important process in image processing. In some applications, image enhancement is required in pre-processing part to detect the important image features and the details of image (see [7,11] for more detail). For instance in [11], enhancement of medical image is needed to help a surgeon to interpret and diagnosis image more accurately. Since generally, medical images are often degraded by noise and external data acquisition.

Several enhancement methods or algorithms are available in some references [3,4,6,10]. This paper is focus on enhancement method which developed based on partial differential equations (PDE). In PDE-based approach, enhancement image can be done by known as shock filtering equation [2,5]. This equation were introduced by Osher and Rudin [5] which assure the edge preserves discontinuity.

Consider an image $I(x,y,t)$ over domain Ω and $\mathfrak{L}$ is a nonlinear elliptic PDE operator, then the shock filtering equation is given as

$$I_t = -F\left(\mathfrak{L}(I)\right)|\nabla I|, \quad (x,y) \in \Omega, \quad t > 0 \tag{1}$$

© Springer Nature Switzerland AG 2019
R. Silhavy (Ed.): CSOC 2019, AISC 984, pp. 83–89, 2019.
https://doi.org/10.1007/978-3-030-19807-7_9

where F is a Lipschitz function which satisfying the following properties

$$\begin{cases} F(\sigma) = 0, & \sigma = 0, \\ \text{sgn}(\sigma)F(\sigma) > 0, & \sigma \neq 0, \end{cases} \tag{2}$$

with $\text{sgn}(\sigma)$ is signum function which takes the sign of real number,

$$\text{sgn}(\sigma) = \begin{cases} 1, & \sigma > 0, \\ 0, & \sigma = 0, \\ -1, & \sigma < 0. \end{cases} \tag{3}$$

Generally in [2,8], (1) is approximated by a simple finite difference method which depend on the sign of function F (a kind of upwind method). Therefore, this method is first order approximation in space and time. However, in this paper, the second order approximation in space and time approach is proposed. Here, time and spatial derivative are discretized using predictor and corrector step. This numerical scheme is known as Mac-Cormack's scheme.

The organization of this paper is given as follow, in Sect. 2, the numerical schemes, upwind and Mac-Cormack's scheme are given. In Sect. 3 the evaluation and discussion of numerical solution of both schemes are elaborated. Finally, the conclusion is given in Sect. 4.

2   Numerical Methods

Before two numerical schemes are elaborated in detail, then the following discrete notations are given. Let the 2-D spatial image domain Ω is discretized into several points. In this case, the points are described as the coordinate of pixels of image. Given an image with size (N_x, N_y), then a set of spatial discrete points is $\mathcal{M} = \{1, 2, \cdots, N_x\} \times \{1, 2, \cdots, N_y\}$. Moreover, another notations for discrete space and time properties are written as follow,

$$\Delta t = \frac{T_f}{N_t}, \qquad t^n = \Delta t \times n, \qquad n \in \mathcal{T} = \{1, 2, \cdots N_t\}, \quad N_t \in \mathbb{Z}^+,$$

$$x_i = i \times \Delta x, \quad y_j = j \times \Delta y, \qquad (i, j) \in \mathcal{M}.$$

In image processing, the number of grids is similar to the sizes of image, thus spatial steps $\Delta x = \Delta y = 1$. Finally, the notation $I_{i,j}^n$ is used to describe the value of a pixel at spatial point (x_i, y_j) and time t^n.

Here, the operator $\mathfrak{L}(I)$ for both numerical schemes is defined as

$$\mathfrak{L}(I) := \frac{I_x^2 I_{xx} + 2I_x I_y I_{xy} + I_y^2 I_{yy}}{|I_{xx} + I_{yy}|^2}. \tag{4}$$

According to [2,5], this operator is better than ordinary Laplace operator. This operator allows the edges are formed from the zero crossings of second directional derivatives.

2.1 Upwind Scheme

In scalar hyperbolic type of PDE, upwind scheme is used to approximate the transport equation with first order approximation in space and time. Here, the discretization of (1) using upwind is given as

$$
I_{i,j}^{n+1} = \begin{cases} I_{i,j}^n - c\sqrt{\left(\dfrac{I_{i+1,j}^n - I_{i,j}^n}{\Delta x}\right)^2 + \left(\dfrac{I_{i,j+1}^n - I_{i,j}^n}{\Delta y}\right)^2}, & \text{if } c < 0, \\[4mm] I_{i,j}^n, \quad \text{if } c = 0, \\[4mm] I_{i,j}^n - c\sqrt{\left(\dfrac{I_{i,j}^n - I_{i-1,j}^n}{\Delta x}\right)^2 + \left(\dfrac{I_{i,j}^n - I_{i,j-1}^n}{\Delta y}\right)^2}, & \text{if } c > 0. \end{cases} \tag{5}
$$

where $c = F(\mathcal{L}(I_{i,j}^n))$. As shown in (5), this scheme depends on the characteristic wave sign of $(F(\mathcal{L}(I)))$. If the sign is positive then backward difference scheme is used, meanwhile the forward difference scheme is implemented if the sign is negative.

2.2 Mac-Cormack's Scheme

In computational fluid dynamics area, Mac-Cormack's scheme is known as a simple and robust scheme for approximating simple Navier-Stokes equation in [1]. This scheme consists of two steps to obtain second order approximation in space and time. Here, the steps are called prediction and correction step.

Prediction step:
   In this step, the prediction value is approximated by forward difference scheme in space,

$$
\left(\frac{\partial I}{\partial t}\right)_{i,j}^n = -F(\mathcal{L}(I_{i,j}^n))\sqrt{\left(\frac{I_{i+1,j}^n - I_{i,j}^n}{\Delta x}\right)^2 + \left(\frac{I_{i,j+1}^n - I_{i,j}^n}{\Delta y}\right)^2}, \tag{6}
$$

then the prediction value is done by discrete time which is given as follow

$$
\overline{I}_{i,j}^n = I_{i,j}^n + \left(\frac{\partial I}{\partial t}\right)_{i,j}^n \Delta t. \tag{7}
$$

Corrector step:
   Next step, the spatial term is discretized using backward scheme using prediction value (7). This scheme is written as

$$
\left(\frac{\overline{\partial I}}{\partial t}\right)_{i,j}^{n+1} = -F(\mathcal{L}(\overline{I}_{i,j}^n))\sqrt{\left(\frac{\overline{I}_{i,j}^n - \overline{I}_{i-1,j}^n}{\Delta x}\right)^2 + \left(\frac{\overline{I}_{i,j}^n - \overline{I}_{i,j-1}^n}{\Delta y}\right)^2}. \tag{8}
$$

Finally, the new value of $I_{i,j}^{n+1}$ is obtained by integrating time evolution using average value of predictor (6) and corrector (8) which given as

$$\left(\frac{\partial I}{\partial t}\right)_{av} = 0.5\left[\left(\frac{\partial I}{\partial t}\right)_{i,j}^{n} + \left(\frac{\overline{\partial I}}{\partial t}\right)_{i,j}^{n+1}\right], \tag{9}$$

$$I_{i,j}^{n+1} = I_{i,j}^{n} + \left(\frac{\partial I}{\partial t}\right)_{av}\Delta t. \tag{10}$$

As shown in [1], Mac-Cormack's scheme is a second order approximation in space and time and analogous to Lax-Wendroff scheme for one-dimensional problem.

3   Results and Discussion

Here, to measure the performance of Mac-Cormack's scheme, a numerical test of deblurring image from the diffusion effect to reduce noise is given. The goal of this test is to see the ability of Mac-Cormack's scheme to improve the quality of image as close as the original image.

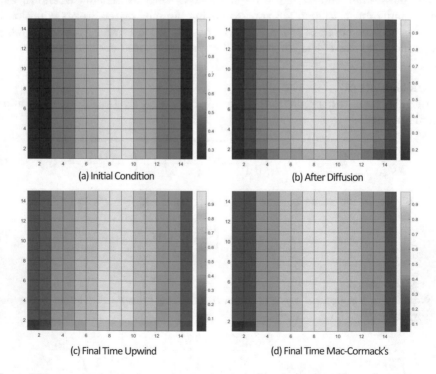

Fig. 1. The measurement of numerical scheme in two-dimensional image, (a) the original image, (b) the result of diffusion effect, (c) the result of upwind scheme, and (d) the result of Mac-Cormack's scheme.

Here, the original image with sizes (16,16) is given as in Fig. 1(a), with the value of pixels in this image is in interval $[0 : 1]$. Further, the original image is blurred using two-dimensional diffusion equation (see [2]) which shown in Fig. 1(b). Finally, in this test, Mac-Cormack's and upwind scheme are used to enhance the image as close as the original.

The results using upwind and Mac-Cormack's numerical scheme are given in Fig. 1(c) and (d) respectively. It is clear that after similar final time of simulation $50t$, with $\Delta t = 10^{-2}$, the result of Mac-Cormack's scheme is shown close to the original image than upwind scheme.

For more clear about this comparison, the observation trough a slice image at $y = 8$ can be seen in Fig. 2. As shown in Fig. 2, the result of Mac-Cormack's scheme is able to approach the original line as close as possible. Meanwhile, the upwind scheme is shown a little bit far from the original line. Moreover, to measure quantitatively, the discrete L^2 or $||\{I_{i,j}\}||_{2,\Delta x}$ norm errors (see [9]) for each numerical schemes are given in Table 1.

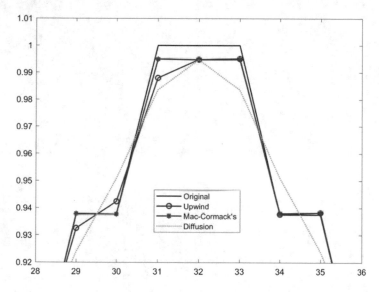

Fig. 2. The comparison results of upwind and Mac-Cormack's scheme in slice image at $y = 8$.

Table 1. The discrete error $||\{I_{i,j}\}||_{2,\Delta x}$ for upwind and Mac-Cormack's scheme.

No	Image sizes	Error of upwind	Error of Mac-Cormack
1	(16,16)	0.64085	0.37166
2	(32,32)	0.63904	0.21817
3	(64,64)	0.79319	0.13762
4	(128,128)	1.06730	0.09127
5	(256,256)	1.48140	0.06626
6	(512,512)	2.07700	0.06640

In Table 1, several simulations using different image sizes are elaborated. It can be seen clearly that, using Mac-Cormack's scheme, the errors are obtained decreasing along the increasing of image sizes. Meanwhile, the numerical error of upwind scheme is shown slightly increasing, but it is not significantly different. Overall, two schemes are shown in a good agreement as the original image where both schemes can preserve the edge discontinuity very well.

The results using Mac-Cormack's scheme for another test can be seen in Fig. 3. Here at first, original image (Fig. 3(a)) is given a Gaussian noise as shown in Fig. 3(b), then to reduce the noise, the diffusion (blurring) effect is

Fig. 3. Results of enhancement using Mac-Cormack's scheme, (a) Original image, (b) image with Gaussian noise, (c) image with diffusion effect to reduce the noise, and (d) the enhanced image using Mac-Cormack's scheme.

implemented (Fig. 3(c)) using diffusion equation. Afterward, the Mac-Cormack's scheme is elaborated to enhance the image as close as the original image (see Fig. 3(d)).

As it can seen in Fig. 3(d), Mac-Cormack's scheme is able to make a blur image becomes sharp image as close to the original image. In this test, the focus is the white area on neck of camera. The results in Fig. 3 show that Mac-Cormack's scheme is capable to sharpen the white area.

4 Conclusion

In this paper, the second order approximation of Shock filtering equation in image enhancement is proposed. The numerical scheme consists of two steps, predictor and corrector step, which is called Mac-Cormack's scheme. In order to evaluate the Mac-Cormack's scheme, a numerical test deblurring image from diffusion effect to original image is elaborated. In numerical simulation, this scheme is shown able to preserve the edge discontinuity as close as the original image. The comparison of Mac-Cormack's and first order upwind scheme is also given. Here, the discrete L^2 norm error of Mac-Cormack's scheme is observed getting smaller along the increasing of image sizes. By using image sizes (512,512), the discrete L^2 norm error of Mac-Cormack's is obtained 0.00640.

References

1. Anderson, J.D., Wendt, J.: Computational Fluid Dynamics, vol. 206. Springer, Heidelberg (1995)
2. Aubert, G., Kornprobst, P.: Mathematical Problems in Image Processing: Partial Differential Equations and the Calculus of Variations, vol. 147. Springer, New York (2006)
3. Iryanto, I., Fristella, F., Gunawan, P.H.: Pendekatan numerik pada model isotropic dan anisotropic diffusion dalam pengolahan citra. Ind. J. Comput. (Indo-JC) 1(2), 83–96 (2016)
4. Lu, J., Healy, D.M., Weaver, J.B.: Contrast enhancement of medical images using multiscale edge representation. Opt. Eng. 33(7), 2151–2162 (1994)
5. Osher, S., Rudin, L.I.: Feature-oriented image enhancement using shock filters. SIAM J. Numer. Anal. 27(4), 919–940 (1990)
6. Perona, P., Malik, J.: Scale-space and edge detection using anisotropic diffusion. IEEE Trans. Pattern Anal. Mach. Intell. 12(7), 629–639 (1990)
7. Rahman, Z.U., Jobson, D.J., Woodell, G.A.: Multi-scale retinex for color image enhancement. In: Proceedings of International Conference on Image Processing, 1996, vol. 3, pp. 1003–1006. IEEE (1996)
8. Remaki, L., Cheriet, M.: Numerical schemes of shock filter models for image enhancement and restoration. J. Math. Imaging Vis. 18(2), 129–143 (2003)
9. Thomas, J.W.: Numerical Partial Differential Equations: Finite Difference Methods, vol. 22. Springer, New York (2013)
10. Wang, Z.: Image inpainting-based edge enhancement using the eikonal equation. In: 2011 IEEE International Conference on Acoustics, Speech and Signal Processing (ICASSP), pp. 1261–1264. IEEE (2011)
11. Yang, Y., Su, Z., Sun, L.: Medical image enhancement algorithm based on wavelet transform. Electron. Lett. 46(2), 120–121 (2010)

Comparison Two of Different Technologies for Outdoor Positioning of Robotic Vehicles

Miroslav Dvorak and Petr Dolezel[✉]

Faculty of Electrical Engineering and Informatics, University of Pardubice,
Pardubice, Czech Republic
{miroslav.dvorak,petr.dolezel}@upce.cz
http://www.upce.cz/

Abstract. This paper aims to compare two different technologies, which can determine the exact position of a robotic vehicle. The first method uses wireless technology and is based on the measurement of the signal strength of the bluetooth beacons. Based on these values, you can calculate the distance from beacons. The second method uses laser light and measurement of the reflected pulses. Based on the reference points of reflection, we can determine the distance. Both methods then use 2D triangulation to determine the position of the robotic vehicle. The exact position of the bluetooth beacons or the reference points must be known for the calculation. The paper also describes experiments with a laser, and the conclusion provides an evaluation of both technologies.

Keywords: Bluetooth · iBeacon · Laser · LiDAR · Photodiode · RSSI · Trilateration

1 Introduction

The requirements for the exact positioning of robotic vehicles are constantly increasing. There are various outdoor positioning technologies such as GPS [1], GALILEO [2] or GLONASS [3]. However, this article deals with the local determination of the position of a robotic vehicle. Examples include drawing lines on a football field or a parking places in front of supermarkets. In addition, the article does not look for a solution for a single robotic vehicle, but for a group of robotic vehicles. Therefore, it solves not only the method of mutual cooperation but also the detection of mutual collisions of vehicles and collisions of vehicles with obstacles, whereas the main goal is to develop a group of vehicles capable to draw patterns on an outdoor surface.

For this purpose, bluetooth wireless technology was considered as a first approach for robot positioning. As a reference point for determining distances, a device called iBeacon [4] was used. These devices periodically transmit their ID. Based on measured signal strength, you can estimate the distance from beacons. Using the 2D triangulation method, the position of the robotic vehicle

© Springer Nature Switzerland AG 2019
R. Silhavy (Ed.): CSOC 2019, AISC 984, pp. 90–98, 2019.
https://doi.org/10.1007/978-3-030-19807-7_10

was calculated. The second considered technology uses a laser beam to measure distance. A laser device sends out a short light pulse that is reflected from the measured object. The reflected pulse comes back and is captured by the optics of the device. The time, that the laser beam takes to complete the distance, is then calculated.

2   IBeacon

The iBeacon is a small and low-power device which periodically transmits its UUID (Universally Unique Identifier) number through the interface of Bluetooth version 4.x. Positioning using an iBeacon is described and tested in the article "An IoT Approach to Positioning of a Robotic Vehicle" [5]. In this article, theory about positioning using bluetooth technology is summarized. Hence, the robot vehicle measures signal strength of iBeacons, called RSSI (Received Signal Strength Indicator) [6], and the distance is calculated according to the strength. Based on these distances, a vehicle is able to estimate its position. According to the experiments presented in the cited paper, the overall accuracy of the position of the robotic vehicle is approximately 0.2m. The mentioned precision of in the position estimation of the robotic vehicle is, however, unacceptable. Therefore, this method is not suitable for further implementations.

3   LiDAR

The use of LiDAR (Light Detection And Ranging) for 2D localization seems more advantageous. In addition to the more precise position of the mobile vehicle, we can obtain the information needed to detect collisions of individual robotic vehicles. For experiments, the LiDAR lite v3 [7] is used. STM32 microcontroller is utilized as a control unit, and LiDAR is connected via the I2C bus.

Since LiDAR applies a method that measures the distance to a target by illuminating the target with pulsed laser light, while measuring the reflected pulses with a sensor, the robotic vehicle needs to be handled with the secondary continuous laser. This laser is used to focus on the target's active reference points. The short flashes of LiDAR are not capable of being detected by photodiodes used as the active reference points.

The mechanical structure consists of a gearbox and an optical sensor engine, and also two lasers attached to the transmission output shaft of the gearbox, as shown in Fig. 1. Information from the optical sensor is used to determine the angle of rotation of the lasers. This information, along with obstacle distance information, is used for collision detection.

4   Active Reference Points of Reflection

The basis for the orientation of the robotic vehicle with laser is the active reference points. Each reference point consists a photodiode matrix with a filter to eliminate sunlight and a control unit, as shown in Fig. 2.

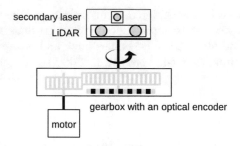

Fig. 1. Part of mechanical structure of robotic vehicle.

When a photodiode is illuminated by the secondary laser beam, the control unit sends the information containing a mark of the reference point and the positions of the photodiode. This information is sent as a UDP (User Datagram Protocol) broadcast using the WiFi interface. then, this information is received by all robotic vehicles. Nevertheless, only one vehicle uses this information for the calculation of the current position.

Fig. 2. Block diagram of active point reference.

The method of controlling robotic vehicles is based on an agent-oriented approach, so the agent periodically sends a request to find a position of a particular vehicle. The already mentioned WiFi interface is used for communication. This vehicle gradually focuses on the available reference points, and calculates its position using 2D triangulation [8] from the obtained responses.

5   Mutual Cooperation of Group of Robotic Vehicles

Common rendering of a pattern on an outdoor surface going to be based on mutual collaboration of group of robotic vehicles. As a considered solution of this problem, each robotic vehicle is fitted with a drawing unit and draws some parts of the pattern.

The required pattern is entered into the main application and the application divides it into individual curves, which are sequentially assigned to the nearest robotic vehicles. Therefore, the emphasis is placed on greater accuracy of positioning of individual vehicles. In the case of large inaccuracies, the curves would not accurately correspond with each other and the resulting shape would be unusable, as shown in Fig. 3.

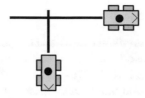

Fig. 3. Two robotic vehicles drawing letter T, incorrect positioning.

6   Obstacle Detection and Collision of Robotic Vehicle

Since LiDAR device detects not only the reference points, but also other objects, obstacle detection procedure can be implemented into the controlling system.

Each robotic vehicle, after receiving a command to move to a new position, begins to monitor its surroundings. As soon as it detects obstacles at a defined minimum distance, the vehicle automatically stops. The agent-oriented system [9] periodically asks for the status of robotic vehicles. Once it receives information about a possible collision, it sends a robotic vehicle command to explore the surroundings in more detail.

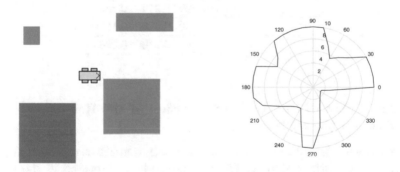

Fig. 4. Environment map

Based on these data, an agent-based system creates an environment map, as shown in Fig. 4. The system then evaluates whether it is a fixed obstacle or collision with another robotic vehicle. According to this, the system suggests another route or waits for the second moving vehicle to leave.

7   Experiments with LiDAR

As the main contribution of this paper, a set of experiments is presented here in order to test the suitability of the use of LiDAR for precise determination of a robotic vehicle position.

7.1   Measurement Accuracy

This experiment is focused on detecting the accuracy of LiDAR measurements. A precision of about 10 cm can be achieved according to the catalog sheet. Data acquisition is performed within a 1–10 m range of distances with an increment of 1m. For each distance, 1000 values are obtained. The results of these measurements are shown in Table 1.

Table 1. Measurement results for individual distances.

Real distance (m)	Error (m)		
	Min	Max	Average
1	0.0300	0.1200	0.0641
2	0.0400	0.0800	0.0673
3	0.0600	0.0900	0.0724
4	0.1100	0.1500	0.1313
5	0.1500	0.1900	0.1763
6	0.1600	0.2000	0.1819
7	0.1500	0.2000	0.1790
8	0.1700	0.2100	0.1888
9	0.1800	0.2200	0.1965
10	0.1800	0.2400	0.2083

7.2   The Effect of the Angle of Rotation of the Reflection Plate on the Distance

This experiment tests the influence of the rotation angle of the reference point reflector plate on the accuracy. The reflection plate is rotated in the range of 0°–80° with steps of 20°. This measurement was repeated for different distances, see Table 2.

It is clear from the measured values that the accuracy of the measurement depends very much on the angle of rotation of the reflection plate. A black reflector plate was used in this experiment. With the increasing distance from the reflection plate and increasing angle of the reflection plate, the distance values tend to be undetectable. The conclusion of this experiment demonstrates the inappropriate use of flat reflection surfaces for reference points. Thus, circular reflection surfaces are used for further experiments.

Changing the shape of the reference point also results in a change in the design of the optical sensor.

The photodiodes are not placed in the matrix, but in individual layers consisting of disks of clear plexiglass, as showed in Fig. 5.

Table 2. Influence of the angle of rotation of the reference point reflector plate

Distance (m)	Angle of rotation (°)	Error (m)		
		Min	Max	Average
1	0	0.0300	0.1200	0.0641
	20	0.0900	0.1700	0.1300
	40	0.1000	0.2200	0.1513
	60	0.1300	0.2500	0.1937
	80	0.1000	0.3300	0.2281
2	0	0.0400	0.0800	0.0673
	20	0.0900	0.2100	0.1213
	40	0.1200	0.2500	0.1436
	60	0.1400	0.2200	0.1540
	80	0.1300	0.3600	0.2257
3	0	0.0600	0.0900	0.0724
	20	0.1000	0.1900	0.1125
	40	0.1300	0.2700	0.1394
	60	0.1400	0.2900	0.1527
	80	0.1600	0.3200	0.1636
4	0	0.1100	0.1500	0.1313
	20	0.1500	0.2400	0.1507
	40	0.1700	0.3200	0.2204
	60	0.2300	0.4500	0.2657
	80	-	-	-
5	0	0.1500	0.1900	0.1763
	20	0.1600	0.3400	0.1833
	40	0.1800	0.3800	0.2256
	60	0.2200	0.4800	0.3257
	80	-	-	-

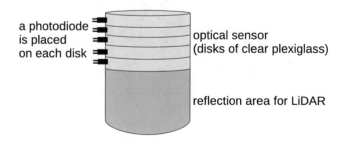

Fig. 5. Construction of circular reference points.

7.3   2D Trilateration

This experiment verifies the accuracy of robotic vehicle positioning using 4 reference points located in the corners of the test surface. The test surface sizes are set to 10 m×10 m. The robot vehicle is placed in a known position, the measurement situation is shown in Fig. 6. For the comparison of both technologies, identical measurements were used to test iBeacons, as presented in [5]. The results of this measurement are shown in Figs. 7 and 8. Apparently, the measurement using LiDAR is more accurate.

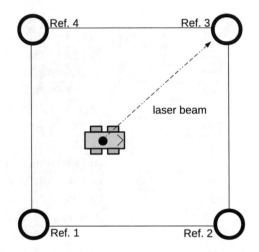

Fig. 6. Layout of experiment using LiDAR.

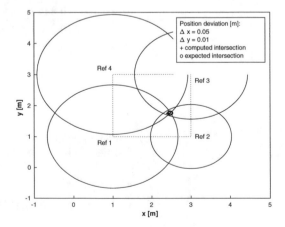

Fig. 7. Experiment using LiDAR.

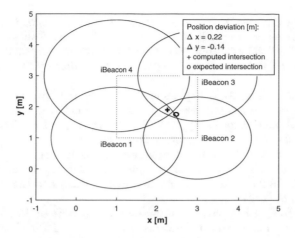

Fig. 8. Experiment using bluetooth technology.

8   Conclusion

The main aim of the article is to compare two different technologies for robot positioning and to determine the more accurate one. For comparison, it is also necessary to take into account the complexity and cost of each individual solution.

Bluetooth technology is less accurate, but the cost of solutions is very low. The iBeacon is commonly available, and the robotic vehicle can be easily equipped with a conventional bluetooth module.

The solution using LiDAR is more accurate, but the circular reference points are not only more expensive, but also more demanding in terms of construction. This is, however, balanced by more precise positioning capability and the ability to use it as the obstacle detection component.

Acknowledgment. The work has been supported by the Funds of University of Pardubice, Czech Republic. This support is very gratefully acknowledged.

References

1. Mulla, A., Baviskar, J., Baviskar, A., Bhovad, A.: GPS assisted Standard Positioning Service for navigation and tracking: review & implementation. In: 2015 International Conference on Pervasive Computing (ICPC), Pune, pp. 1–6 (2015). https://doi.org/10.1109/PERVASIVE.2015.7087165
2. Kwasniak, D.L.: Single point positioning using GPS, Galileo and BeiDou system. In: 2018 Baltic Geodetic Congress (BGC Geomatics), Olsztyn, pp. 310–315 (2018).https://doi.org/10.1109/BGC-Geomatics.2018.00065

3. Kwaśniak, D., Cellmer, S., Nowel, K.: Precise positioning using the modified ambiguity function approach with combination of GPS and Galileo observaions. In: 2018 25th Saint Petersburg International Conference on Integrated Navigation Systems (ICINS), St. Petersburg, pp. 1–6 (2018). https://doi.org/10.23919/ICINS.2018.8405841
4. Li, X., Xu, D., Wang, X., Muhammad, R.: Design and implementation of indoor positioning system based on iBeacon. In: 2016 International Conference on Audio, Language and Image Processing (ICALIP), Shanghai, pp. 126–130 (2016). https://doi.org/10.1109/ICALIP.201trilateration6.7846648
5. Dvorak, M., Dolezel, P.: An IoT approach to positioning of a robotic vehicle. In: Silhavy, R. (eds) Software Engineering and Algorithms in Intelligent Systems. CSOC2018 2018. Advances in Intelligent Systems and Computing, vol. 763. Springer, Cham (2019)
6. Mahapatra, R.K., Shet, N.S.V.: Experimental analysis of RSSI-based distance estimation for wireless sensor networks. In: 2016 IEEE Distributed Computing, VLSI, Electrical Circuits and Robotics (DISCOVER), Mangalore, pp. 211–215 (2016).https://doi.org/10.1109/DISCOVER.2016.7806221
7. Catapang, A.N., Ramos, M.: Obstacle detection using a 2D LIDAR system for an Autonomous Vehicle. In: 2016 6th IEEE International Conference on Control System, Computing and Engineering (ICCSCE), Batu Ferringhi, pp. 441–445 (2016). https://doi.org/10.1109/ICCSCE.2016.7893614
8. Hereman, W., Murphy, W.S.: Determination of a position in three dimensions using trilateration and approximate distances. https://inside.mines.edu/~whereman/papers/Murphy-Hereman-Trilateration-MCS-07-1995.pdf. Accessed 03 Jan 2018
9. Xiao, N.-F., Nahavandi, S.: Multi-agent model for robotic assembly system. In: Proceedings of the 5th Biannual World Automation Congress, Orlando, FL, pp. 495–500 (2002). https://doi.org/10.1109/WAC.2002.1049486

An Effective Hardware-Based Bidirectional Security Aware M-Payment System by Using Biometric Authentication

Jerrin Yomas[1(✉)] and N. Chitra Kiran[2]

[1] Visvesvaraya Technological University, Belagavi, Karnataka, India
jerrinyomas@gmail.com
[2] Department of ECE, Alliance College of Engineering and Design
Alliance University, Bengaluru, Karnataka, India
chitrakiran.n@alliance.edu.in

Abstract. The increased use of the internet all over the world has witnessed the emerging services of electronic payment systems. The use of portable electronic gadgets with high-speed internet and mobile networking mechanisms have offered the user to access and perform a financial transaction. The technological advancement in wireless technology has enabled the mobile devices as the prime component in a digital economy where the user can execute all the transactions. However, the attackers are finding the various ways to identify those transaction details and is leading to security concern for both the mobile application developers and also the users of it. Thus, this paper considers all the security concerns of mobile payment systems which fails to offer secure customer verification during the transaction. This paper introduces a hardware-based bidirectional secure payment system by using a biometric authentication mechanism at both the ends of the seller and buyer. This system eliminates the physical requirement of cash for all the transactions and resolves the identity needs during each transaction. The outcomes suggest that the system fulfills all the secure payment system and it takes approximately less than 30 s of time to complete the entire process of the transaction.

Keywords: Security · Hardware components · Bidirectional system · Payment system

1 Introduction

The financial transactions over the mobile network in a secure manner can be referred to as "mobile payment system". The m-payment system has gained a lot of interest in today's world as it offers financial transactions from one individual/organization to other in a secure manner [1]. Hence, the financial corporates, business stakeholders and users are fond of m-payment and it also yields opportunistic features with simple and ease of financial transactions. The m-payments were also targeting a particular group of peoples by giving a lot of discounts, cash back and rewards for each transaction and with these users are liking m-payment [2]. With all these favoring initiatives in m-payment system lot of applications were introduced in recent times. Some of the m-payment applications

© Springer Nature Switzerland AG 2019
R. Silhavy (Ed.): CSOC 2019, AISC 984, pp. 99–108, 2019.
https://doi.org/10.1007/978-3-030-19807-7_11

are Phone pay, Google pay, PayTM, etc. From the analysis, it is been also reported that the privacy, trust, and security are always been a major concern in growing e-commerce systems as the use of m-payment is growing in a rapid manner where the attacks through the internet are common [3, 4]. However, the latest m-payment were advantageous but lags with disadvantages like less effective user authentication mechanism. Hence, security is always been a major concern in the m-payment system. This paper focuses on these security concerns and introduces a hardware-based bidirectional m-payment system by using biometric authentication. The categorization of the paper is performed with different sections like the background of m-payment system is discussed in Sect. 2, review of existing researches in securing m-payment in Sect. 3, problem addressed for securing m-payment in Sect. 4, proposed solution to overcome the problem addressed in Sect. 5 and the algorithm implemented in the proposed system (Sect. 6), analysis of results accomplished in Sect. 7 and the conclusion in Sect. 8.

2   Background

Since last five decades, m-payment owing a booming advancement with favorable features. The entities involved in an m-payment system are given in Fig. 1, which has four major units like a buyer, seller, seller bank and buyer bank. In this system, both the seller and buyer are connected via the internet, mobile network etc. But, in m-payment should compromise with seller, buyer and gateway entities to perform transaction clearance among buyer and seller [5]. The connection among all these entities can be performed by using a wired, wireless or cellular network service provider. The communication among buyer, seller, and gateway takes place over the wired connections of banks while its communication is made secure by using some of the security protocols like Transport layer, Socket layer security etc. [6, 7].

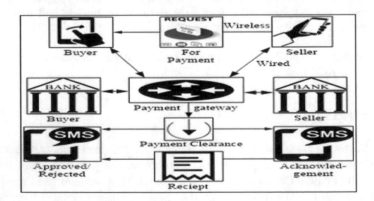

Fig. 1. The basic mobile payment system

In Fig. 1, the seller requests for payment to the buyer in wireless connection and also requests bank to deposit the requested amount from buyer to his account through the payment gateway. The gateway builds the communication among both the banks of seller and buyer via a wired or wireless manner. After receiving the request from the seller, the buyer requests bank to deduct the money from his/her account and credit into sellers account via payment gateway [8]. Once the payment is cleared from both the banks, the gateway sends a notification to the buyer about payment approval and rejection and also sends an acknowledgment to seller mobile phone via SMS. Finally, the seller sends a payment receipt to the buyer. The classification of the m-payment system can be performed by considering different parameters. The globally accepted m-payment can be remote or proximity payment. The remote payment system offers users to perform the payment by using mobile phones accessing backend systems of m-payment over mobile networks. The proximity payment systems facilitate users to perform payment via mobile phones and short rage communication mechanisms. Following are the some of the technologies which are implemented at user end [9, 10].

2.1 Short Message Service

This kind of m-payment service offers mobile phones and other networked devices for exchanging in the short messages in the form of the text of 160 characters and the SMS service is available at a wide range for both business and individual prospects.

2.2 Unstructured Service Delivery

This kind of service uses transaction-oriented mechanism related to the GSM module and has the ability to build the standard to have serviced over GSM networks.

2.3 GPRS Based Service

The mobile data over GPRS plays a major role in this service that used mobile phones over the Internet and its communication services such as email and World Wide Web access.

2.4 Phone-Based Application

The applications developed through Java platform can be used in this service where GSM based mobiles and wireless systems help in achieving transaction over code division multiple access.

2.5 Near-Field Communication (NFC) System

This system uses wireless communication of shorter range and performs a transaction via the combination of the contactless Radio-Frequency Identification (RFID) smart card and the mobile phone.15

2.6   Dual Chip

In this, the phones exhibit 2-slots for SIM card and payment chip each which permits payment application provider to build an application within the chip card without interfacing with an owner of the SIM card.

2.7   SIM-Based Application

The applications using SIM in GSM based mobile phones as a smart card with protecting the information via cryptographic keys and algorithms and keys. These applications can be considered as more secure in comparison with user applications exist in the mobile phone.

2.8   Mobile Wallet

This exhibits a software within the mobile phone having user bank account details which allow the user to perform payments via the mobile phone.

3   Existing Researches Towards Securing M-Payment

The recent researches were more focused on improving the security level of the m-payment system to a greater extent. One of such research was observed in Abughazalah et al. [11] which gives one-time-password (OTP) based system. The payment system with QR code mechanism is observed in Zhang et al. [12] which yields better security in terms of sensitive applications. Focusing towards offline e-payment system the work of Fan et al. [13] have introduced a renewal protocol which helps in exchanging the old currency with the new one. A payment system for retail shops was introduced in Chaudhry et al. [14] by solving different anonymity and security issues faced by users. Another QR code-based technology is presented in Subpratatsavee and Kuacharoen [15] for storing, transmitting and recognition of data with information security which becomes more convenient from all the user prospective. The discussion on e-payment system with mobile phones was given in Ortiz-Yepes [16] which is a profitable and least time-consuming. Similar concept with QR model is introduced in Teran et al. [17] that offers a visual cryptographic mechanism yielding highly feasible and secure authentication for m-payment. However, the attackers can find the various ways to hack these internet-based transactions. Towards this concern Yang and Lin [18] have introduced anonymity in cloud-based m-payment. This [18] system is efficient and secure in cloud-based transaction. Similarly, Kang and Xu [19] gave an anonymity aware secure payment system to protect user privacy. An authenticated pseudonym system for user's identity is presented in Qin et al. [20] which helps to track the transactions of all the authenticated users. The work of Barkhondari et al. [21] discussed different factors causing security concerns in m-payment. A work considering the analysis of anonymity in m-payment Breaken [22] introduced a payment system for visually impaired and blind persons. A recent survey work of Yomus and Chaitra [23] gives current state of art in m-payment security systems. Numerous works towards considering m-payment security are found in Chaitra et al. [24–28] which has a greater impact on research in the m-payment system. A recent work of Kang and Nyang [29] presented privacy preserve e-payment system for passage mass transactions. Similarly, Cao and Zhu [30] have presented a secure e-cash transaction mechanism for ride hailing services and are a reusable, flexible system.

4   Problem Statement

The security of m-payment is always the biggest concern in today's genre as the attackers are developing an advanced hacking technique to attack the payment systems happening over the internet. The countries with less use of smartphones are facing a problem with the m-payment system. From the research survey of existing researches, it is observed that most of the researches were considered the factors of security over the network or mobile but no much was considered with authentication at both ends receiver and sender. Also, very rare researches have incorporated the hardware-based approaches using biometric features of the user. Thus, there is a need for an authenticated m-payment at both ends. The problem statement is *"to design hardware based bidirectional authenticated m-payment system using biometric authentication"*. The following section gives the proposed solution to this stated problem.

5   Proposed Solution

The problems addressed above were considered while introducing the proposed solution which considers the biometric feature to achieve secure transaction. The proposed solution is aimed to meet the needs of payment systems. The architectural model of proposed hardware based biometric system is given in Fig. 2, which consists of both seller and buyer modules having a different set of hardware components. The seller module composed of Bluetooth, low powered 32bit ARM Cortex M3 processor board, fingerprint sensor, a power supply unit, GSM 900 module and Graphical display while the buyer module also consists of same low powered 32bit Cortex M3 processor board, Bluetooth module, fingerprint sensor, Power supply unit, and TFT display. The ARM Cortex M3 is selected as it is cost efficient and effective embedded system for real time applications. This Arduino board is having 32-bit ARM Cortex-M3 central-processing unit of type LPC1768, four hardware serial ports, extended serial peripheral interface (SPI) support, and 512-kB static random-access memory (SRAM). The available Bluetooth component is used for peer2peer communication. A capacitive fingerprint sensor (R303/305) is selected from Fingerprint Cards. This also stores fingerprint data on external flash memory for later verification.

From Fig. 2, the buyer will authenticate with his/her fingerprint and then waits for account details of the seller. The account details of the seller can be processed only after seller authentication through the fingerprint. Once the authentication is done, the buyer will get the details of seller account on Graphical display via Bluetooth. Then will enter the required amount to be paid for seller via the graphical display and press the submit button of the module. Then the amount from buyer account will be

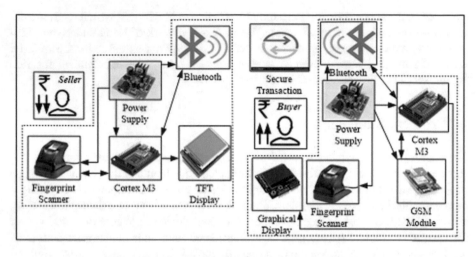

Fig. 2. Architectural model of the proposed solution

transferred to the seller account. On successful transaction, the seller will get an SMS to his mobile about amount confirmation through the GSM module. From this, it can be observed that the module is able to provide the simple and effective authentication mechanism at both the ends of the seller and buyer module with biometric features.

6 Algorithm Implementation

The implementation of an algorithm for both seller and buyer module is given in this section.

For the seller module, both the Bluetooth and fingerprint (fp) were initialized through serial communication protocols like UART0 and UART3 respectively (L-1). Later, the LCD initialized for a graphical display (L-2) and read the message of module initialization (L-3). Then, training of fingerprint (fp) is conducted for authentication purpose. During training, the sample of fingerprints (fp) is collected to $fp1$, $fp2$....fp_n times and are stored in the RAM of CPU. Further verification of fingerprint (fp). Is conducted by matching the samples collected (L-4 to L-5). In case matching is found then it will pass the account details of the seller to the buyer via Bluetooth (L-6). If matching is not found then repeat step L-4.

Similarly, in the buyer module, both the Bluetooth and fingerprint (fp) were initialized through serial communication protocols like UART0 and UART3 respectively (L-1). Later, the LCD initialized for TFT display (L-2) and read the message of module initialization (L-3). Then, training of fingerprint (fp) is conducted for authentication purpose. During training, the sample of fingerprints (fp) is collected to $fp1$, $fp2$....fp_n times and are stored in the RAM of CPU. Further verification of fingerprint (fp). Is conducted by matching the samples collected (L-4). In case matching is found then it receives the account details of seller via Bluetooth (L-5). The buyer will be having a facility of entering the amount to be paid and submit using TFT screen. On successful

---

Algorithm implementation for hardware based bidirectional payment system

---

Input: Seller and Buyer module
Output: Authenticated payment transaction
Start:
Module-1
L-1: Initialize → UART0 and UART3
L-2: Initialize →LCD
L-3: Read→ LCD
L-4: Training →*fp* samples
        L-4.1: TrainingSet→*fp1*
        L-4.2: Store→*fp1*
        L-4.3: TrainingSet→*fp2...fp$_n$*
        L-4.4: Store→*fp2...fp$_n$*
        L-4.5: Create →Template *fp1, fp2..., fp$_n$*
        L-4.6: Validate→*fp* in database
  if (matching)
    move → L-5
  else
    move→ L-4
L-5: Display →message & transfer Account details→Buyer module
L-6: Repeat L-4
End
Module-2
L-1: Initialize → UART0 and UART3, GSM module
L-2: Initialize →LCD
L-3: Read→ LCD
L-4: Training →*fp* samples
        L-4.1: TrainingSet→*fp1*
        L-4.2: Store→*fp1*
        L-4.3: TrainingSet→*fp2...fp$_n$*
        L-4.4: Store→*fp2...fp$_n$*
        L-4.5: Create →Template *fp1, fp2.., fp$_n$*
        L-4.6: Validate→*fp* in database
  if (matching)
    move → L-5
  else
    move→ L-4
L-5: Display →Enter amount & Submit
L-6: Send→ SMS via GSM module→Seller mobile
L-7: Repeat L-4
Stop:

transaction, the GSM module sends a confirmation SMS to seller mobile phone with GSM server (L-6). Similarly, if matching is not found then repeat step L-4. From this, it is observed that both the ends of the module are authenticated through a biometric parameter (fingerprint) of both seller and buyer, which yields highly secure m-payment. The following section gives an analysis of accomplished.

7   Results and Analysis

The coding part of the proposed hardware based bidirectional m-payment with the biometric parameter is performed by using Embedded Kiel software. The experimental setup of the hardware model is given in Fig. 3.

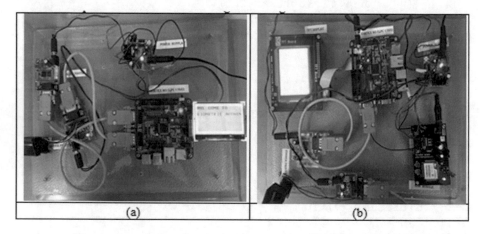

Fig. 3. Experimental set of (a) Seller Module and (b) Buyer Module

Both the module communicates with each other via Bluetooth devices and the hardware components are interfaced with each other. After giving supply to the modules, it starts functioning by displaying message into respective modules. However, authentication of each module is carried out by taking samples of fingerprints of seller and buyer and the GSM module is used to give payment confirmation to the seller. The biometric feature like a fingerprint is unique for each individual and hence the payment through the proposed model is more effective and secure. The performance analysis of this hardware model is performed by considering processing time and it takes only approximately 30 s of time to complete the entire transaction process.

8   Conclusion

Aiming with designing hardware based bidirectional security aware m-payment system using the biometric feature, this paper uses fingerprint for authentication. The proposed model incorporates the hardware components like low powered ARM Cortex M3 is used for real time implementation, GSM module, power unit, Bluetooth module is used to build communication among both seller and buyer modules. In order to execute the model, embedded Kiel software is adapted for programming. From verification and validation of the modules, it is observed that the proposed model is able to secure the m-payment with fingerprint at greater accuracy. As each human being is having unique fingerprint feature, the m-payment model with this feature offers higher security than existing ones. The performance is measured in terms of processing time and is noted approximately 30 s.

This model can be incorporated for futuristic researches with different biometric features to enhance security level. Also, the system can be considered for real-time application where processing time is must for each transaction.

References

1. Raina, V.K.: Overview of mobile payment: technologies and security. In: Electronic Payment Systems for Competitive Advantage in E-Commerce, pp. 186–222. IGI Global (2014)
2. Carr, M.: Framework for mobile payment systems in India. In: Mobile and Ubiquitous Commerce: Advanced E-Business Methods, pp. 237–254. IGI Global (2009)
3. Carr, M., et al.: Machine learning techniques applied to profile mobile banking users in India. Int. J. Inf. Syst. Serv. Sect. (IJISSS) 5(1), 82–92 (2013)
4. Wirtz, B.W., Ullrich, S.: Mobile marketing in multi-channel-marketing. In: Erfolgsfaktoren des Mobile Marketing, pp. 165–181. Springer, Heidelberg (2008)
5. Yu, H.-C., Hsi, K.-H., Kuo, P.-J.: Electronic payment systems: an analysis and comparison of types. Technol. Soc. 24(3), 331–347 (2002)
6. Kungpisdan, S.: Design and Analysis of Secure Mobile Payment Systems. PhD dissertation, Faculty of Information Technology, Monash University (2005)
7. Kshetri, N., Acharya, S.: Mobile payments in emerging markets. IT Prof. 14(4), 9–13 (2012)
8. Niranjanamurthy, M., et al.: M-commerce: security challenges issues and recommended secure payment method. Int. J. Manage IT Eng. 2(8), 374–393 (2012)
9. Halonen, T.: A System for Secure Mobile Payment Transactions. Master's thesis, Department of Computer Science, Helsinki University of Technology (2002)
10. van der Heijden, H.: Factors affecting the successful introduction of mobile payment systems. In: Proceedings 15th Bled e-Commerce Conference, pp. 430–443 (2002)
11. Abughazalah, S., Markantonakis, K., Mayes, K.: Secure mobile payment on NFC-enabled mobile phones formally analyzed using CasperFDR. In: Proceedings of the 13th IEEE International Conference on Trust, Security and Privacy in Computing and Communications (TrustCom 2014), pp. 422–431. IEEE, Beijing, September 2014
12. Zhang, B., Ren, K., Xing, G., Fu, X., Wang, C.: SBVLC: secure barcode-based visible light communication for smartphones. In: Proceedings of the 33rd IEEE Conference on Computer Communications (IEEE INFOCOM 2014), pp. 2661–2669, Toronto, Canada, May 2014
13. Fan, C.-I., Sun, W.-Z., Hau, H.-T.: Date attachable offline electron-ic cash scheme. Sci. World J. 2014, 19 (2014)
14. Chaudhry, S.A., Farash, M.S., Naqvi, H., Sher, M.: A secure and efficient authenticated encryption for electronic payment systems using elliptic curve cryptography. Electron. Commer. Res. 16(1), 113–139 (2015)

15. Subpratatsavee, P., Kuacharoen, P.: Internet banking transaction authentication using a mobile one-time password and QR code. Adv. Sci. Lett. **21**(10), 3189–3193 (2015)
16. Ortiz-Yepes, D.A.: A review of technical approaches to realizing near-field communication mobile payments. IEEE Secur. Priv. **14**(4), 54–62 (2016)
17. Terán, L., Horst, C., Fausto Valencia, B., Rodriguez, P.: Public electronic payments: a case study of the electronic cash system in Ecuador. In: 2016 Third International Conference on edemocracy & eGovernment (ICEDEG), pp. 65–70. IEEE (2016)
18. Yang, J.-H., Lin, P.-Y.: A mobile payment mechanism with anonymity for cloud computing. J. Syst. Softw. **116**, 69–74 (2016)
19. Kang, B., Xu, D.: Secure electronic cash scheme with anonymity revocation. Mob. Inf. Syst. **2016**, 10 (2016)
20. Qin, Z., Sun, J., Wahaballa, A., Zheng, W., Xiong, H., Qin, Z.: A secure and privacy-preserving mobile wallet with outsourced verification in cloud computing. Comput. Stand. Interfaces **54**, 55–60 (2017)
21. Barkhordari, M., Nourollah, Z., Mashayekhi, H., Mashayekhi, Y., Ahangar, M.S.: Factors influencing the adoption of e-payment systems: an empirical study on Iranian customers. Inf. Syst. e-Business Manage. **15**(1), 89–116 (2017)
22. Braeken, A.: An improved e-payment system and its extension to a payment system for visually impaired and blind people with user anonymity. Wireless Pers. Commun. **96**(1), 563–581 (2017)
23. Yomas, J., Chitra Kiran, N.: Critical analysis on the evolution in the e-payment system, security risk, threats and vulnerability. Commun. Appl. Electron. (CAE) (2018)
24. Kiran, C.N., Narendra Kumar, G.: Modelling efficient process oriented architecture for secure mobile commerce using hybrid routing protocol in mobile adhoc network. Int. J. Comput. Sci. Issues (IJCSI) **9**(1), 311 (2012)
25. Chitra Kiran, N., Narendra Kumar, G.: A robust client verification in cloud enabled m-commerce using gaining protocol. Int. J. Comput. Sci. Issues **8**(6) (2012)
26. Kiran, N.C., Narendra Kumar, G.: Reliable OSPM schema for secure transaction using mobile agent in micropayment system. In: 2013 Fourth International Conference on Computing, Communications, and Networking Technologies (ICCCNT), IEEE (2013)
27. Kiran, N.C., Narendra Kumar, G.: Building robust m-commerce payment system on offline wireless network. In: 2011 IEEE 5th International Conference on Advanced Networks and Telecommunication Systems (ANTS), IEEE (2011)
28. Chitra Kiran, N., Narendra Kumar, G.: Implication of secure micropayment system using process oriented structural design by Hash chain in mobile network. IJCSI Int. J. Comput. Sci. Issues **9**(1), 329 (2012)
29. Kang, J., Nyang, D.: A privacy-preserving mobile payment system for mass transit. IEEE Trans. Intell. Transp. Syst. **PP**(99), 1–14 (2017)
30. Cao, C., Zhu, X.: Practical secure transaction for privacy-preserving ride-hailing services. Secur. Commun. Netw. **2018**, 8 (2018)

A Study of DDoS Reflection Attack on Internet of Things in IPv4/IPv6 Networks

Marek Šimon and Ladislav Huraj[(✉)]

Department of Applied Informatics, University of SS. Cyril and Methodius,
Trnava, Slovakia
{marek.simon, ladislav.huraj}@ucm.sk

Abstract. Today, many smart devices support IPv6 communication. IPv6 provides a larger address space capable to address up to 3.4×10^{38} devices. This is the reason why IoT devices have started addressing and networking in IPv6. However, a large number of IoT devices can be misused by hackers. A preferred method of attackers in recent years is the DDoS attack carried from IoT devices.

The main aim of this manuscript is to compare the behavior of commonly available IoT devices under a real DDoS reflection attack led in IPv4 and IPv6 networks and to analyse the potential of such attack.

Keywords: IPv6 networks · DDoS reflection attack · IoT devices

1 Introduction

Nowadays, DDoS attacks have been considered as significant threats in network security even in new IPv6 networks. The new IPv6 protocol has not only brought more addresses, but many changes in the communication have been implemented. Since the IPv6 supports an address demand of 2^{128} it is appropriate for the Internet of Things as well. Although at present the IPv6 packets account for less than 2% of all Internet traffic, their usage is increasing [1]. Due to the huge number of IPv6 addresses, a much larger attack is possible, and because many new devices and networks can support IPv6, but security tools do not, it is an attractive target with high potential for attackers.

DDoS attacks are driven by a large group of computers that overwhelm the victim machine with unreasonable traffic that is unable to handle. Sometimes attacks are led by attacked computers by unsuspecting users who are abused for such criminal activity. Moreover, hundreds of thousands of devices connected to the Internet, such as webcams, routers, set-top boxes or smart fridges can be misused for these attacks. The Internet of Things is still on the rise and it can be expected that similar attacks will increase. But in many IoT devices there is a lack of security from their manufacturers and implementation of IPv6 into them might aggravate the situation [2–5].

In the manuscript, the comparison between IPv4 and IPv6 is evaluated for the special case where a UDP-based DDoS reflection attack using IoT devices is led in the IPv4 as well as in the IPv6 network and the potential of such attack is inspected. Four IoT devices supporting IPv4/IPv6 protocol were tested for this purpose and the real

R. Silhavy (Ed.): CSOC 2019, AISC 984, pp. 109–118, 2019.
https://doi.org/10.1007/978-3-030-19807-7_12

behavior of the devices is described as well as the differences for both networks are analyzed.

The rest of the article is structured as follows: a short background about IoT, IPv6 and DDoS attack is described in Sect. 2. The device profiles and network scenario of the proposed tests is explained in Sect. 3. The performance as well as the experimental results are described in Sect. 4. Finally, the conclusion and future work is stated in Sect. 5.

2   Background

2.1   Internet of Things

The Internet of Things (IoT) is made up of many small smart resource-constrained devices continuously connected to the untrusted and unreliable Internet; each device is uniquely identifiable and they can interact and exchange data. Currently, there are billions of devices connected to the Internet, and the trend suggests that the number will continue to rise enhanced by phenomena like smart homes, smart cities or industrial IoT.

The safety of these devices is still a big question, most IoT devices are cheaply made actuators and sensors which are likely to be very insecure. But even when they are secured with mechanisms like authentication or encryption, they are still exposed to various attacks from the Internet [6–8].

We tested four heterogenous IoT devices with the ability to communicate in IPv4 as well as in IPv6 networks in the experiments for a special kind of DDoS attack. Each tested device is a specific IoT device representing a whole group of IoT devices from the point of view of devices' specificity of the purpose they serve and of the degree of processing power. The IoT devices are representatives of: (i) single board computer, (ii) virtual assistant, (iii) surveillance IP camera, and (iv) network printer.

2.2   IPv6

IPv6 (Internet Protocol Version 6) is the upcoming protocol for communication on the current Internet or computer networks created by the Internet. IPv6 has replaced the IPv4 protocol. It brings in particular a massive extension of the address space and for example, also the ability to allocate their own IPv6 addresses to all devices. It also improves the ability to transfer high-speed data.

Although IPv6 deployment brings many improvements over the IPv4 protocol, many vulnerabilities persist, and even the IPv6 protocol also brings many new problems that have not previously occurred [9, 10].

There are many general comparisons of IPv4 and IPv6 protocols; e.g. Table 1 [11] which shows the main differences between these protocols. However, differences often appear only when they are actually deployed in specific situations. Our experiments are focused on a special case using ICMP and ICMPv4 messages for DDoS reflective attacks and on comparison of the achieved results for both protocols.

Table 1. Short comparisons of IPv4 and IPv6 protocols [11]

Group	IPv4	IPv6
Addressing	32 bits	128 bits
Fault isolation	ICMP	ICMPv6
Address resolution	ARP	ICMPv6 NS/NA
IPsec support	Optional	Recommended
Fragmentation	Both in hosts and routers	Only in hosts

2.3 DDoS Attack

DDoS attacks are the stage of DoS (Denial of Service) attacks that have been known since the 80s of the 20th century. The major difference between these types of attacks is that the DDoS attack is essentially distributed - it comes from more than one source, making it also more effective and dangerous. The main goal of the attack is to consume the computational resources and network bandwidth of the target system and to cause unavailability of the system. The attack is characterized by a larger number of computers or other devices, e.g. IoT devices, trying to overwhelm the victim of attack. The attack is often conducted without the knowledge of the owners of the attacking computers [12].

Distributed DoS attacks have several advantages compared to simple DoS attacks. First, a larger number of attacking devices can generate more network traffic than a single computer, and thus cause a greater load on the victim system. Second, attacking multiple attackers are more difficult to be traced and filtered out of the network traffic. Moreover, the attacker's scalability can also be easily increased by adding other attacking devices if the victim increases the transmission capacity of his or her Internet connection or the computing resources.

A special case of DDoS is a DDoS reflection attack using resources that belong to another, third party. The attack is based on sending bogus requests on a large number of innocent devices to respond to these requests. Invalid requests have a spoofed victim address as the source address and the response is sent to the victim, not back to the attacker. Consequently, the victim is overwhelmed by the responses to these requests. The vulnerable service chosen is the one that replies with a message much larger than the one received [13]. In our experiment, the UDP-based DDoS reflection attack was applied.

3 Device Profiles and Network Scenario

IoT devices are non-standard by their nature; therefore, there is no prototype of IoT device that could be considered as standard for the IoT environment. In the testing of our experiment, the four most popular IoT devices were used, namely IP Camera D-Link DCS-4703 3-Megapixel, network printer Kyocera FS-C5150DN, Raspberry Pi and Google Home Assistant. Based on the classification in [8], the IoT devices used in the experiment are categorized into three groups on the grounds of their specific functionality characterizing analogous IoT devices as well as on their variations between the IoT devices, as shown in Table 2.

Table 2. Devices used in the experiment

Group	IoT device	Function(s)
I	Raspberry Pi	Embedded computing devices
I	Google Home Assistant	Embedded computing devices
II	IP Camera D-Link DCS-4703	Surveillance camera
III	Kyocera FS-C5150DN	Network printer

Raspberry Pi as a single board computer represents a small computer system. It is a cheap, small, education-oriented and portable size of computer board that allows a user to develop and run several applications. Raspberry Pi works in the same way as a standard PC, needing a keyboard for command entry, a power supply and a display unit; it contains essential (memory, processor, graphics chip) and other optional devices (various interfaces and connectors for peripherals) and uses an operating system [14].

The Google Home Assistant is a device constructed as a Virtual Assistant to allow for controlling other IoT devices in a home. It interacts via a special respective app or by voice. The user is allowed to control playback of videos or photos, to listen to music or receive news updates entirely by voice; other services, both in-house and third-party, are integrated. Such devices are highly interesting for attackers since infection of this assistant can lead to infection of other devices connected to it. Additionally, a security update of a Virtual Assistant is usually rare, making it vulnerable to exploitation. Moreover, a frequent problem of IoT devices is that the default password to the device access is retained [15].

Google Home Assistant can be with Raspberry Pi and some other devices (e.g. Intel Galileo, Intel Edison, Arduino, Amazon Echo) included into group of embedded computing devices (ECDs) that read data from sensors, receive and send instructions to actuators, and interpret sensor data. Compared to Raspberry Pi, Google Home Assistant has become a very popular device in ordinary households and is usually directly connected to the Internet [16].

IP Camera D-Link DCS-4703 is 3-megapixel surveillance camera. This category of IoT devices represents devices not only serving their special functions but allowing for several extensions via available apps. Nowadays, lots of companies provide surveillance cameras to provide remote monitoring services. These cameras generally offer a web interface for configuration of the camera by the owner, send and receive data and watch the video captured. Security of IP cameras appears to be a serious issue in light of their widespread use [17].

The network printer Kyocera FS-C5150DN is the last IoT device used. It acts for IoT devices strictly providing one functionality, interacting through a specific respective app; on the other hand, this kind of device is commonly used in many offices or homes and the devices feature web-based interfaces through which owners can remotely check the status of the printer, configure or reboot it.

After defining the device profiles that were going to be analysed, a network environment was created to support them. The four devices and a switch were wired, connected by 100 Mbps Ethernet interface for both cases, i.e. for IPv4 as well as the IPv6 scenario where Google Home Assistant communicated via WiFi 802.11n networking.

Figure 1 depicts the experiment environment used to generate the tests. The Packet generator was used to generate the UDP packets of each time frame, starting from 100 till 600 μs with steps of 50 μs to see the best packet frequency. For this purpose, the packet manipulation tool for computer networks Scapy was used. Scapy as an open source packet manipulation environment written in Python language is designed to build custom packets to communicate with other devices. Using such kind of software, it is possible to construct packets by modifying field values which are convenient to the user even if these values are not foreseen. This fact allows the assessment of the behavior of devices and network services when they receive an unexpected packet [18].

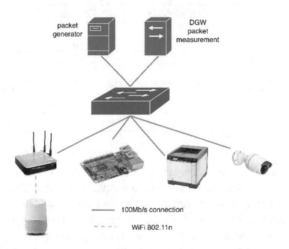

Fig. 1. Devices connection and the testing IPv4/IPv6 network topology

The second machine – DGW packet measurement machine – worked as a Default Gateway for IoT devices and all communications from IoT devices to spoofed addresses outside the testing network went through this DGW. Moreover, no communication among IoT devices was considered during the experiment. A router advertiser daemon for IPv6 Stateless Address Autoconfiguration (SLAAC) was installed on this machine.

All IoT devices were used for DDoS reflection attack without being infected or otherwise modified and the sent packets were reflected by them as ICMP messages against the target network. A successful DDoS attack is available even with small amount of nodes, as can be seen from our previous research in this field [19, 20].

4 Performance and Results

A DRDoS attack was performed using the tool Scapy. We generated the UDP flood attack for 105 s separately for each IoT device; a UDP datagram with a spoofed source IP address and with a nonexistent destination port was sent for different time periods. Each IoT device generated for every UDP datagram a reflected error ICMP message *Destination Unreachable* Code Field *Port unreachable* regarding a limit of its resources. Since the spoofed source IP address was in the UDP datagram, the IoT device resent the message to target machine. The analysis of IoT behaviour as well as the comparison of IPv4 and IPv6 networks for DDoS reflection attack is based on two metrics: the amount of packets reflected from a particular IoT device to the target machine per second; and the average of all reflected sent bytes, Table 3.

Table 3. Sent reflected bytes per second – the average amount

	Raspberry Pi	Google Home Assistant	IP Camera D-Link DCS-4703	Network printer Kyocera
Reflected bytes in IPv4	4239.28	4398.68	4126.61	4164.78
Reflected bytes in IPv6	8806.96	8690.68	8844.44	8861.74
Ratio IPv6/IPv4	2.077	1.976	2.143	2.128

It should be noted that in IPv4, the size of the UDP request sent by the attacker consists of a 1 byte payload UDP datagram (a simple number) and the UDP header is 8 bytes long; subsequently it is on the network layer encapsulated into the 28 byte IP datagram; after encapsulation into the Ethernet frame on the data link layer, the whole frame has the size of 43 bytes. The size of the outgoing message from the reflector (IoT device) to the victim machine consists of the 29 bytes payload ICMP packet with an error message and of 8 bytes ICMP header with added 20 bytes for IP packet encapsulation. Consequently, the encapsulation into the Ethernet frame on the data link layer is done and finally, the whole leaving frame is 71 bytes long.

The situation is different in IPv6. The UDP payload is 1 byte, the UDP header is 8 bytes, followed by 40 bytes of IPv6 header and by 14 bytes of Ethernet header on the data link layer; the whole frame is 63 bytes long. The IoT device replies with an ICMPv6 message *Destination Unreachable Message* where 8 bytes is the payload and then the ICMPv6 message can contain as much of invoking packet as possible without the ICMPv6 packet exceeding the minimum IPv6 Maximum Transmission Unit (MTU) [21], in our case 49 bytes. After encapsulation into the Ethernet frame on the data link layer, the whole leaving frame has the size of 111 bytes which is 1.76 times bigger than incoming frame and 1.56 time bigger than the ICMPv4 leaving frame.

As can be seen from Fig. 2, DDoS reflective attacks are possible using IPv4 as well as IPv6 for the proposed topology. But application attacks in IPv6 networks show some differences. The risk associated with the use of an IoT device for the attack in the new

IP version approximately doubled, as shown in Table 3. The main reason for the increase is caused by the modified packet header. On the other hand, the increase does not depend only on the size of the IPv6 header; in our case the IPv6 header increased only 1.56 times. The increase is also caused by the ability of IoT devices in IPv6 networks to generate the ICMP error messages quicker compared with IPv4 networks. Other influences may be considered as negligible due to the connection of all examined devices to a 100 Mbps switch in the tested network topology.

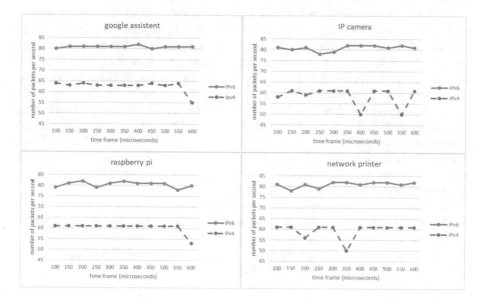

Fig. 2. Particular IoT devices: median values of the reflected packets per second

The increase is most significant for the IP Camera, as shown in Table 3. For example, Fig. 3 illustrates the best selected packet frequency of 550 μs, where the IP Camera reflected the highest volume of ICMP messages.

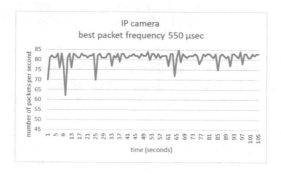

Fig. 3. IP Camera D-Link: best packet frequency 550 μs in IPv6

The next experiment dealt with our previous research [22] testing the ability of IoT devices to form directed DDoS reflective attacks on an individual target machine in IPv4 to overwhelm the machine. The results of the IPv6 network experiment confirmed the results found in IPv4 networks. If the IoT devices are flooded with the UDP packets only to one exact spoofed IP address directed to a victim, the amount of reflected packages is very low (does not exceed value 5 packets per second) and so it is not possible to perform such a DDoS attack to overwhelm an individual machine; it can be exploited for an attack on the whole network and its connectivity.

On the other hand, while in the IPv4 network the directed DDoS reflective attack is meaningful mostly for special cases where the attack is targeted at a large company with lots of IP addresses (i.e. data centres), the situation in IPv6 networks is slightly different. Usually, a client in data centres has assigned a subnet mask class C in IPv4 networks where it is possible to address only up to 254 hosts per subnet. An IoT device reflecting all 254 IPv4 addresses in the directed DDoS reflective attack is not loaded with such a small amount of IP addresses and the amount of reflected packets is small as well. In contrast, the typically allocated prefix in IPv6 networks is 64 bits for the network and the rest of the 64 bits is for the host portion; and therefore it is possible to address 2^{64} hosts per subnet. Hence the IoT device could be loaded to its limits by reflecting packets with direct IPv6 addresses up to 2^{64} to the victim network and so it can overburden the connectivity of the network.

Generally, the IoT devices can be implemented as reflectors into a DDoS reflection attack for both IPv4 as well as IPv6 networks. It should be noted that only freely available equipment as well as Open Source software or Free software (GPLv2+) were used for the implementation of the DDoS attack and that no IoT devices had to be compromised, hacked or however modified.

5   Conclusions and Future Work

The aim of this manuscript was to design, implement and test a real DDoS reflective attack on IoT devices in the IPv4/IPv6 networks. A range of IoT devices was used to perform the DDoS reflective attack: IP Camera D-Link, network printer Kyocera, Raspberry Pi and Google Home Assistant. The comparison between attacks led in the IPv4 network as well as in IPv6 network was analysed. Consequently, the experimental outcomes demonstrate that the DDoS reflection attack potential involving IoT devices exists in both networks.

From the experiments presented in the manuscript, it can be seen that risk of DDoS attack employing IoT devices is in IPv6 networks more efficient and more overwhelming than in IPv4 networks. That is the reason why new IoT technologies should be developed with a view to possible DDoS reflection attacks with emphasis on countermeasures.

Future research will also focus on scaling of DDoS reflective attacks and on involving a larger variety of IoT devices. In addition, other aspects of security in IPv6 networks as well as countermeasures to discovered vulnerabilities can be investigated.

Acknowledgements. The work was supported by the grant VEGA 1/0145/18 *Optimization of network security by computational intelligence.*

References

1. Cusack, B., Tian, Z., Kyaw, A.K.: Identifying DOS and DDOS attack origin: IP traceback methods comparison and evaluation for IoT. In: Interoperability, Safety and Security in IoT, pp. 127–138. Springer, Cham (2016)
2. Srivastava, S., Pal, N.: Smart cities: the support for Internet of Things (IoT). Int. J. Comput. Appl. Eng. Sci. **6**(1), 5, 5–7 (2016)
3. Ölvecký, M., Gabriška, D.: Motion capture as an extension of web-based simulation. Appl. Mech. Mater. **513**, 827–833 (2014)
4. Horváthová, D., Siládi, V., Lacková, E.: Phobia treatment with the help of virtual reality. In: 13th International Scientific Conference on Informatics, pp. 114–119. IEEE (2015)
5. Hosťovecký, M., Novák, M., Horváthová, Z.: Problem-based learning: serious game in science education. In: ICEL 2017-Proceedings of the 12th International Conference on e-Learning. ACPI 2017, pp. 303–310 (2017)
6. Raza, S., Wallgren, L., Voigt, T.: SVELTE: real-time intrusion detection in the Internet of Things. Ad Hoc Netw. **11**(8), 2661–2674 (2013)
7. Halenar, I., Juhasova, B., Juhas, M.: Proposal of communication standardization of industrial networks in Industry 4.0. In: IEEE 20th Jubilee International Conference on Intelligent Engineering Systems (INES), pp. 119–124 (2016)
8. Habibi, J., Midi, D., Mudgerikar, A., Bertino, E.: Heimdall: mitigating the Internet of Insecure Things. IEEE Internet Things J. **4**(4), 968–978 (2017)
9. Dirgová Luptáková, I., Pospíchal, J.: Community cut-off attack on malicious networks. In: Conference on Creativity in Intelligent Technologies and Data Science, pp. 697–708. Springer, Cham (2017)
10. Pishva, D.: IoT: their conveniences, security challenges and possible solutions. Adv. Sci. Technol. Eng. Syst. J. **2**(3), 1211–1217 (2017)
11. Saad, R., et al.: Design & deployment of testbed based on ICMPv6 flooding attack. J. Theor. Appl. Inf. Technol. **64**(3) (2014)
12. Singh, S., Gyanchandani, M.: Analysis of botnet behavior using queuing theory. Int. J. Comput. Sci. Commun. **1**(2), 239–241 (2010)
13. Mansfield-Devine, S.: The growth and evolution of DDoS. Netw. Secur. **2015**(10), 13–20 (2015)
14. Maksimović, M., Vujović, V., Davidović, N., Milošević, V., Perišić, B.: Raspberry Pi as Internet of Things hardware: performances and constraints. In: 1st International Conference on Electrical, Electronic and Computing Engineering - IcETRAN 2014, pp. ELI1.6.1-6 (2014)
15. Koopman, M.: Preventing Ransomware on the Internet of Things (2017). https://pdfs.semanticscholar.org/419f/8d22ba3eea1f4d9fbcd6b3b68d294504d276.pdf
16. Bastos, D., Shackleton, M., El-Moussa, F.: Internet of Things: a survey of technologies and security risks in smart home and city environments. In: International Conference Living in the Internet of Things: Cybersecurity of the IoT – 2018, London, UK (2018). https://doi.org/10.1049/cp.2018.0030
17. Tekeoglu, A., Tosun, A.S.: Investigating security and privacy of a cloud-based wireless IP camera: NetCam. In: IEEE 24th International Conference on Computer Communication and Networks (ICCCN), USA, pp. 1–6 (2015)

18. Kobayashi, T.H., et al.: Using a packet manipulation tool for security analysis of industrial network protocols. In: 2007 IEEE Conference on Emerging Technologies and Factory Automation, ETFA, IEEE (2007)
19. Šimon, M., Huraj, L., Čerňanský, M.: Performance evaluations of IPTables firewall solutions under DDoS attacks. J. Appl. Math. Stat. Inf. 11(2), 35–45 (2015)
20. Šimon, M., Huraj, L., Hosťovecký, M.: IPv6 network DDoS attack with P2P grid. In: Creativity in Intelligent, Technologies and Data Science, pp. 407–415. Springer, Cham (2015)
21. Conta, A., Gupta, M.: Internet control message protocol (ICMPv6) for the internet protocol version 6 (IPv6) specification. Request for Comments 4443 (2015)
22. Huraj, L., Šimon, M., Horák, T.: IoT measuring of UDP-based distributed reflective DoS attack. In: IEEE 16th International Symposium on Intelligent Systems and Informatics (SISY 2018), pp. 209–214. IEEE, Serbia (2018)

Perspective on Slovakia's Freelancers in Sharing Economy – Case Study

Michal Beno[✉]

VSM/City University of Seattle, Panonska Cesta 17, 851 04 Bratislava, Slovakia
michal.beno@vsm-student.sk

Abstract. In this paper, we provide an overview of freelancers from a Slovakian perspective. Our purpose specifically is to provide an insight into the field of freelancing in a sharing economy and to offer general guidelines for beginner freelancers. The paper reviews survey data on users of freelancers in Slovakia on the basis of literature reviews and survey data. It illustrates that freelancers comprise a skilled, specialised workforce that enables businesses to reduce their dependence on full-time employees. It also reviews the freelancers' barriers to entry, risk and financial requirements, while they enhance their skills and improve their business agility, flexibility and efficiency. Freelancers are an important input that enables innovative and entrepreneurial economies. Possible barriers to freelancing are discussed, paying attention to demands on freelancers, e.g. lack of stable income and boundaries between work and privacy, irreplaceability and getting paid on time. The results indicate that freelancing in Slovakia is on the increase, and most freelancers are specialists, predominantly university-educated, highly-qualified professionals, using modern technology with smaller numbers of clients. The four key features for success in freelancing are a good reputation, expertise, reliability and contacts.

Keywords: Sharing economy · Freelancer · Slovakian perspective

1 Introduction

The Industrial Revolution brought employees from their homes to the factories. With Information and Communication Technology (ICT), we have reached full circle and the reverse is possible, with employees now able to move back to their homes [1].

Generally, the Internet had its genesis as a tool for sharing information, and its earliest origins served only scientific and military communities. Progressively, it became the World Wide Web, including the power of networking. The sharing economy uses this tool to match people who want to share assets and services such as mobility, food, education, home and office, leisure/travel and other.

The spread of globalisation and the development of modern innovative technologies, social changes and, associated with them, the increase in collective environmental awareness, have augmented the interests in mobile and alternative forms of working in recent years [2–4]. The population is exploring products and services through ICTs, e.g. desktops, laptops, tablets, smartphones and even virtual reality devices.

© Springer Nature Switzerland AG 2019
R. Silhavy (Ed.): CSOC 2019, AISC 984, pp. 119–130, 2019.
https://doi.org/10.1007/978-3-030-19807-7_13

Using these technological innovations, more and more organisations have started to redesign their work approach. We feel strongly that, central to this new approach, is the fact that employees are required to organise their work flexibly.

Selloni [5] emphasised that collaborative consumption, collaborative economy, on-demand economy, peer-to-peer economy, zero-marginal cost economy, and crowd-based capitalism are examples of different interpretations interconnected with the notion of a sharing economy.

Sharing is a natural, pro-social behaviour and has always been a sign of solidarity, cooperation and mutual aid [6]. Nowadays, we recognise the networked information economy. We understand the sharing economy as a system built around the sharing of human, physical and intellectual resources of goods and services by different people and organisations, for example Airbnb, Lyft, Uber, Amazon, Ebay, HomeAway, and Freelancer.

Freelancers are usually defined as those genuinely in business on their own account, working alone or with partners or co-directors, in skilled non-manual occupations. Other characteristics of freelancers such as being enablers of innovation, entrepreneurship and risk-management have also come to the fore [7].

This paper is intended to provide a concise survey of some recent theoretical and empirical research on the role and value of freelancers in Slovakia. This evidence gives an insight into the functions that freelancers serve in the modern economy. In the past, the freelancer has been viewed as a cheap substitute for full-time employees, but nowadays the freelancer reflects a modern organisational form such as independent contractor, moonlighter, diversified worker, business owner, temporary worker, co-worker [8] and many others. In this environment these workers are increasingly skilled, more expensive and also create employment.

In the following section, we briefly outline the methodology which we used in our research project. The third section gives a brief overview of the concept of a shared economy. The fourth section provides an account of freelancing in the modern digital economy. In the fifth section, a case study is presented to analyse the results of an e-mail questionnaire survey. The paper closes with our conclusions.

2   Methodology

We carried out a two-step research project to study freelancers in Slovakia. The first step was a literature review on the extent and nature of a sharing economy and free-lancers, including examples of best practice as well as factors contributing to the effectiveness of freelancers. The second was a questionnaire survey where we e-mailed questionnaires to all registered freelancers who use the following web pages: www.flexipraca.sk, www.free-lance.sk, www.freelanceri.sk, and www.startitup.sk. The data used in this study were collected from November to December 2017, with a probability sample of 310 after a response rate of 52.2%.

Electronic surveys offer the means to carry out large-scale data collection other than through organisations at the centre of power in society [9]. Technology provides an inexpensive mechanism for conducting surveys online, instead of through the postal service [10, 11] and one in which costs per response decrease instead of increase significantly as sample size increases [12]. Electronic surveys are becoming increasingly common [13], and research comparing electronic vs. postal surveys is starting to confirm that electronic survey content results may be no different from postal survey results, yet provide the distinct advantages of speedy distribution and response cycles [14, 15].

One can divide the collection of survey data via computers into three main categories, based upon the type of technology used for distribution and for collecting the data, namely point of contact, e-mail-based and web-based. In this paper, we use the second option.

The rapidly expanding body of literature on electronic survey techniques reflects a growing concern among researchers over the methodological issues associated with their use [9, 16–20]. The e-mail questionnaire contained several types of questions for respondents to answer. Some questions were also open-ended, which allowed respondents to submit their own answers.

3   The Rise of Sharing Economy

We divide the evolution of the sharing economy into three phases: (1) communication (e.g. Yahoo, Aol), (2) Web 2.0 and social media (e.g. Google, Facebook, Twitter) and (3) the sharing economy. Sharing-economy workers drive you home, deliver groceries and office lunches or rent your home.

According to Stephany [21], sharing economy can be a confusing term. In practice, various definitions of this term are found, e.g. gig economy or on-demand economy, collaborative economy, access economy, peer-to-peer economy [5, 22]. Stephany [23, pp. 9] has provided a concise definition with five main limbs: "The sharing economy is the **value** in taking **underutilised assets** and making them **accessible online** to a community, leading to a **reduced need for ownership** of those assets."

We believe that the sharing economy will experience significant growth in the future. According to a PricewaterhouseCoopers report [24], five key sharing sectors (travel, car sharing, finance, staffing, and music and video streaming) have the potential to increase global revenues from roughly $15 billion today to around $335 billion by 2025. China is leading the sharing-economy revolution with 600 million people involved in the sharing economy; the transaction volume of China's sharing economy topped $500 billion in 2016 [25]. An incomplete list of the known sharing models in transport, accommodation, goods, services, finance and media/entertainment is given in Table 1 (Author's own compilation).

Table 1. Main Areas of sharing economy.

Transport	Accommodation	Goods	Services	Finance	Media/Entertainment
Uber, BlaBla Car, Shared Parking, Lyft, Ola	Airbnb, InstantOffices, Housetrip, HomeAway, Wimdu	Patagonia, Ebay, Etsy, trademe	UpWork, TaskRabbit, Helplinge, Freelancer.com, GOGET, fiverr	Auxmoney, Cashare, Bondorra, TransferWise, LendingClub, bitcoin	Spotify, Netflix, Amazon Prime, Aldi Life, Joox

Sharing-economy flag-bearers like Uber essentially provide average cars in a premium way, but own no cars themselves [26]. On its site, Airbnb makes everyday apartments look luxurious in order to drive higher booking rates [27]. Homejoy gives the population easy, cheap access to products and services that would otherwise go unused, free of the burden of ownership [28] (see the Eckhardt and Bardhi [29] headline saying "The Sharing Economy Isn't About Sharing at All"). In 2017, Airbnb had 150 million users, an average of 500,000 stays per night, and in 2015 it operated in 65,000 cities in 191 countries [30]. In 2017, Uber recorded 50 million rides and 7 million drivers in 450 cities [31].

According to Lindak and Hayek Foundation [32], the sharing economy in Slovakia is currently facing a difficult time. Kovanda emphasises that we should prepare ourselves for Uberisation with the key benefit of a shared economy which is the social dimension [33] according to the research of Alan Krueger [34].

Through the technologies, there is a more effective link between supply and demand. This will lead to more efficient use of economic resources - both labour and capital - which will also mean a decline in unemployment, a saving in transaction costs and lower prices. The result of this will be greater economic efficiency and lower depletion, leading to less wastage of natural and economic resources. Along with Uberisation there will be a decrease in information asymmetry - the concept of information asymmetry has become a key theoretical argument for the regulation of various species. With its decline, there will be pressure to reduce regulations, which Kovanda stresses as the main advantages of the shared economy [33].

4   Occupation Freelancer

The number of independent professionals (freelancers) in the EU rose from 7.7 million to 9.6 million between 2008 and 2015 [35]. Freelance workers, also called free agents, temporary/contingent workers or e-lancers [36–38] follow a type of non-standard work arrangement that comprises a large percentage of the workforce [36, 37, 39]. Malone and a colleague [40] coined the term "e-lancer", or an electronically connected freelancer, to describe the person engaged in this system of working. In other words, he/she is a person who carries out work via electronics.

Several criteria are relevant to determining freelance status and to distinguishing freelancers from other labour market factors. According to Smallbone and Kitching [41], freelance status attaches to workers performing particular kinds of work, in particular end-user relationships. Hence, individuals might be freelancers in one relationship and employees in another; indeed multiple jobholders may occupy several statuses simultaneously.

Generally, there is no typical freelancer because of differences in personal, work and organisational characteristics, all of which tend to influence business processes and outcomes [41]. Globally the workplace is evolving. Freelancers are widely recognised as having an important impact on today's economy [7]. The modern innovation and dynamic economy has provided the opportunity for the rise of freelancers.

Freelancers are typically defined on the basis of their specific occupations and skills [7] and as self-employed workers and directors of limited companies without employees in the primary or secondary paid work role, on either a full-time or part-time basis [42]. These workers are highly skilled and provide professional services to existing businesses, enabling the latter to be: innovative, flexible and agile, able to manage entrepreneurial risk, and capable of prospering despite greater market uncertainty [7, pp. 6]. We understand this work form as independent contractors who are usually hired by individuals or companies to perform specific tasks.

These workers work for various clients, and the relationship between the client and the freelancer is often settled in a contract of service which outlines specific terms. This is the main difference from regular employment in which the employer/employee relationship is determined in a contract of service. The freelancer is not an employee, since he/she is not dependent on the employer and has more autonomy than regular employees. But freelancers are responsible for paying their own taxes and insurances, which means that they also lack the benefits enjoyed by employees [41]. Burke [43] emphasises that freelancers are often paid for their productivity rather for the time input, and for example do not get a regular wage. From an entrepreneurial viewpoint, the main distinguishing aspect of freelancers from small business owners is that the former do not hire other workers. The next defining factor is short-term and task-based, because contracts are set for a limited time, and once they have been completed, the freelancer must find another job.

According to the data from Horowitz et al. [44], the main benefits for the freelancer are a flexible schedule, diversity of projects, maintaining a more balanced life, and freedom from politics. Unstable income and the need to be constantly looking for work are among the disadvantages.

To be able to perform the work of a freelancer in Slovakia, the worker must register the particular activity in the trade licence. The Trade Licence Office will register all trade activities in the trade licence certificate, including the business name and business account. A fee has to be paid for each trade. The Tax Office will send the freelancer a notice with the tax identification number, which has to be stated on all invoices [45].

Slovakia's freelance status according to our results can be summarised as shown in Table 2.

Table 2. Freelance status.

Freelance status	
Work status	Self-employed freelancers or partners mostly without employees
Occupation	Creative, managerial, professional, scientific, technical occupations only
Number of clients	Workers with single client per contract
Contract duration	Contract of any duration depends on customer
Primary/Secondary work role	Primary or secondary paid work roles; full-time or part-time basis

5 Results

There are many freelancers in Slovakia, and the numbers will continue to grow. This modern occupation can be seen as a reversion to the system of the old crafts. In the past it was the smith, carpenter or potter who worked at home, but now the freelancer works at home and supports himself and his family with his craft. According to our results, the first five years are important, as this is a crucial period for the freelancer to obtain sufficient work and to establish a reliable network of clients and references so as to have enough supply and demand.

Why is it important that Slovakia's citizens should be working in a modern way? The economic implications of a more flexible workforce are significant. This kind of workforce can pursue a meaningful independent life anytime and anywhere. It also reflects the change of culture and social structure in the conversion from an agrarian to an industrial society and from an industrial to an information society, including the effects of this on social structures e.g. civil rights, worktime, and workforce participation.

In Slovakia, it is no longer an exception to work from home, mostly in the form of freelancing. On the basis of our survey results, we firstly summarised some advice for people interested in this kind of work. The first requirement is to create a portfolio by stating the information about your personality, experience, knowledge and abilities on the Internet and elsewhere. Most of the respondents emphasised the need to be truthful. One of the respondents expressed it as follows: "People are looking everywhere. Remember, once you put something on the Internet, the more popular you become, the less likely it is that potential customers will not know all about you … this means not only a good reputation but also the danger involved in giving the wrong information, which could make people suspect you of cheating, failure to honour the terms of the contract, making mistakes…". The next requirement is the fee structure for the work the freelancer does. On one hand, the freelancer expects a fair reward for the work done, and on the other hand it is good to have an overview of how much it costs in general: "One design project has a different value than designing prints for multiple promotional items". As advised in the third chapter, a contract is non plus ultra in this work environment. Contracts, invoices or any other documentation relating to the work must be saved, preferably with a back-up on an external hard disk. The freelancer has to be reliable and trustworthy (this is the key to professionalism). All the interviewees stressed that in this work form they had to be constantly learning, had to broaden their general outlook, expand their knowledge and share with others in order to acquire contacts.

The survey defined freelancers as individuals who have engaged in contract work in the past 12 months. The impact of freelancers on the economy is very important as this workforce adds money to economic growth through their work every year. A total of 32% of the freelancers reported an increased demand for their services in the three past years, compared to 15% who reported a decrease. With more demand comes more work, and 45% of freelancers expect to work more hours, compared with 10% who expect to work less. A total of 97% of the respondents emphasised that additional work is a plus point.

Freelancers covered by the survey data are mostly specialists with higher qualifications and they work primarily on a desktop computer or a notebook. About 40% of Slovakia's freelancers are women, 58% of the respondents have a university education, 78% speak English, 25% speak German and 10% Russian; 75% normally work from home, and 81% respond to the demand within one business day.

The survey found that the most common reasons for going freelance were "flexibility and freedom" (86%), and not "to earn extra money" (45%). Interestingly, 70% of the respondents began freelancing by choice and as the main source of income. Some comments were: "I have more control over when and where I work and what I do for my work", "I can work anytime/anywhere", "I set my schedule and work on opportunities that interest and challenge me".

This form of work also has barriers, which are identified as a lack of stable income (65%), disappearance of the boundary between work and privacy (48%), irreplaceability (46%) and difficulty in finding work (37%). Some other comments were: "tired and exhausted", "it is hard, but very nice", "it is possible to work more and have less restrictions". Most of the interviewees answered that if it were easier to find offers of jobs, their income would be more consistent (48%). It is not just finding a job that can be challenging, but also getting paid on time, and 48% of responses pointed this out as the next barrier for freelancing. A total of 49% of the freelancers described the payment behaviour of customers as very good, and 33% as good because they paid according to the terms of the agreed payment conditions. The respondents highlighted some methods for avoiding the problem of unpaid contracts, as displayed in Table 3.

Table 3. Methods for avoiding unpaid contracts (N = 310).

Method	Total share in %
Intuition	58.06
Payment in advance, proforma invoice	29.03
Search and colleagues interviewing	14.51
Documentation and recovery of claims	6.45
No answer	14.51

The leading responses in relation to this problem were: "I use a company for recovering claims", "In the invoice I also include non-payment penalties", "After payment deadline, I usually send a payment reminder, then I call and try to find a solution… if that does not help I contact my lawyer".

But technology is making it easier to search for work. All the interviewees agreed that the Internet had made it easier to find freelance work, e.g. on websites (74%), Facebook (45%), online portfolios (25%), blogs (19%) and Twitter (19%). In the words of one respondent: "there are more ways to find a job, to make contacts, to share with others, to connect with others… thanks to the modern equipment and its ability to network… social networking has drastically changed the dynamics of networking", "a professional freelancer is behind a computer every morning". In order to appreciate the freelancer's contribution with regard to work, we analysed the respondents' perception of the technology tools, as depicted in Table 4.

Table 4. Technology tools used for work (N = 310).

Tool	Total share in %
Smartphone and apps	65.16
Gmail or Google Apps email	58.06
Disk Google	43.55
Google Calendar	41.94
DropBox	37.10
DataBox	35.48
Evernote	19.35
Toggl	16.13
Fakturoid	12.90
Basecamp	6.45
Todoist	4.83

The four key features to be successful in the freelance world are a good reputation (60%), expertise (58%), reliability (52%) and contacts (46%). Some interviewees also noted: "Luck", "resistance to stress", "mentoring", "other work form options not considered by me". A total of 95% of the freelancers are happier, 90% started the business without any capital or own family savings, 80% work on weekends and holidays.

More than half of the customers are from Slovakia (55%), from Czech Republic (31%), others are from the EU (8%) or the rest of the world (6%). A total of 60% of respondents have up to 20 clients per year, which is for regular professional work on short- and long-term projects. The dominant sources of freelancer contacts are the social links, e.g. reputation, personal recommendations, contacts, use of social networks and networking.

This modern way of working seems to work well for Slovakia's modern workforce. This is why we have developed an interest in the decision of freelancers to choose this occupation. Here we list some of the open-ended responses to this question: "I needed additional income for my family", "I prefer to work anytime/anywhere... and to be my own boss", "Self-determination to set my own schedule and work on projects that interest and challenge me", "After some time in my job, I realised that the company did not care about their customers as I thought they should so I decided to do it by myself", "After nine years of working as an employee, I set up a creative studio together with my ex-colleague. Now we determine our direction on how and where we go, but at the same time we are responsible for our work. It is best for a creative person to manage himself".

6   Conclusion

Globalisation, lifestyle, technological progress, demographics, development of knowledge-based economy, and increased demands for a specialised skilled workforce lead people to choose a more flexible and business-like approach to work that is less bound to a specific time and place.

The world of work is changing profoundly. The ILO [46] estimated that global unemployment figures reached 201 million in 2014. Providing more jobs seems to be an encouraging challenge. In Europe we observe that most people work for someone else, but it seems that this could change soon, as our survey results indicate. What was marginal in the past has become commonplace today.

Workers in the creative industries are often labelled as freelancers. These workers pursue gainful employment on their own behalf, under conditions laid down by national law [47], which is the opposite of being an employee. Generally, we see freelancers as falling somewhere between dependent employees and self-employed workers. In short, freelancers are often self-employed knowledge or creative workers providing specialised services for a limited time-frame and working under a service contract; these workers are responsible for the own taxes, insurances and health care.

The typical career of today's workforce is when a worker is employed, self-employed or between jobs. Work is no longer linked to organisations but focuses on individuals on the basis of a whole range of different types of contracts - freelancer, temporary worker, part-time worker, contract worker, contractor, etc.

In relation to the survey, we have collected useful advice from the respondents, which can be summarised as follows: create your own portfolio, remember everybody is watching you, the price of the work, real expectations, contracts, archive documents, finishing contract on time, always learning.

According to our results, not surprisingly, with more flexibility it is easier to achieve a work-life balance. A total of 86% of the respondents consider freedom and flexibility as the greatest advantage of freelancing. We believe that the people of the future will have jobs, but not the traditional ones. Robin Chase, co-founder of Zipcar, said: "My father had one job in his lifetime, I will have six jobs in my lifetime, and my children will have six jobs at the same time" [48].

The main entrepreneurial aspect of freelancing is commonly considered to be the risk and uncertainty associated with job and income security [43]. The main barriers arising from the survey data are lack of stable income (65%), disappearance of the boundary between work and private life (48%), irreplaceability (46%) and difficulty in finding work (37%). According to the answers of our survey, Slovakia's freelancers are mostly specialists, higher qualified, predominantly high-qualified professionals, use modern technology, have a smaller number of clients, and mostly have a university education. To avoid the problem of obtaining payment, they rely on intuition and payment in advance. The four key features for being successful in the freelance world are a good reputation, expertise, reliability and contacts.

We believe that public policies must respond flexibly to the changes required by people who are trying to organise their working lives, and this must be done through a modern approach of investing in skills, decreasing bureaucracy, modernising education systems, adapting tax rules, and creating physical and digital infrastructure. Freelancers are nevertheless not sitting and waiting for change.

Freelance work seems to be becoming a more prominent career trend. Top free-lancers work together on important contracts for previously unthinkable rewards. Slovakia's freelancers have started to collaborate more, share experiences, publish, lecture, educate and support their colleagues. The ethos of friendly colleagues and cooperation among the Americans, French and British that Slovakia's citizen admired for so many years, and which are still missing in a number of countries according to our results, have finally become established in this country.

References

1. Simitis, S.: The juridification of labour relations. Comp. Labour Law **93**, 93–142 (1986)
2. Chung, H.: Explaining the provision of flexitime in companies across Europe (in the pre-and post-crisis Europe), Research gate. https://www.researchgate.net/profile/Heejung_Chung/publication/283667243_Working_paper_for_Work_autonomy_flexibility_and_work-life_balance_The_provision_of_flexitime_across_Europe/links/56430e0108aeacfd8938a905.pdf. Accessed 14 Feb 2018
3. Manyika, J., Bughin, J., Woetzel, J., Lund, S., Stamenov, K., Dhingra, D.: Digital globalization: the new era of global flows, McKinsey Global Institute. https://www.mckinsey.com/ ~ /media/McKinsey/Business%20Functions/McKinsey%20Digital/Our%20Insights/Digital%20globalization%20The%20new%20era%20of%20global%20flows/MGI-Digital-globalization-Executive-summary.ashx. Accessed 14 Feb 2018
4. UNDP: Human Development Report 2015, Work for Human Development. http://hdr.undp.org/sites/default/files/2015_human_development_report.pdf. Accessed 14 Feb 2018
5. Selloni, D.: CoDesign for Public-Interest Services, New Forms of Economies: Sharing Economy, Collaborative Consumption, Peer-to-Peer Economy. Springer International Publishing AG, XXVI, pp. 15–26 (2017). https://doi.org/10.1007/978-3-319-53243-1_2
6. Benkler, Y.: The Wealth of Networks: How Social Production Transforms Markets and Freedom. Yale University Press, New Haven and London (2006)
7. Burke, A.: The Role of Freelancers in the 21st Century British Economy. PCG Report. PCG, London (2012)
8. Deloitte: The workplace of the future. How digital technology and the sharing economy are changing the Swiss workforce, Deloitte. https://www2.deloitte.com/content/dam/Deloitte/ch/Documents/consumer-business/ch-cb-en-the-workplace-of-the-future.pdf. Accessed 14 Feb 2018
9. Couper, M.P.: Web-based surveys: a review of issues and approaches. Public Opin. Q. **64**, 464–494 (2000)
10. Sheehan, K.B., Hoy, M.B.: Using e-mail to survey internet users in the United States: methodology and assessment. J. Comput. Mediated Commun. **4**(3), JCMC435 (1999)
11. Weible, R., Wallace, J.: The impact of the internet on data collection. Mark. Res. **10**(3), 19–23 (1998)
12. Watt, J.H.: Internet systems for evaluation research. In: Gay, G., Bennington (eds.) Information technologies in evaluation: social, moral epistemological and practical implications, San Francisco: Josey-Bass, no. 84, pp. 23–44 (1999)
13. Lazar, J., Preece, J.: Designing and implementing web-based surveys. J. Comput. Inf. Syst. **xxxix**(4), 63–67 (1999)
14. Yun, G.W., Trumbo, C.W.: Comparative response to a survey executed by post, e-mail, & web form. J. Comput. Mediated Commun. **6**(1), JCMC613 (2000)
15. Swoboda, S.J., Muehlberger, N., Weitkunat, R., Schneeweiss, S.: Web-based surveys by direct mailing: an innovative way of collecting data. Soc. Sci. Comput. Rev. **15**(3), 242–255 (1997)
16. Dillman, D.A.: Mail and Telephone Surveys. Wiley, New York (1978)
17. Dillman, D.A.: The design and administration of mail surveys. Annual Rev. Soc. **17**(1), 225–249 (1991)

18. Fink, A.: The Survey Handbook, vol. 1. Sage, Thousand Oaks (1995)
19. Fowler, F.J.: Improving Survey Questions: Design and Evaluation, vol. 38. Sage Publications, Thousand Oaks (1995)
20. Krosnick, J.A.: Survey research. Annu. Rev. Psychol. **50**, 537–567 (1999)
21. Stephany, A.: Alex Stephany: How to understand the Sharing Economy. https://www.lsnglobal.com/opinion/article/16302/alex-stephany-how-to-understand-the-sharing-economy. Accessed 15 Feb 2018
22. Steinmetz, K.: Exclusive; See How Big the Gig Economy is, Time. http://time.com/4169532/sharing-economy-poll/. Accessed 15 Feb 2018
23. Stephany, A.: The Business Sharing: Making it in the New Sharing Economy. Palgrave Macmillan, Hampshire (2015)
24. PricewaterhouseCoopers: The Sharing Economy. Consumer Intelligence Series, PWC. https://www.pwc.fr/fr/assets/files/pdf/2015/05/pwc_etude_sharing_economy.pdf. Accessed 15 Feb 2018
25. Pennington, J.: The number that makes China the world's largest sharing economy, World Economic Forum. https://www.weforum.org/agenda/2017/06/china-sharing-economy-in-numbers/. Accessed 15 Feb 2018
26. Goodwin, T.: The Battle Is For The Customer Interface, TechCrunch. https://techcrunch.com/2015/03/03/in-the-age-of-disintermediation-the-battle-is-all-for-the-customer-interface/. Accessed 15 Feb 2018
27. Crook, J., Escher, A.: A Brief History of Airbnb. TechCrunch, TechCrunch. https://techcrunch.com/gallery/a-brief-history-of-airbnb/slide/3/. Accessed 15 Feb 2018
28. Bloomberg: The Sharing Economy, Bloomberg.com, Bloombergbrief. https://newsletters.briefs.bloomberg.com/repo/uploadsb/pdf/false_false/bloombergbriefs/4vz1acbgfrxz8uwan9_0_1019.pdf. Accessed 16 Feb 2018
29. Eckhardt, G.M., Bardhi, F.: The Sharing Economy isn't About Sharing at All, Harvard Business Review. https://hbr.org/2015/01/the-sharing-economy-isnt-about-sharing-at-all. Accessed 15 Feb 2018
30. Smith, C.: 100 Airbnb Statistics and Facts (November 2017)|By the Numbers, DMR. https://expandedramblings.com/index.php/airbnb-statistics/. Accessed 15 Feb 2018
31. Smith, C.: 90 Amazing Uber Statistics, Demographics and Facts (January 2018), DMR. https://expandedramblings.com/index.php/uber-statistics/. Accessed 15 Feb 2018
32. Lindak, M., Hayek Foundation, F.A.: Slovakia Better than the Czech Republic in Regulating Sharing Economy, 4liberty.eu. http://4liberty.eu/slovakia-better-than-the-czech-republic-in-regulating-sharing-economy/. Accessed 15 Feb 2018
33. Brejčák, P.: Lukáš Kovanda: Pripravme sa na japonizáciu, izraelizáciu a uberizáciu, Etrend.sk, Trend.sk. https://www.etrend.sk/ekonomika/lukas-kovanda-pripravme-sa-na-japonizaciu-izraelizaciu-a-uberizaciu.html. Accessed 15 Feb 2018
34. Cramer, J., Krueger, A.B.: Disruptive change in the taxi business: the case of Uber. Am. Econ. Rev. **106**, 177–182 (2016)
35. EFIP: Independent professionals driving Europe's employment revolution. http://www.efip.org/node/21. Accessed 15 Feb 2018
36. Cappelli, P., Keller, J.R.: A study of the extent and potential causes of alternative employment arrangements. Ind. Labor Relat. Rev. **66**(4), 874–901 (2013)
37. Cappelli, P., Keller, J.R.: Classifying work in the new economy. Acad. Manag. Rev. **38**(4), 575–596 (2013)
38. Kochan, T.A., Litwin, A.S.: The future of human capital: an employment relations perspective. In: Burton-Jones, A., Spender, J.C. (eds.) The Oxford Handbook of Human Capital, pp. 647–670. Oxford University Press, New York (2011)

39. Cascio, W.F., Boudreau, J. W.: Talent management of nonstandard employees. CEO Publication, G15-19(666), Chap. 28. Oxford Handbook of Talent Management, Los Angeles (2015)
40. Malone, T.W.: The future of work: How the new order of business will shape your organization, your management style, and your life. Harvard Business School Press, Boston (2004)
41. Smallbone, D., Kitching, J.: Are freelancers a neglected form of small business? J. Small Bus. Enterp. Dev. **19**(1), 74–91 (2012)
42. Kitching, J., Smallbone, D.: Defining and Estimating the size of the UK Freelance Workforce, Eprints Kingston University. http://eprints.kingston.ac.uk/3880/1/Kitching-J-3880.pdf. Accessed 15 02 2018
43. Burke, A.: The entrepreneurship enabling role for freelancers, theory with evidence from the construction industry. Int. Rev. Entrepreneurship **9**(3), 1–28 (2011)
44. Horowitz, S., Buchanan, S., Alexandris, M., Anteby, M., Rothman, N., Syman, S., Vural, L.: The Rise Of The Freelance Class, A New Constituency of Workers Building a Social Safety Net. http://www.people.hbs.edu/manteby/RiseoftheFreelanceClass.pdf. Accessed 15 Feb 2018
45. Titans.sk.: Ako sa stať IT FREELANCEROM, Titans.sk. https://www.titans.sk/upload/ako_sa_stat_it_freelancerom.pdf. Accessed 14 Feb 2018
46. ILO: World Employment Social Outlook, The changing nature of jobs, ILO. http://www.ilo.org/wcmsp5/groups/public/—dgreports/—dcomm/—publ/documents/publication/wcms_368626.pdf. Accessed 15 Feb 2018
47. EUR-Lex: Self-employment — equal treatment between men and women, Directive 2010/41/EU, EUR-Lex - em0035 - EN - EUR-Lex. http://eur-lex.europa.eu/legal-content/EN/TXT/?uri=LEGISSUM:em0035. Accessed 16 Feb 2018
48. Adams, T.: My father had one job in his life, I've had six in mine, my kids will have six at the same time, The Guardian, Guardian News and Media. https://www.theguardian.com/society/2015/nov/29/future-of-work-gig-sharing-economy-juggling-jobs. Accessed 16 Feb 2018

Framework Based on a Fuzzy Inference Model for the Selection of University Program Applicants

Omar L. Loaiza⊕, Nemias Saboya$^{(\boxtimes)}$⊕, José Bustamante⊕, and Juan J. Soria⊕

Universidad Peruana Unión, Carretera Central Km. 19.5,
Lurigancho-Chosica/Lima, Peru
{omarlj, saboya, joseb, jesussoria}@upeu.edu.pe

Abstract. This study consisted in the construction of an integrated framework based on a fuzzy inference model for evaluating the competencies in the entry profile of applicants to university programs in order to evaluate the process and make it more efficient. This framework was designed to systematize the method for evaluating the applicant profiles based on specific requirements and to improve performance indicators in the evaluation process. This architecture framework has two components: (a) the evaluation system and (b) the fuzzy model; where the system collects input information from knowledge assessment tests and applicant interviews, which are processed by a web service and delivered as parameters to libraries embedded in the fuzzy model, which was designed in Matlab SDK, and as a result the level of the applicant is returned. Furthermore, the system can be configured to add linguistic variables to the fuzzy model and set membership functions, as well as other modifications. This study is important because it facilitates the evaluators and improves the efficiency of the process to evaluate competencies in a profile as well as the efficiency of results which serve as input for process improvements. Finally, the results of the implementation of the framework based on indicators improved time efficiency (80.3%), use of resources (45.7%) and increased evaluator satisfaction (25.7%) with a value of $p < 0.05$ for all cases.

Keywords: Framework · Enrollment profile · Fuzzy inference model

1 Introduction

1.1 Context for the Research Study

Currently in the context of technological convergence and digital transformation, information technologies intervene daily in the lives of people and organizations in a wide range of contexts [1], making them in many cases essential and catalytic tools for greater productivity and decision making in different fields [2].

In organizations, processes become the engine to achieve their objectives so it is important to manage them properly [3]; these may be influenced by considerations present in their business environment [4]; in this sense, information systems and other

© Springer Nature Switzerland AG 2019
R. Silhavy (Ed.): CSOC 2019, AISC 984, pp. 131–140, 2019.
https://doi.org/10.1007/978-3-030-19807-7_14

information technologies must provide technological support [5] to equip them with competitive capacity.

Today, information systems are increasingly complex as a result of the digital transformation process [1]; where it is no longer just about transactional operations contained in components of a software architecture [6] and [7], but rather with its integration with artificial intelligence models [8] which form an abstract representation of the complex reality within the context of organizations.

Thus there are research projects where IT projects are integrated with fuzzy models in different fields, such as in the educational field where Chin and Baba [9] found a way to integrate a diffuse model with software to determine the learning profile of language students and improve the evaluation process; or as the case of Mitra and Das [10] where they also helped assess learning skills through software and a fuzzy model. Here it is also pertinent to cite Alonso and Magdalena [11] who showed the generation of fuzzy rules from a Java programming environment.

The university context of this research study includes internal considerations of the institution and external from the Peruvian normative framework regarding the applicant selection processes. These processes may vary from one university to another [12], but usually consist of a knowledge test as a mandatory element, and optionally can include skill and attitude exams [13]; such that resources and methods are necessary to make the evaluation process effective and reliable.

Traditional methods of evaluation have limitations because they only use a few evaluation rules due to the limitations of the evaluators who are faced with a wide range of aspects and evaluation rules that must be observed [12]. That is why they must be integrated with innovative and intelligent alternatives [14].

In this line of thought, the objective of this research study is to build a framework based on a fuzzy inference model for the evaluation of the profile of new students in unviersitary study programs. [14] in order to make this process more efficient in terms of time, use of resources and level of participant satisfaction.

This research study is relevant and important because it allows universities to make their evaluation processes more effective and reliable; as well as show evidence of compliance with basic quality conditions that are regulated by Peruvian Law 30220 and the quality model of the National System of Evaluation and Accreditation of Educational Quality (SINEACE) in Peru.

1.2   Evaluation of the Applicant Profile

The evaluations of the admission process, according to the experience reported by universities in Peru, are basically oriented to measure the level of knowledge [13]. However, thanks to the educational reform established by Peruvian Law 30220 and SINEACE, the universities are looking for mechanisms that strengthen this process and provide favorable results for the future student. As part of these mechanisms, some universities no longer perform a common exam for all applicants but rather have started using differentiated exams according to the major.

The applicant profile "are characteristics (skills, abilities, qualities and values) that guide the admission process to a study program" for this reason it requires tools that perform the operational tasks and facilitate compliance with these characteristics [13].

Saboya *et al.* conducted a study considering the characteristics of the applicant profile mentioned above through a method called PED (see Fig. 1), this method is different because the evaluation of the applicants is differentiated and it establishes levels to recognize the level obtained in each of the applicant's competencies [14]. Everything mentioned above was an input for the elaboration of the fuzzy inference model [14].

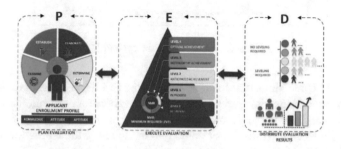

Fig. 1. Structure of the applicant profile evaluation method [14].

The method is developed in three phases, the first one covers the planning of the evaluation of the applicant profile, which consists in establishing percentage weights for the competencies of the profile according to the subjects that contribute the most to the field of study, later they elaborate scales and levels in function of the weights and the number of questions elaborated by each subject and finally establish the minimum level that the applicant must reach to be admitted to the study program, this can be varied as needed. The second phase consists in the execution solely of the evaluation that is done manually, from the development of the evaluation to the results of the same and does not have a tool that supports the whole process and the third is the dissemination of the results according to the specifications given in the planning phase.

1.3 Fuzzy Evaluation Model

Fuzzy inference systems (FIS) are also known as systems based on fuzzy rules, fuzzy models, diffuse expert system or diffuse associative memory and they support decision making [15]. This is based on the concepts of fuzzy set theory, fuzzy IF-THEN rules and fuzzy reasoning. The FIS use the declassifications "IF… THEN…" with connectors ("AND" or "OR") present in the declaration of the rules as mentioned by Sivanandam [16].

A basic FIS can take fuzzy inputs or crisp inputs, but the outputs it produces are almost always fuzzy sets. When the FIS is used as a controller, a clear output is needed. Therefore, in this case, the defuzzification method is adopted to extract a crisp value that best represents a diffuse set.

1.3.1 Construction and Function of the Inference System

Antao [15] mentions that the fuzzy inference system consists of a fuzzification interface, a rule base, a database, a decision-making unit and a defuzzification interface. An FIS with the aforementioned elements is described in Fig. 2.

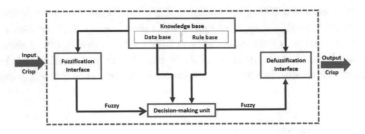

Fig. 2. Fuzzy inference system [17]

Paluszek [17] mentions that the operation of the FIS has a clear input that becomes fuzzy using the fuzzification method. After the fuzzification, the rule base is formed. The rule base and the database are known together as the knowledge base. Defuzzification is used to convert the fuzzy value into the value of the real world which is the output. The rules of the diffuse inference model based in the PED method [14], designed in Matlab SDK show the input and output variables which have the following form: If (C1CAPAS is C1N2LPRE) and (C2COBASI is C2N3LDEST) and (C3COBALET is C3N0INI) then the student profile (PERIngresante is PFNivel2). The intervals for the input variables are [21; 24>, [26; 31>, [0; 4> and the values of the output interval are [50; 60>, which gives a fuzzy degree of belonging from 0.40 to 0.60 and that means that the profile of the student shows anticipated achievement. If it is executed in Matlab SDK you will get an entry of [22; 27; 3] whose output is 53.2 which means that it is in level 2 of expected achievement shown in Fig. 3.

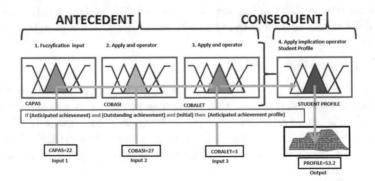

Fig. 3. Rule of inputs and outputs for the fuzzy inference model

2   Methodology

This research study was carried out in 4 stages: (1) in the first stage, the applicant profile evaluation method was analized to elicit functional requirements that the framework should satisfy; (2) in the second, the functionalities derived from the requirements and functionalities with process actors were validated; then, (3) the integration with the fuzzy inference model was made [14] to accept inputs and extract results; and, finally, (4) their efficiency was validated through KPIs of satisfaction, time efficiency and results in the use of resources according to the budget allocated for the process of selecting applicants in a pilot university.

Returning to give additional depth to the methodological sequence, in the first stage the topic of study was how the incoming profile evaluation method works, which proceeded to raise functional requirements that were then transformed into Use cases and documented in the UML language and in a specification document (SRS).

For the second stage, prototypes were prepared that were validated by the relevant process actors who would interact with the framework.

Regarding the third stage, the key step was the integration of the framework with Matlab SDK's fuzzy logic libraries, providing them with the parameters that the membership functions (CAPAS, COBACI and COBALET) from the knowledge dimension of the applicant profile.

Finally, with regard to the fourth stage, the results of the implemented framework were validated, assessing their effectiveness through the KPI calculation of the admission process. Figure 4 expresses graphically what was expressed previously.

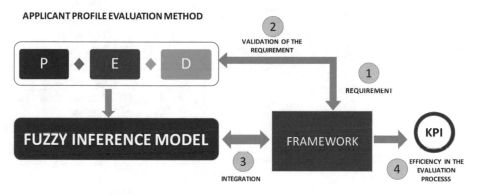

Fig. 4. Methodological sketch of the research study

3 Results

3.1 Requirements and Validation of the Framework

The elicitation of requirements is evidenced in a matrix (see Table 1), which is the result of the review of documents and interviews with the actors of the evaluation process by means of the profile evaluation method in the stages of: planning, execution and diffusion [14]. The requirements are grouped into two categories: (1) those that have to do with the configuration of the fuzzy model (FR-01, FR-02, FR-03 and FR-04) from the software and have to do with the planning stage of the PED; and, (2) those that have to do with the input of information and presentation of results (FR-05, FR-06, FR-07 and FR-08) that are aligned to stage 2 and 3 of the PED. These requirements were approved by the actors of the process through the revision of some prototypes and the sequencing of the activities of the PED.

Table 1. Requirement matrix of the framework

Stage of the process		Functional requirements		Use case	
Plan evaluation	Weigh	FR-01	Register competencies	UC-01	Register competencies
		FR-02	Adjust the information of the areas (subjects) to be evaluated	UC-02	Register thematic areas
	Establish			UC-03	Establish weights for courses in the course plan
	Elaborate	FR-03	The system must allow for the management of the evaluation dimensions	UC-04	Register dimension
				UC-05	Register membership function
				UC-06	Readjust parameters for the membership function
	Determine	FR-04	Establish the minimum level to enter	UC-07	Register minimum level to enter
Execution		FR-05	The system must generate documentation for the process	UC-08	Open the process
				UC-09	Generate instruments
				UC-10	Generate evaluation template
				UC-11	Generate interview template
		FR-06	Record evaluation information	UC-12	Register knowledge exam
				UC-13	Register the interview
		FR-07	The system should evaluate the applicant profile	UC-14	Evaluate the profile manually
				UC-15	Evaluate profile by batches
Emit results		FR-08	The system must publish the evaluated results	UC-16	Consolidate results
				UC-17	Publish results

3.2 Framework Architecture

The construction of the framework was based on the design of an architecture guided by architectural drivers such as functional requirements and defined use cases.

The architecture is basically divided into two large components that are integrated: the first one is the Evaluation System for the applicant profile with graphic interfaces oriented for users to manage and configure the functionality of the fuzzy model and the evaluation of the applicant profile; and the second component is a web service that contains the operations necessary for the interaction with the fuzzy inference model and its execution in the Matlab SDK from a JAVA application, which provides a method *evaluarCompetencias* that receives the parameters CAPAS, COBACI and COBALET necessary for the execution of the fuzzy inference model and will return the processed information that will be used by the first component. Unlike the framework of other advanced fuzzy inference systems that propose operations from an object-oriented programming language [18], this proposed framework uses design patterns such as Data Access Object (DAO), Interface, Singleton, MVC and N-Layers for its implementation, with the Java language.

Below in Fig. 5. the logical view of the architecture of the applicant profile evaluation system is presented, which consists of 4 layers: the presentation layer that contains components such as *VistaEvalConocimiento*, *VistaEvalEntrevista* and *VistaEvalFinalModel* that take care of receiving the information from user requests (UC-12, UC-13, UC-14 and UC-15) and showing results (UC-16 and UC-17), the business layer contains components with methods of evaluation and configuration of the fuzzy model and that are exposed by means of *interfaces*, the data layer contains components that are responsible for storing objects in the relational database and the integration layer where the components *ClienteModelDifusoImpl* and its interface *ClienteModelDifuso* allow for the integration of the aforementioned services.

Regarding the configuration of the framework from within the system, similar to the evaluation part, the web service is also used to provide one of its interfaces where the desired configuration can be stipulated for use in the fuzzy model with a parameter which accesses a method of *ModelDifuso.fis* by means of a *SimpleFactoryConfiguracion* class and the request is carried out in one of the UC-04, UC-05 or UC-06.

Likewise, the component diagram (see Fig. 6) is presented in order to understand the ecosystem of the framework in relation to the fuzzy inference model. There are three main components: an evaluation component for the applicant profile, which represents the main user interfaces of the framework; a second component that represents the web service that contains a javabuilder.jar library that allows Java programmers to incorporate Java classes created from Matlab SDK functions in their own Java applications and the ModelDiffuse.jar subcomponent that contains a type of interface type *ModelDifuso.class* and the classes *ModelDifusoRemote.class* and *ModelDifusoMCRfactory* necessary to call and execute the fuzzy model (compiled file with extension *.fis*) in Matlab SDK, this component is the bridge of communication between the user interfaces and Matlab SDK, since it exposes the methods *evalCompetencias* and *ConfigurarModel*, the first will receive the CAPAS, COBACI and COBALET parameters, which are sent from the user interfaces and are sent to the Matlab SDK application for the execution of the fuzzy inference model using the ModelDifuso.jar component and the javabuilder.jar library, once executed, the result is

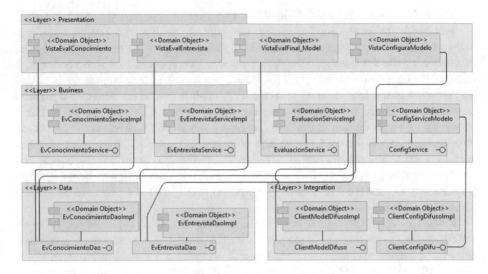

Fig. 5. Framework architecture

captured after process in the fuzzy motor (see Fig. 3) and this is interpreted to determine the applicant's level, which is sent to the graphic user interface for the end user to
make decisions; the other method (*ConfigurarModel*) allows for the configuration of
the fuzzy model. Finally we have the component that represents the application in
Matlab SDK that contains the fuzzy inference model (.fis) and the Matlab SDK that is
the collection of specific native libraries of the platform needed to execute the functions
of Matlab SDK.

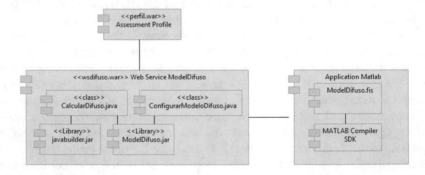

Fig. 6. Integration of the fuzzy inference model with the framework

4   Discussion

The efficiency of the framework in the process of selection of applicants was related to
3 KPIs, the first depending on the time required to evaluate the exams and interviews,
the second depending on the resources used and finally the level of satisfaction of

process actors. For these results, the information of one application cycle was used and the process actors were surveyed by means of a sample to show to what degree they were satisfied with the process using the framework.

The results of efficiency as a function of the time required for the evaluation process was improved by 80.3%, comparing before and after, as well as the efficiency of resources which was at 45.7%, and finally an increase was noted in the satisfaction of the process actors by 25.7% with respect to before and after the implementation of the framework, all these results were statistically significant in favor of the efficiency of the framework with a $p < 0.05$. The results of the framework were efficient and relate to the results of Chin and Baba [9] who integrated a fuzzy model with software for language students as well as improved their evaluation process and those of Mitra and Das [10] where software integrated into a fuzzy model helped to assess learning skills.

5 Conclusions

The framework has a high growth potential in the development of its functions, since it is integrated with a fuzzy inference model, where both are derived from the method of evaluation for the applicant profile (PED), which is flexible, a characteristic that has also been incorporated into the architecture of the framework software. This is due to the configuration capabilities implemented in the framework to introduce changes in the fuzzy inference model contained in the usage cases found in the functional requirements FR-03 and FR-04. In addition, the study concludes that the developed framework is efficient regarding execution time of the evaluation process, use of resources and the level of satisfaction of the process actors.

References

1. Carell, A., Kim, L., Dirk, P.: Using design thinking for requirements engineering in the context of digitalization and digital transformation: a motivation and an experience report. In: Gruhn, V., Striemer, R. (eds.) The Essence of Software Engineering. Springer, Cham (2018)
2. Pagani, M., Pardo, C.: The impact of digital technology on relationships in a business network. Ind. Mark. Manag. 67(September 2016), 185–192 (2017)
3. Plakoutsi, A., Papadogianni, G., Glykas, M.: Performance measurement in business process, workflow and human resource management. In: Glykas, M. (ed.) Business Process Management - Theory and Applications, p. 474. Springer, Londres (2013)
4. Thompson, A., Peteraf, M., Gamble, J., Strickland, A.: Administración Estratégica, 18th edn. McGrawHill, México DF (2012)
5. Laudon, K.C., Laudon, J.P.: Sistemas de Información Gerencial, 12th edn. Pearson Education, México (2012)
6. Sommerville, I.: Ingeniería De Software (2011)
7. Hasselbring, W.: Software architecture: past, present, future. In: Gruhn, V., Striemer, R. (eds.) The Essence of Software Engineering (2018)
8. Pressman, R.: Software Engineering: A Practitioner's Approach, 8th edn. McGraw-Hill, New York (2014)

9. Chin, F.M., Baba, F.A.: Assessment of English proficiency by fuzzy logic approach. In: International Educational Technology Conference, pp. 355–359 (2008)
10. Mitra, M., Das, A.: A fuzzy logic approach to assess web learner's joint skills. Int. J. Mod. Educ. Comput. Sci. Mod. Educ. Comput. Sci. **9**(9), 14–21 (2015)
11. Alonso, J., Magdalena, L.: Generating understandable and accurate fuzzy rule-based systems in a java environment. Fuzzy Logic Appl. **6857**, 212–219 (2011)
12. Davidovitch, N., Soen, D.: Predicting academic success using admission profiles. J. Int. Educ. Res. **11**(3), 125–142 (2015)
13. SINEACE: Modelo de Acreditación para Programas de Estudios de Institutos y Escuelas de Educación Superior, p. 36 (2016)
14. Saboya, N., Loaiza, O., Soria, Q., Bustamante, J.: Fuzzy logic model for the selection of applicants to university study programs according to enrollment profile. In: Advances in Intelligent Systems and Computing 311 ICT Innovations, pp. 121–133 (2018)
15. Antão, R., Mota, A., Escadas, R., José, M., Machado, T.: Nonlinear physical science type-2 fuzzy logic uncertain systems' modeling and control with contributions by (2017)
16. Sivanandam, S.N., Sumathi, S., Deepa, S.N.: Introduction to fuzzy logic using MATLAB (2007)
17. Paluszek, M., Thomas, S.: MATLAB Machine Learning (2017)
18. Perronne, J.M., Petitjean, C., Thiry, L., Hassenforder, M.: A framework for advanced fuzzy logic inference systems. IFAC Proc. **35**(1), 175–180 (2002)

Collaborative Data Mining in Agriculture for Prediction of Soil Moisture and Temperature

Carmen Ana Anton[✉], Oliviu Matei, and Anca Avram

Electric, Electronic and Computer Engineering Department,
Technical University of Cluj-Napoca, North University Center Baia Mare,
Baia Mare, Romania
{carmen.anton, oliviu.matei}@cunbm.utcluj.ro,
anca.avram@ieee.org

Abstract. Climate change affects agriculture in many ways. Reducing the vulnerability of agricultural systems to climate change and enhancing their capacity to adapt would generate better results with fewer losses. Under the conditions, according to The United Nations Food and Agriculture Organization, the world has to produce 70% more food in 2050 than it produced in 2006, to feed the growing population, it is obvious that any innovative ideas that help agriculture are optimal and needed. An option for increasing efficiency of agriculture is a data mining process that can predict climate conditions and humidity of soil. Determining the optimal time for planting and harvesting could be based on predictions from a data mining process. In this scenario, the application of collaborative data mining techniques, would offer solution for the cases in which one sources do not poses useful data for mining, and the process uses date from another sources correlated.

Keywords: Machine learning · Data mining · Collaborative data mining · Soil moisture prediction · Prediction in agriculture

1 Introduction

1.1 Data Mining in Agriculture

Today, knowledge of soil characteristics, weather forecasts for an optimum period and plant production capacity are essential conditions for smart agriculture, which could facilitate and improve their success in the area. The implementation of a weather forecasting system could simplify and optimize the decisions farmers take in sowing, harvesting etc. Weather forecasting is an action that is based on a large amount of data and allows for the timing of parameters such as temperature at different levels, soil humidity, precipitation, atmospheric pressure and wind.

In areas such as agriculture, aviation, transport, everyday life, the weather forecast can provide useful information that can be used to make profits for companies, avoid disasters, and adapt life to every day. Due to the fact that at the end of the 21st century

© Springer Nature Switzerland AG 2019
R. Silhavy (Ed.): CSOC 2019, AISC 984, pp. 141–151, 2019.
https://doi.org/10.1007/978-3-030-19807-7_15

(2090–2099), compared to the period 1980–1999, in the "Guide for Adaptation to the Effects of Climate Change"[1] by the National Meteorological Administration in Romania, was foretells that is expected in summer a growth for the temperatures and a decrease of pluviometric level with 10–30% and in winter an increase of precipitation with 5–10%, agriculture will have to suffer and must action and modifie strategies trying to overcome the climate change.

In agriculture, as in other areas, data mining methods could be applied, resulting in process and methods that can lead to efficient and performing results. The method of extraction of knowledge, as shown by Lika in [1], considers that the following stages:

1. Understand the scope and the problem formulation;
2. Create or collect data sets for the study;
3. Cleans and preprocessing data;
4. Reduction and projection the data (finding useful features to represent the data depending on the goal of the task);
5. Match the goals of the extracting process with a particular data mining method;
6. Apply data mining algorithms and methods;
7. Extract the patterns;
8. Analyzes the knowledge discovered;
9. Apply what was discovered (integration of operational systems).

Figure 1 illustrates a schematic representation of the concept of data mining.

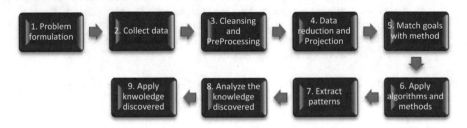

Fig. 1. Representation of the process of data mining

1.2   Related Work

As is mentioned by Moyle in [2], collaborative data mining is a framework in which the data mining effort is distributed to several collaborating agents - people or software. The ultimate outcome of the collaborative effort is to provide solutions to the problem that is considered better than what would have been offered from individual agent actions. If a certain level of collaboration is required for the human form, in the case of

---

[1] http://www.meteoromania.ro/anm/images/clima/SSCGhidASC.pdf.

applications or other uses of the notion it is necessary to study the mechanism and the procedures necessary for its operation Moyle in [2], propose three interpretations of the notion collaborative data mining:

(1) Several software programs or agents that apply data processing algorithms to solve the same problem;
(2) People using modern collaboration techniques to apply Data Mining to a defined problem;
(3) Data Mining in activities of human collaboration.

Many researches have approached the notion of collaboration in terms of collaboration between individuals, an interesting approach would be in the field of predictability of phenomena or actions in order to improve them. The questions that arise are:

– How many levels of data should be used in a collaborative mining data?
– What type of data can we use?
– What is the accuracy of the results?

According to the annual report from the ITU/UNESCO Broadband Commission for Sustainable Development[2], the access to the Internet has grown to a 3.4 billion people or 45.9% of the world's population who are estimated to have been online at the end of 2016 (a year-on-year increase of some 180 million people). The Internet has started to be used in all aspects of the life, from home, to industry, health and equipment maintenance. Therefore, much research has been invested in determine these usage patterns, such as the ones reported by Di Orio et al. in [3] and Matei in [4]. Both use data mining to fulfill this objective. The aim is to determine an expected behavior of the device or user based on some known sensors outputs. Matei et al. in [5] proposed a very powerful methodology for mining data in such a hyper-connected world. Moreover, in [6] Matei et al. applies the concept of collaborative network to exploring the data from several home appliances. He suggests that using the data from two correlated data sources influences positively the accuracy of the data mining. On the other hand, if the data sources are not or low correlated, the accuracy decreases.

Preliminary research to establish and define the concept of collaborative data mining has been carried out by Matei et al. in [6] which show that the notion can be applied in many areas, but for each case the parameters involved in the process need to be specified and adapted.

In [7], Matei et al., presents some preliminary discovery on a system that gave in real time, predictive soil moisture for agriculture. The paper presents a system based on data collected from several weather stations in Transylvania Plateau and predicts soil moisture for the next day, in real time. The system has been tested on 10 stations and demonstrates very high prediction accuracy, allowing the use of the agricultural process to obtain reliable and basic information for future decisions in the field.

Matei in [8] shows that information can be obtained from different systems, making it possible to use them in a complex data extraction process and to determine the degree of precision of the obtained results.

---

[2] https://www.itu.int/dms_pub/itu-s/opb/pol/S-POL-BROADBAND.18-2017-PDF-E.pdf.

According to the Food and Agriculture Organization (FAO), an important aspect in achieving success in the agricultural sector is rapid adaptability to climate change. In document "Tracking adaptation in agricultural sectors"[3] published by the organization in 2017, states that, the Paris Agreement adopted in 2015, marks the importance of monitoring and learning from adaptation actions and recommends that periodic assessment of overall progress towards adaptation to climate change is recommended.

In [9], Matei et al. proposed a definition for an architecture and a framework to be able to manage data analysis on the basis of their complexity according to the computational resources available at each stage. The article contains details about the flow of data in IoT and present a new architecture based on the diversity and complexity of the various obtained data flow stages, and experiments based on a new architectural pattern emerging important scientific findings. The results of the experiments and research underlying the article have led to conclusions that support a layered system of data architecture used in exploitation starting from the bottom level that is based on simple calculations and reaching a superior layer based on complex algorithms. Each layer is presented with the features and benefits of use.

The collaborative data mining process has been experimented in a field of home appliances by Matei et al. in [6], part of which was translated and applied in the agricultural field by Matei et al. in [7], representing a start of research confirming the results obtained in the first case. As stated by Matei in [7], the quality and quantity of the data used in the process is required and determined an improvement of the accuracy of predictions.

2   Methods

Following the steps from Fig. 1, the problem addressed in this article concerns the efficiency of predictions for soil humidity and air temperature in Transylvanian Plateau.

2.1   Understand the Scope and the Problem Formulation

The question for the research is as follows: Which predictions are better: those based on the data from the meteorological station in question, or which also use cumulative data from another station from the nearest perimeter?

2.2   Create or Collect Data Sets for the Study

The study collected information from the meteorological stations in the Transylvanian Plateau during the period 2008–2015. The process uses data from weather stations (type of files: dtf, hobo, csv) and contains information such as: timestamp, three temperatures of soil at three depths: −10 cm, −30 cm, −50 cm; soil moisture.

Recorded periods and station locations allowed a selection of an observation area and a continuous and project-related time span. Data was imported with the HoboWare

---

[3] http://www.fao.org/climate-change/resources/publications/en/.

application, which is a software package that allows users to filter and extract data. These were imported from *.dtf files collected from meteorological stations and exported to files with *.csv extension to access the collected information.

2.3 Cleans and Preprocessing Data

Three issues of data cleaning and preprocessing have been addressed in this report, namely: establishing the common data range, preprocessing data, files concatenation.

The step of establishing the common data range, it was obtained by selecting stations that for a continuous period of time had complete and relevant data. From this point of view some stations were removed from the study, leaving only those who fulfilled the condition, namely: Cojocna, Filipisu Mare, Luduş, Silivaş, Triteni. Data from January 2010 to April 2010 were used in the process. The air distances between the mentioned localities, calculated with DistanceFromTo, are situated in the range of 14 km and 57 km: the shortest distance being between Ludus and Triteni, the largest distance between Cojocna and Filipisu.

Preprocessing task involved obtaining data that allows independent access to the following columns: date, temperature to −10 cm, temperature to −30 cm, temperature to −50 cm, soil humidity. In this scope the csv files have been converted to xlsx files.

2.4 Reduction and Projection the Data

Finding useful features to the data is the step that decides what information is relevant to the task. In this case, important data for the process are the temperatures and humidity of each calendar day. This led to a data reduction operation existing in xls files (each containing 17274 lines, records being made every 10 min) so that there is an average for each day with the corresponding standard deviation. The operation was performed using a SQL command applied to a database in Microsoft Access. Table 1 illustrates the data from Cojocna2010ian_aprZILE.xlsx file after using the SQL command.

Table 1. The structure for the final data

Data	AT_10	SDT_10	AT_30	SDT_30	AT_50	SDT_50	UMID	SD_UMID
01.01.2010	4,407	0,086	2,871	0,245	2,381	0,978	0,317	0,002
02.01.2010	4,513	0,079	3,826	0,181	3,689	0,165	0,315	0,001

The columns prefixed with "A" (UMID, AT_10, AT_30, AT_50) refer to the average values of soil moisture, respectively soil temperature at −10, −30 and −50 cm. and the columns prefixed with "SD" (SDUMID, SDT_10, SDT_30, SDT_50)) refer to the standard deviations over the days for the corresponding sensors. The unit for temperature and humidity is Celsius degrees.

2.5   Match the Goals of the Extracting Process with a Particular Data Mining Method

Most studies have addressed areas where data mining was implemented various user options such examples by Camarinha-Matos and Afsarmanesh in [10], Badrul et al. in [11] or Billsus and Pazzani in [12], but an approach to collaboration in terms of predictive believe is appropriate in terms of meteorological issues treated by Folorunsho and Adeyemo in [13].

We can deduce that: products or events that have the same behaviors or characteristics and they are correlated with values in a certain range, can have predictable behavior. Going forward with the hypothesis, we can say that in the absence of data for specific products or events, we can use the data correlated with a similar item from which performed the analysis. Collaborative data mining could define situations in which products or events that have the same behaviors or characteristics and they are correlated with values in a certain range, can have predictable behavior, and can be causing an algorithm and methodology for extracting relevant information for the user.

Going forward with this reasoning, the processes that have been approached have been implemented for two concepts for predicting soil moisture, namely:

- standalone (predictions based on data from the current station)
- collaborative (predictions based on information from other stations).

The algorithms used were: k-Nearest Neighbor model (KNN), Local polynomial regression (LPR), Neural Net model (NN) and Support Vector Machine (SVM).

An earlier stage of the algorithm application was that of establishing the correlations between weather stations. A process was performed using the RapidMiner Studio application (version 7.4). The correlations regarding the recorded moisture were established, and the result are: Cojocna-Filipisu Mare (0,519 °C), Cojocna-Ludus (0,505 °C), Cojocna-Silivas (0,716 °C), Cojocna-Triteni (0,613 °C), Filipisu Mare-Ludus (0,366 °C), Filipisu Mare-Silivas (0,788 °C), Filipisu Mare-Triteni (0,791 °C), Ludus-Silivas (0,354 °C), Ludus-Triteni (0,387 °C), Silivas-Triteni (0,855 °C).

The already presented information has a huge importance when a machine learning algorithm is used. It has a higher prediction result for correlated stations and lower for not correlated stations.

2.6   Apply Data Mining Algorithms and Methods

The two steps of applying the mining algorithms led to the following results that is describing the standalone stage and collaborative mining.

In the standalone stage, the results obtained, applying the algorithms LPR, KNN, NN, SVM, can be seen in the list below, showing the observed meteorological stations and the predictive values. The values of prediction of the humidity for the next day is:

- Cojocna: 0.7670 °C (LPR), 0.3100 °C (KNN), 0.3280 °C (NN), 0.3380 °C (SVM);
- Filipisu Mare: 0.767 °C (LPR), 0.526 °C (KNN), 0.509 °C (NN), 0.4400 °C (SVM);
- Ludus: 0.4400 °C (LPR), 0.5000 °C (KNN), 0.7070 °C (NN), 0.7330 °C (SVM);
- Silivas: 0.8530 °C (LPR), 0.2840 °C (KNN), 0.4480 °C (NN), 0.5000 °C (SVM);
- Triteni: 0.5950 °C (LPR), 0.4740 °C (KNN), 0.5340 °C (NN), 0.4830 °C (SVM);

In the **collaborative stage**, for each algorithm, a process was applied to the weather stations participating in the experiment, obtaining the results found in the following tables with the corresponding conclusions.

Tabels contain the values of prediction of the humidity for the next day, treated in two modes: standalone and collaborative: for standalone stage, the values are found on the first line and for collaborative stage, was applied to pairs, and the results are in the cells. The unit of measurement for the values in the tables is Celsius degrees.

1. *KNN - k-nearest neighbor*

In this case, in Table 2, we see that 11 of the 20 values are increasing compared to the prediction obtained in the standalone variant, which represents percent of 55% which is in favor for the application of a collaborative data mining.

Table 2. Prediction for the meteo station in collaborative mining

Standalone	0.310	0.526	0.500	0.284	0.474
	Collaborative windowing step size = 1				
KNN	Cojocna	Filipisu	Ludus	Silivas	Triteni
Cojocna		0.500	**0.507** ↑	0.284	**0.483** ↑
Filipisu	**0.328** ↑		**0.517** ↑	**0.388** ↑	0.431
Ludus	**0.336** ↑	0.509		**0.328** ↑	0.457
Silivas	**0.328** ↑	0.500	0.491		**0.491** ↑
Triteni	0.284	0.526	**0.534** ↑	**0.397** ↑	

2. *LPR - local polynomial regression*

In this case 10 of the 20 values are increasing compared to the prediction obtained in the standalone variant, which represents percent of 50%. This percentage is in favor of collaborative data mining, but we consider this is not enough to pick this algorithm. The results are in Table 3.

Table 3. Prediction for the meteo station in collaborative mining

Standalone	0.328	0.509	0.707	0.448	0.534
	Collaborative (prediction is read on the column of the weather station)				
LPR	Cojocna	Filipisu	Ludus	Silivas	Triteni
Cojocna		0.500	0.612	0.552 ↑	0.483
Filipisu	0.379 ↑		0.810 ↑	0.509 ↑	0.414
Ludus	0.448 ↑	0.457		0.509 ↑	0.448
Silivas	0.474 ↑	0.491	0.724 ↑		0.466
Triteni	0.422 ↑	0.457	0.647	0.509 ↑	

3. *NN - neural networks;*

In the Table 4, for the algorithm of Neural Networks, only 4 out of 20 values are increasing, what represents a percentage 20%.

Table 4. Prediction for the meteo station in collaborative mining

Standalone	0.328	0.509	0.707	0.448	0.534
	Collaborative (prediction is read on the column of the weather station)				
NN	Cojocna	Filipisu	Ludus	Silivas	Triteni
Cojocna		0.509	0.741 ↑	0.328	0.431
Filipisu	0.310		0.681	0.397	0.405
Ludus	0.319	0.483		0.474 ↑	0.422
Silivas	0.397 ↑	0.422	0.681		0.388
Triteni	0.310	0.500	0.698	0.517 ↑	

4. *SVM - support vector machine;*

For the Support Vector Machine from 20 values 8 are increasing which represents percent of 40%. This fact is presented in Table 5.

Table 5. Prediction for the meteo station in collaborative mining

Standalone	0.388	0.440	0.733	0.500	0.483
	Collaborative (prediction is read on the column of the weather station)				
SVM	Cojocna	Filipisu	Ludus	Silivas	Triteni
Cojocna		0.509 ↑	0.750 ↑	0.474	0.440
Filipisu	0.336		0.698	0.483	0.414
Ludus	0.345	0.491 ↑		0.509 ↑	0.543 ↑
Silivas	0.353	0.457 ↑	0.707		0.457
Triteni	0.336	0.466 ↑	0.733	0.509 ↑	

2.7   Extract the Patterns

In various meteorological studies proposed by Rashid et al. in [14], Moghadam and Ravanmehr in [15], Charles et al. in [16], or Comeau et al. in [17], models were found in the researched data, applying a wide variety of methodologies and methods of extraction. In this research, the result obtained by comparing the applied methods both on a standalone and collaborative basis led to the results that reveal a good percentage of 55% for KNN algorithm and 50% for LPR algorithm. The results reveal that a collaborative approach can have better outcomes if we are using the right technique.

2.8   Analysis of the Results

From the previous sections it turns out that choosing the highest values for each station and considering the collaborative working model, there is a significant increase in the accuracy of the predictions, so from 20 set values, 14 are increasing 4 in decreasing and 2 stagnating, which represents 70% of the collaborative values are higher than those obtained with the standalone model.

3   Results

Due to the fact the algorithm with the highest percentage of increasing values for data extraction in collaborative mode was KNN, a change in process parameters and a tracking of the evolution of the values were attempted. After testing different data mining algorithms on the newly introduced data, we observe the following:

- KNN was the algorithm with highest percentage, in collaborative mode;
- modifying the step size attribute from the windowing operator (from 1 to 3) gives significantly improved values for the stations that are correlated.

During the tests we change and track the parameters evolution.
We have shown that collaborative data exploitation means better results when

- the data is sufficient and correlated;
- the algorithm used is set to optimal parameters.

Using data from different sources could increase the accuracy of the results of the machine learning process. Ultimately, the goal pursued, could be achieved, defining the notion "collaborative data mining", such as: obtain predictive information for a source without data available, using data form a complete and correlatted source with initial one.

4   Conclusions

In the process of weather stations, a new architectural data analysis is addressed. The concept takes into account the characteristics of the data, and its source for the whole process. The levels approached, locally, contextually and in collaboration provide a new perspective on the steps necessary for preprocessing and analyzing data, finding multiple solutions to obtain more accurate predictions.

The concept has been developed, tested and validated on real-time data in exper-imental cases (weather stations form Transylvanian Plateau), with the development of a collaborative data mining process usable in the extensive predictive process of some decisions or phenomena.

The multiple sources considered in the collaborative data mining process for weather stations reveal that a large amount of data needs a wider analysis and more complex preprocessing.

From a brief side of view, the steps followed in the experiments are preprocessing, training, testing, improvement. The implementation for the collaborative data mining and follow-up of these phases was done, in parallel, from two points of view:

– to obtain predictions for a source using only its stand-alone data;
– based on data from a source correlated with the one observed (collaborative).

The results obtained were compared, established the parameters to take into account for obtaining a better precision and better results. In initial steps, the algorithms (trained model) tested was four: k-Nearest Neighbor model (KNN), Generate Weight (LPR), Neural Net model (NN) and Support Vector Machine (SVM).

The basic idea is that the most of the cases obtained better accuracy for the collaborative process than one with data from stand-alone sources, but very important facts are that the data used and their volume is a major attribute. The conclusions that have been reached so far indicate that the collaborative data mining technique involves an extensive process of analyzing the participating data and their sources, requiring a correlation between sources and high accuracy for the data used.

A future line that deserves attention and development is the one of approaching a collaborative data mining process with three sources and eventually generalizing it.

References

1. Lika, B., Kolomvatsos, K., Hadjiefthymiades, S.: Facing the cold start problem in recommender systems. Expert Syst. Appl. **41**, 2065–2073 (2014)
2. Moyle, S.: "Collaborative Data Mining. In: Maimon, O., Rokach, L. (eds.) Data Mining and Knowledge Discovery Handbook. Springer, Boston, MA (2005)
3. Di Orio, G., Matei, O., Scholze, S., Stokic, D., Barata, J., Cenedese, C.: A platform to support the product servitization. IJACSA **7**(2), 392–400 (2016)
4. Matei, O.: Preliminary results of the analysis of field data from ovens. Carpathian J. Electr. Eng. **8**(1), 7–12 (2014)
5. Matei, O., Nagorny, K., Stoebener, K.: Applying data mining in the context of Industrial Internet. Int. J. Adv. Comput. Sci. Appl. **1**(7), 621–626 (2016)
6. Matei, O., Di Orio, G., Jassbi, J., Barata, J., Cenedese, C.: Collaborative data mining for intelligent home appliances. In: Working Conference on Virtual Enterprises, pp. 313–323. Springer, Cham (2016)
7. Matei, O., Rusu, T., Petrovan, A., Mihuţ, G.: A data mining system for real time soil moisture prediction. Procedia Eng. **181**(837–44), 31 (2017)
8. Matei, O., Rusu, T., Bozga, A., Pop-Sitar, P., Anton, C.: Context-aware data mining: embedding external data sources in a machine learning process. In: International Conference on Hybrid Artificial Intelligence Systems, pp. 415–426. Springer, Cham (2017)
9. Matei, O., Anton, C., Scholze, S.: Multi-layered data mining architecture in the context of Internet of Things. In: IEEE 15th International Conference on Industrial Informatics (INDIN) (2017)
10. Camarinha-Matos, L., Afsarmanesh, H.: Collaboration forms in collaborative networks: reference modeling, pp. 51–56. Springer US (2008)
11. Badrul, S., et al.: Item-based collaborative filtering recommendation algorithms. In: Proceedings of the 10th International Conference on World Wide Web. ACM (2001)

12. Billsus, D., Pazzani, M.: Learning collaborative information filters. In: Proceedings of the International Conference on Machine Learning. Morgan Kaufmann Publishers. Madison (1998)
13. Folorunsho, O., Adeyem, A.B.: Application of data mining techniques in weather prediction and climate change studies. Int. J. Inf. Eng. Electron. Bus. 4(1), 51 (2012)
14. Rashid, R.A., Nohuddin, P.R., Zainol, Z.: Association rule mining using time series data for malaysia climate variability prediction. In: Badioze Zaman, H., et al. (eds.) Advances in Visual Informatics. IVIC 2017. Lecture Notes in Computer Science, vol. 10645. Springer, Cham (2017)
15. Niazalizadeh Moghadam, A., Ravanmehr, R.: Multi-agent distributed data mining approach for classifying meteorology data: case study on Iran's synoptic weather stations. Int. J. Environ. Sci. Technol. 15(1), 149–158 (2018)
16. Namen, A., Charles, A., Rodrigues, P.: Comparison of data mining models applied to a surface meteorological station. RBRH 22, 58–67 (2017)
17. Comeau, D., Zhao, Z., Giannakis, D., Majda, A.: Data-driven prediction strategies for low-frequency patterns of North Pacific climate variability. Clim. Dyn. 48(5–6), 1855–1872 (2016)

Novel Anti-Collision Algorithm in RFID Tag Identification Process

Musaddak Maher Abdulzahra$^{(\boxtimes)}$

Al Mustaqbal University College, Hilla, Babil, Iraq
MusaddaqMahir@mustaqbal-college.edu.iq

Abstract. Radio Frequency Identification (RFID) system is a new communication technology identifying one or more objects simultaneously in favor of using electromagnetic waves. In RFID systems, tags and readers communicate together over a shared wireless channel, and their signals may collide during transmission process and collision problem may occur in these systems. When two or more tags want to communicate with the reader, this collision problem may occur. This collision problem will waste the time for identification and it is also energy consuming as well and consequently reduces efficiency of the tag identification process. Therefore, it is required to attempt minimizing collision occurrence and decrease possibility of collision by using RFID anti-collision algorithms. In this project a novel Anti-collision Algorithm was suggested, which is called Enhance Dynamic Tree Slotted Aloha (EDTSA) by combining Tree slotted ALOHA and Advance Dynamic Framed Slotted Aloha Anti-Collision Algorithms and using Cubic Spline-based tag estimation technique. We have designed this algorithm by using C# programming language in visual studio 2008. The final results has simulated and compared with other algorithms. Simulated results also have illustrated that proposed algorithm can improve the efficiency of RFID systems.

Keywords: RFID · Dynamic Tree · Collision · Frame · Algorithm ·
Aloha based · Passive tag · Binary tree · Tree slotted ALOHA algorithm

1 Introduction

Despite its familiarity to barcode system, RFID is known as a new technology, which is used widely for identify many objects at the same time. empirical evidence indicate obvious similarities between RFID and barcode. However, RFID has been able to solve serious problems that barcode system faced, which enable RFID to enhance the performance and efficiency of identification process. In particular, barcode disadvantages compared to RFID include less memory capacity, weakness in identifying objects simultaneously, identifying objects from short distance, and straight identifying [1].

RFID system is mainly relying on electromagnetic signals than intends to recognize objects automatically and very fast. This system is able to identify many objects at the same time.

Sometimes during identification process signals may create collision. Signals that collied cannot be read and identify by reader. As a result, objects in this situation would

© Springer Nature Switzerland AG 2019
R. Silhavy (Ed.): CSOC 2019, AISC 984, pp. 152–169, 2019.
https://doi.org/10.1007/978-3-030-19807-7_16

not be identified, which make system to repeat the entire identification process. In fact, collision has contribution to wasting time and energy.

Generally, anti-collision algorithms are divided into two main groups namely Tree-based and ALOHA-based [2]. According to Tree-based algorithms, once the collision occurs all tags at interrogation zone will be divided into two subgroups. Thereafter, the process will repeat similar to the first one. In this phase if collision occurs again then the each of the subgroups will be divided into two new subgroups, which depends on tags' ID or random number. This process will continue until reader identify all the tags correctly.

On the other hand, in ALOHA-based anti-collision algorithm, FSA (Framed Slotted ALOHA) represents good performance. Based on this algorithm, reader sends fixed frame size, which includes time slots as well. Once the slots were received, tags start to transmit their data to the reader via randomly taken slots that need to be big enough for sending IDs to the reader. Once the first frame finished and yet there is unread tags, reader has to send next frame size to the tags. In the second frame tags will take another slot randomly to send back their data to the reader again.

As mentioned before, collision will waste both time and energy, which in turn will affect efficiency of the identification process. Hence, developing a perfect and anti-collision algorithm really matters. Some anti-collision algorithms have already been introduced. These algorithms have many disadvantages and they are not sufficient and reliable enough [3]. The main objective of present study is to focus on tag collision problem in RFID systems by developing a new identification method or new anti-collision algorithm to minimize wasting time and energy and improve the performance of RFID system.

2 Literature Review

RFID as a wireless technology has been used widely that is able to transmit electro-magnetic signals to recognize and track objects. RFID usually deals with one tag or more. We have to attach each tag to one object with unique ID that could be readable. As discussed, RFID is relying mainly based on broadcasting and electronic magnetic waves. Although this systems are similar to barcode systems but they have very fundamental differences. In the 1800s, the Electromagnetic theory was expanded and developed for the first time.

Michael Faraday realized that radio waves and lights are components of electro-magnetic energy. In 1896, Guglielmo Marconi began experimenting for sending signals via electromagnetic waves. He made a transmitter to send electromagnetic waves and designed a receiver to detect and receive radio waves. Marconi had managed to transmit signal between two places without using any wire.

Finally, in December 1901, Marconi succeeded to send radio signal across Atlantic without using any wire [4]. Alexanderson demonstrated transmission of radio signals and discovered new communication technology and started modern radio communication technology and also illustrated radio frequency identification (RFID).

In 1935, Sir Robert Alexander Watson-Watt who was Scottish physicist and discovered the radar to detect objects and represented this new technology. Radar systems can send radio waves and these radio waves osculated with objects. By the reflection of these waves, then, radar can detect and locate each object and determine the speed of it [3].

In 1935, during second World War, radar used by Germans to distinguish their planes from very far distance. They had a serious problem to distinguish their own planes from enemies' planes when they are flying in very far distance, so they have used Radio Frequency Identification (RFID) technology to distinguish their own and enemy's planes from very long distance. Hence, passive RFID systems have used by Germans for the first time. They used RFID systems to alert the radar systems. In 1989, American company, named Dallas North Turnpike, followed these RFID applications.

In this decade, 1980 s, the full usage of RFID applications was happened, all of these applications have identified objects but they have used in different field. Each application has worked on special situations to achieve special goals for instance industrial applications. These applications could be used in transportation. They have worked in toll collection, and drivers could pay toll with smart cards. They also created some applications in agriculture to track the animals and could find their location. These applications also have used in business [4].

In 1990s, usage of RFID systems was spread in the large scale and North America made many RFID tags to use in navigation systems. According to the reports in this decade, North America installed more than 3 million RFID tags and made the 1990 s decade as very important between other prior decades in terms of RFID technology. In that decade, first open highway electronic tolling and traffic management system were developed by some of American States such as Oklahoma, Georgia, Kansas, and Houston. Additionally, European countries were interested in using RFID systems such as access control and toll collection, as well as they were interested in using this technology in commercial applications [5].

In 1990s, this particular technology was developed more quickly than before and smart cards were extensively used. Between 1992 and 1995 many standards were developed for contactless smart cards. In 1999, some books that investigated RFID systems was published that Klavs Finkenzeller wrote the first RFID books. In the 21st century, by increasing people interest in telematics, mobile commerce, object tracking, and wireless communication systems, the necessity of RFID technology usage is enhanced and the usage of these technologies will rise quickly [6].

The major features of current RFID system is to identify the objects without need to contact between tag and reader and also no need line of sight between reader and tag. In current RFID systems tags have read and write capability and readers have cluster reading ability [7].

In these days, Because of the new technologies in the world, RFID tags size became smaller than before. Nowadays, the companies are able to produce small tags in stick label form. These kind of tags can be easily attached to materials, and articles, and be able to make them manageable.

One of the important problems in industry is in asset management and tracking. Previously companies were using barcodes to track their assets. They had many problems by using barcode systems for tracking, problems such as damage of labels or losing labels. To solve these problems we can use RFID technology and replace barcode labels with RFID tags [8].

2.1 Advantages of RFID Systems vs. Barcodes

Barcode is a traditional way to identify objects, but RFID tag is modern method to identify them. These two technologies have many differences in various point of view, and you can see some of these differences in Table 1. Some of these differences is presented in detail as follows [9].

- Read/Write Data: in RFID systems each tag is attached to one object and these tags can store information about that special object. The readers are able to write or read these information from long distance.
- Storage Capacity: RFID tags are able to store a large amount of information about the tagged object for example: object manufacturing time, date, and location or information about the content of tagged object and also we can change their information, but barcodes cannot store large amount of data and when barcode label is printed we cannot change it's data. In comparison between tags and barcodes, it can find out passive tags and active tags respectively are able to store data 30 times and 100 times more than capacity of barcodes [10].
- Environmental Information: In RFID systems, active tags can have any kind of sensor. So these tags are able to obtain real-time information about the carrier object via using these sensors. There are several kind of sensors, such as temperature and vibration sensors.
- Simultaneous Identification: Barcodes work based on traditional identification systems. To guarantee successful identification in these systems, all items must be read individually. They also must be read manually, while RFID systems provide simultaneous identification for large number of tags that placed in reader's read range [11].
- Resilience to Forgery: In barcode systems, every person can duplicate each barcode by using scanner and printer. We can scan our target barcode and then must print the scanned file, so the barcodes may be replicated easily. In RFID systems, on the contrary, unskilled people are not able to copy any kind of tags. This means that, copying these tags is not trivial for ordinary people [12].
- Life Span: Sales man and woman have to scan barcodes very fast, so we must place these barcodes labels in a accessible area of object and we cannot place them on every where on object, so during transportation or handling they may get damaged and their life time is short, but about RFID tags we can say that they have a greater shelf life because they can be placed in a safe place and we can attach them to the safest place on the object because there is no need to scan them [13].

Table 1. Comparison between RFID and barcode.

System	Barcode	RFID
Data transmission	Optical	Electromagnetic
Memory/Size	Up to 100 byte	Up to 256 kb
Tag writeable	No	Possible
Position of scann/Reader	Line-of-sight	No line-of-sight
Read range	Up to several meters (line of sight)	Up to several meters
Access security	Low	High
Anti-collision	No possible	Possible

As shown in Fig. 1 in 2005 almost 6 million RFID tags were used, which increased to more than 80 billion in 2010 and it is also expected that it will reach a number of 10 trillion in 2015. So we have to decrease possibility of collision in these system and use them in many application [9].

3   Components of RFID Systems

In RFID systems, Tags have a unique ID and they attach to the objects and known as essential parts of RFID systems, which carry very important information about the objects. The most important activity that each tag conducts is to send data to the reader continuously. Figure 2 illustrates a basic RFID system.

As discussed, Tags are integral parts of any RFID system that are attached to the objects and being used for labeling. The tags are very simple devices and simplest component in RFID systems. The tags composed by an antenna, and a basic and simple electronic circuit, which has a memory.

Nevertheless, there are three types of tags namely passive, active and semi-passive tags. In particular, passive tags are very simple and inexpensive devices, which do not have any batteries.

The reader signals are supplied their energy when reader starts to work many signals emitted by it and the tags will receive the signals. These signals supply their energy. This energy can activate the tag's electrical circuit, then it will produce a signal called a response signal.

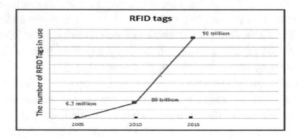

Fig. 1. Number of tags 2005–2015 [9]

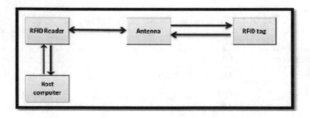

Fig. 2. Basic RFID system

On the other hand, active tags cannot work without power source, so need energy source for working like batteries. They embed batteries in their body. Each tag has it's own battery, hence embed a memory and a microprocessor in their circuits to read and write, rewrite, and erase information from an external device, thereby their cost increases. However, semi-passive tags characterization and their work manner are placed between passive tags and active tags. These tags have built-in batteries to supply energy integrated circuit. Table 2 illustrates main differences between active and passive tags.

Table 2. Active and passive tags

	Active	Passive
Tag power source	Internal to tag	Energy transferred from the reader
Availability of tag power	Continuous	Only when found in the field of the reader
Required signal strength	Low	High
Tag readability	Able to provide signal from far distance up to 100 m	Only in covered region by reader around 3 m
Available signal strength	High	Low
Communication range	Long range	Short range
Sensor capability	Ability to monitor continuously monitor sensor input	Monitor sensor input when tag is powered from the reader
Data storage	Large	Small
Shelf life	Less than 5 years equal to battery	Very high, work over a life time
Cost	Expensive	Cheap

When reader starts to work, its emitted electromagnetic waves are used to activate these types of tags. Then, these tags generate their response by using the energy of their batteries, so they work faster than circuitry activation that used in passive tags [10].

RFID Reader: Practically, readers are the key parts of RFID systems that represent RFID readers, RFID interrogators, and RFID scanners. Readers mainly receive data and modify them into readable and usable information for computer. Consistent with tags' types, there are two groups of readers. The readers normally supply tags' energy in a passive RFID system that do not have independent source of energy while active tags have their one source of energy. Figure 3 shows components of one RFID system [13].

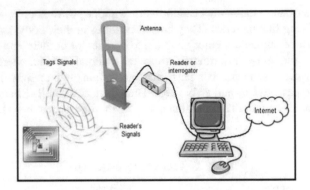

Fig. 3. Components of RFID systems.

RFID Antenna: RFID antenna is used to connect readers and tags. In addition it can be used also as a channel for moving data in a round trip. In fact, antennas are the main player in converting electric current into electromagnetic waves and contrariwise. Antenna also transmit these waves in a special pattern and intensity [13].

4 Anti-Collision Algorithms

RFID system also includes at least one reader. This reader must communicate with all tags located in integration zone and collect their information. This reader can identify an object through RF wireless communications with corresponding tag. In tag identification process, first reader sends a signal to all the tags and then all the tags in integration zone send back their unique ID number to the reader and in the next step, The reader has to recognize all tags very quickly. Therefore reader's signal and tag's signal are transmitted between the reader and the tag in both directions and they may collide because tags and readers communicate together over a shared wireless channel you can see the cause of collisions in the next figure.

Thus when collision occurs in RFID systems (Fig. 4), the reader is unable to identify all the tags correctly, so reader must repeat again this identification process. Repeating the identification process is time consuming and it can waste energy.

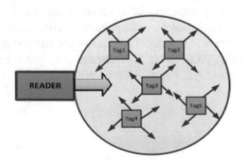

Fig. 4. The cause of collision

Therefore, we have to attempt to minimize collision occurrence, and we can solve this problem by using anti-collision algorithms.

The anti-collision algorithms can increase identification process speed, and they also can decrease the possibility of collision, and consequently, these algorithms can enhance the efficiency of RFID performance [10].

We can categorize these anti-collision algorithms into two various groups: ALOHA-based and Tree-based anti-collision algorithms.

Basic ALOHA

In Basic ALOHA algorithm, reader usually begins the first level by sending query to all tags in interrogation zone than can receive the signals. The tags will send back their data to the reader by taking random time slot.

Thus, if tags send back their data simultaneously then the likelihood of collision occurence may increase. Having many tags with huge amount of data also can increase collision occurence and affect process efficiency as well. As a result, it seems that this approach cannot be reliable. Figure 5 illustrates basic ALOHA algorithm [5].

In this method it can observe three various situations:

- Readable: it happened if just one tag transfers its ID.
- Collision: it happens if more than one tag send their ID to the reader simultaneously.
- Idle: it happens if tags do not send data or send it incomplete.

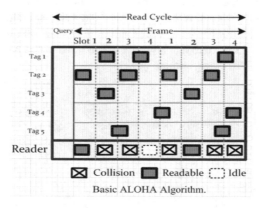

Fig. 5. Basic Aloha algorithm

Slotted ALOHA Algorithm

SA Algorithm is mainly based on Pure Aloha algorithm. Although these two algorithms are similar but the identification time divided into some slots, which are considered as intervals for transmitting the tags' data. It's up to tags to transmit their data in this slot or the next one. This algorithm is shown in Fig. 6.

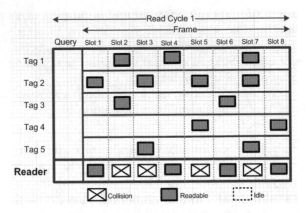

Fig. 6. Slotted ALOHA algorithm

Framed Slotted ALOHA Algorithm

Framed Slotted ALOHA (FSA) is introduced to improve the efficiency and performance of Slotted ALOHA algorithm. The size of these frames is fixed and this algorithm uses frames to group and sort the slots. In this algorithm tags have to send their data only once at each frame [4].

In this identification process, each tag has to choose a random time slot and it will send its' data to the reader. In FSA algorithm each tag can send its' data only once in any frame. Hence, they cannot retransmit their data, when collision happens in two or more tags, in that frame for the second time and they have to wait to resend their ID's to the next frame, Fig. 7 illustrate this algorithm. FSA algorithm has better performance than SA and Pure ALOHA algorithms [7].

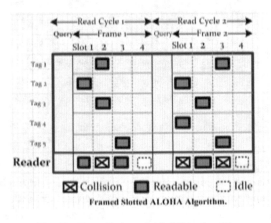

Fig. 7. Framed slotted ALOHA algorithm

Dynamic Framed Slotted ALOHA Algorithm

IN 1983, Schoute developed Dynamic Framed Slotted ALOHA (DFSA) algorithm. Based on DFSA, previous frame slots' information such as readable slots, collision slots, and idle slots would determine frame size of the new frame size [13].

However, DFSA is designed to identify tags based on minimal frame size, which is either four or two. In the first round if reader cannot identify any tag then reader automatically increases the frame size. Once the frame size was determined then the process will be repeated to identify a tag then it will stopped. Thereafter, a new read cycle will be started with minimum frame size to strat identification process [13].

According to Schoute, DFSA will generate its best performance if the frame size and unidentified tags are equal. However. Sometimes we have to increase frame size due to tags' limited memory capacity. The DFSA algorithm in short manner is shown in Fig. 8 [11]. DFSA identification process contains five steps:

Step 1: Read cycle starts with a primary frame size.

Step 2: According to distribution of the tags in current frame, reader has to compute C_0, C_1, C_K.

Step 3: Reader must estimate the number of all tags by using one of tag estimation methods.

Step 4: Reader has to estimate the frame size for next read cycle by using one of the frame size estimation methods.

Step 5: At the end of each read cycle, reader checks all of the tags are identified or not and then decides to stop operation or repeats all of these steps from step1 till all the tags are identified [4].

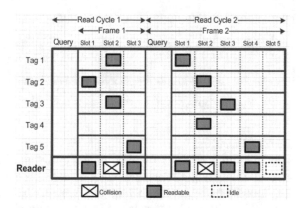

Fig. 8. Dynamic framed slotted ALOHA algorithm

Tree Based Anti-collision Algorithms

Tree-based code of standard behavior tags can be recognized by means of readers in the recognition structure which consists of a lot of scanning sequences. In reading sequence, reader attempts to recognize groups of tags. When we say groups, our purpose is tags which transmits identification at one reading sequence. Collision takes place in sending while groups consist of two or some tags. If collision takes place, the groups are split by means of tag identification or haphazard binomial in two small groups.

Thus small groups attempt to be recognized by means of reader in one structure, this operation keeps untill one group consists one tag so reader can read this tag. Figure 9 shows an Tree-based identification methods. This figure consists of three scctions A, B, and C. Part A illustrates Tree-based identification process, and part B and C shows two type of this identification method, so In part B you can see Binary tree protocol and in part C you can see Query tree protocol [5].

- **Idle sequence:** In Idle sequence there is not any tag so reader cannot identify any tag in this sequence.
- **Readable sequence:** Only one label sends identification, so tag recognition acted very well.
- **Impact sequence:** Two or some labels send identification to a reader so collision takes place and reader is not able to recognize all the tags, and identification process is not successful.

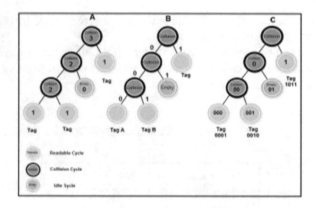

Fig. 9. Different kind of Tree algorithms

Tree Slotted ALOHA Algorithm

As mentioned earlier, there are two main groups of anti-collision algorithm namely ALOHA based and TREE based. Combining these two algorithms will generate another group called Tree Slotted ALOHA (TSA), which is very beneficial algorithm. This is because in ALOHA algorithm if tags failed to collide to each other in a frame they will be collided in next one. However, TSA will solve any collision once it happens, which save time and energy. In case of collision in a certain slot, tags that are involved in the following read cycle that particular slot retransmit the data to the reader, Fig. 10 illustrates Tree Slotted ALOHA Algorithm [10].

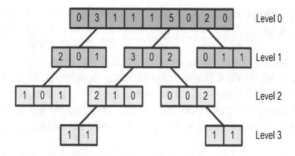

Fig. 10. Tree slotted ALOHA algorithm

Later than each read cycle in TSA algorithm, the tags number is estimated by applying of Chebyshev's inequality tag estimation method. The size of frame for level $i + 1$ is given by

$$N(i+1) = \left\lceil \frac{n(i) - C_1(i)}{C_k(i)} \right\rceil$$

5   Proposed Anti-Collision Algorithm

I have proposed new anti-collision algorithm which works based on the TSA to solve the problems and increase the performance of the tag identification process. Figure 11 shows the classification of anti-collision algorithms and presents the position of the suggested algorithm too.

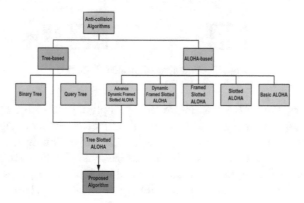

Fig. 11. Classification of anti-collision algorithms

New presented algorithm combines the strength points of FSA-based and tree-based anti-collision algorithms to achieve a better performance in tag identification process. Present study intends to combine dynamic frame slotted aloha and tree based anti-collision algorithms to make Dynamic Tree Slotted ALOHA and by using Cubic Spline tag estimation method we made Enhanced Dynamic Tree Slotted ALOHA (EDTSA).

In this algorithm, first the reader produces the initial frame size by sending a query to all tags. Similar to all FSA-based anti-collision algorithms, tags choose a slot randomly. Tags send their IDs back to the reader at the selected slot. After first read cycle, if any collision occurs, depending on the number of collided slots and by means of the Cubic Spline-based tag estimation technique, the numbers of unread tags are estimated accurately.

In accordance with the estimated number of unread tags the next frame size is also determined. The main idea of this algorithm is to solve the collision as same as tree-based algorithms. Thus, at the next frame just the tags, which involve at the first collided slot are considered and other unread tags have to wait until all these tags are identified successfully.

Based on the exact number of tags involved at the first collided slot, after know the first collided slot, the frame size for other collided slots is recalculate to identify the optimum frame size that in effect it will increase the efficiency of the system as well. Figure 12 demonstrated a sample of proposed algorithm.

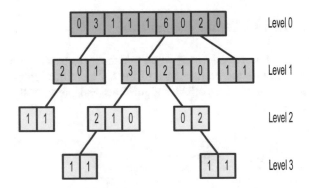

Fig. 12. Example of proposed method

In conclusion, the DTSA was improved by applying the Cubic Spline-based tag estimation technique to increase the accuracy of estimated number of tags and we made EDTSA. We produced this algorithm by using C# programming language in visual studio software and It also determine the frame size dynamically to reduce the required time slot to identify the tag that make faster identification process, which save the time and reduce the energy consumption in RFID tag verification system.

Implementation of Proposed Algorithm

The proposed anti-collision algorithm was implemented by using C# programming language in visual studio 2008. You can see how to compare these results with other algorithms and how to calculate the efficiency of proposed algorithm and compare it

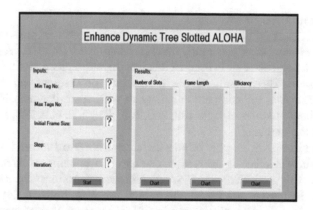

Fig. 13. Program interface

with other algorithms. Figure 13 shows the interface of program and User has to insert numbers and push start button.

In this part you can see the results of the proposed anti-collision algorithm based on different number of tags. The performance of proposed algorithm is compared with previous algorithms such as: Frame Slotted ALOHA (FSA), Dynamic Frame Slotted ALOHA (DFSA), Advance Dynamic Frame Slotted ALOHA (ADFSA), Tree Slotted ALOHA (TSA) algorithms. Three main factors of proposed algorithm were compared with other algorithm. There are three different diagrams for these three factors namely number of slots, frame length, and efficiency.

6 Evaluation of Proposed Algorithm

Some simulations were conducted to evaluate suggested algorithm. In the simulation process N is considered to be frame size and n is the number of tags and C_K represents the number of collision slots. Tag estimation method can estimate the number of tags based on these numbers and then based on number of unread tags our algorithm chooses the next frame size.

The program was tested many times and in this part the simulated results will be elaborated. There are many diagrams, which can be used to illustrate the results and compare them with results of other algorithms. These are the common diagrams, which used to compare RFID systems:

- Number of Slots need to identify all the tags.
- Frame Length used to identify all the tags.
- Efficiency of RFID System.

Number of Slots Used to Identify
The main motivation of this diagram is to focus on number of slots, which used to identify different number of tags in my system and compare simulated result with other algorithms. The number of slots used in this identification process will be appeared in output of program. These results also will be shown on diagram and user can record these simulated results and compare this diagram with other diagrams. The program is running with these numbers of tags: 100, 200, 300, 400, 500, 600, 700, 800, 900, 1000, and record all the outputs.

In this experiment, the number of tags increases from 0 to 1000 and you can see our simulated result with other algorithms in one diagram to compare them accurately. It also shows the EDFSA, BT, and DTSA algorithms offer almost the same result and they are better than FSA and DFSA algorithms. As it is shown in Fig. 10 proposed algorithm used less number of slots to identify different number of tags.

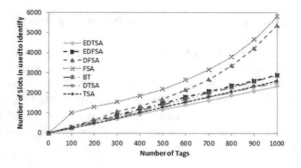

Fig. 14. Number of slots used

Figure 14 shows that proposed algorithm (EDTSA) can identify all the tags with less number of slots and work faster than other algorithms, thus this algorithm can increase the efficiency of RFID systems.

Frame Length used to Identify Tags in RFID Systems
The results were compared with simulated result of famous algorithms. Figure 11 shows amount of frame length used in other algorithms to identify same number of tags. This figure shows frame length used for identification in several algorithms such as: FSA, DFSA, EDSA and so on. We have compared our proposed algorithm with other algorithms based on frame length used in their identification process. Each algorithm can identify all the tags by using less frame length is work better than other algorithms.

Figure 15 shows amount of frame length used in other algorithms to identify same number of tags in comparison with proposed algorithm. This figure shows frame length used for identification in several algorithms such as: FSA, DFSA, EDSA. Figure 11 illustrates all algorithms together; therefore by using this figure it is possible to compare these algorithms easily.

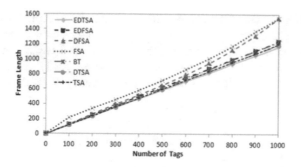

Fig. 15. Frame length used to identify

As shown, for small number of tags the results of these algorithms EDFSA, DTSA, and TSA are very close to results of the proposed algorithm (EDTSA). Although, they work very similar but by increasing the number of tags, the proposed algorithm (EDTSA) have used less frame length than other algorithms. Therefore, this figure shows that the proposed algorithm is able to determine the best frame size for next read cycle. It also illustrates that our proposed algorithm can identify all the tags faster than other algorithms.

It is demonstrated that proposed algorithm increases tag identification process speed and decrease energy consumption. Consequently this algorithm can improve the efficiency of RFID systems.

System Efficiency
The efficiency of RFID systems can be calculated by using:

$$\text{SEr} = \text{Rid}/\text{Rtot}$$

In this formula Rid is the amount of rounds in identification process, and Rid is equal to the number of tags. Rtot shows the total number of all rounds, and Rtot must be equal to number of slots which system used in identification process to identify all the tags. Therefore, system efficiency can be measured by:

$$\text{System Efficiency} = \text{Number of tags} \,/\, \text{number of slots}$$

In terms of time (time system efficiency), the system efficiency is:

$$\text{SEt} \quad = \quad \text{Tid}$$

In this formula Tid is equal with amount of time spent to identify all the tags in RFID system, and Ttot is also equal with execution time [4]. Figure 16 shows efficiency diagram for several algorithms such as TSA, BS, and QTI and our proposed algorithm (EDTSA). This figure represents that all the algorithms together in one diagram. In this figure, system efficiency in different algorithms is provided based on different number of tags. Therefore, taking this figure will assist comparing the efficiency of the proposed algorithm with other algorithms easily. Based on next figure you can see that my algorithm can improve the efficiency of RFID systems.

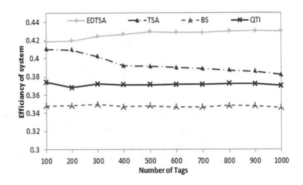

Fig. 16. System efficiency

7 Conclusion

In this project, the first Dynamic Tree Slotted ALOHA algorithm was created, which estimates number of tags by applying the Cubic Spline-based tag estimation technique. It also can increase the accuracy of estimated number of tags, and determining the frame size dynamically that in turn reduces the required time slot to identify the number of tags.

The proposed algorithm was implemented and it was called Enhanced Dynamic Tree Slotted ALOHA (EDTSA). The results were simulated and compared with other algorithms. It also was demonstrated that the efficiency of this algorithm is better than other algorithms and Consequently this algorithm can increase the performance and efficiency of RFID systems.

References

1. Shakiba, M.: Fitted dynamic framed slotted ALOHA anti-collision algorithm in RFID systems. In: ICIMU 2011 Proceedings of the 5th International Conference on Information Technology & Multimedia, November 2011
2. Lecture Notes in Computer Science (2006)
3. Lee, C.-W.: An enhanced dynamic framed slotted ALOHA algorithm for RFID tag identification. In: The Second Annual International Conference on Mobile and Ubiquitous Systems Networking and Services (2005)
4. Maselli, G., Petrioli, C., Vicari, C.: Dynamic tag estimation for optimizing tree slotted aloha in RFID networks. In: Proceedings of the 11th International Symposium on Modeling Analysis and Simulation of Wireless and Mobile systems MSWiM 2008 (2008)

5. Park, C.W., Ahn, J.H., Lee, T.-J.: RFID identification protocol with optimal frame size for varying slot time. Int. J. Inf. Electron. Eng. **4**(2), 87–91 (2014)
6. EPCRadio-Frequency Identification Protocols Generation-2 UHF RFID Protocol for Communications at 860 MHz–960 MHz. Version 2.0.0, EPCglobal, November 2013
7. Sample, A.P., Yeager, D.J., Powledge, P.S., Mamishev, A.V.: Design of an RFID based battery-free programmable sensing platform. IEEE Trans. Instrum. Measur. **57**(11), 2608–2615 (2008)
8. Wickramasinghe, A., Ranasinghe, D.C.: Ambulatory monitoring using passive computational RFID sensors. IEEE Trans. Sens. J. **15**(10), 5859–5869 (2015)
9. Khandelwal, G., Lee, K., Yener, A., Serbetli, S.: ASAP: a MAC protocol for dense and time constrained RFID systems. EURASIP J. Wireless Commun. Networking **2007**(2), 1–13 (2007)
10. Kawakita, Y., Mitsugi, J.: Anti-collision performance of gen2 air protocol in random error communication link. In: Proceedings of SAINT-W (2006)
11. Sheng, B., Tan, C.C., Li, Q., Mao, W.: Finding popular categoried for RFID tags. In: Proceedings of ACM Mobihoc (2008)
12. Di Marco, P., Alesii, R., Santucci, F., Fischione, C.: An UWB-enhanced identification procedure for large-scale passive RFID systems. In: Proceedings of the IEEE ICUWB 2014, September 2014
13. Vizziello, A., Savazzi, P.: Efficient RFID tag identification exploiting hybrid UHF-UWB tags and compressive sensing. IEEE Sens. J. **16**(12), 4932–4939 (2016)

An Empirical Assessment of Error Masking Semantic Metric

Dalila Amara[1(✉)] and Latifa Rabai[1,2]

[1] Université de Tunis, Institut Supérieur de Gestion, SMART Lab,
2000 Le Bardo, Tunisia
dalilaa.amara@gmail.com, latifa.rabai@gmail.com
[2] College of Business, University of Buraimi, Al Buraimi P.C. 512, Sultanate of Oman

Abstract. Semantic metrics are quantitative measures of software quality attributes based on the program functionality not only to the syntax. Different semantic metrics are proposed in literature and most of them are successfully used to assess internal quality attributes like complexity and cohesion. Among these metrics, a recent semantic suite for software testing is proposed to monitor software reliability. The purpose of this suite is to quantify an aspect of software testing and reliability that is fault tolerance by assessing the program redundancy. One of these metrics namely error masking is proposed to reflect the program non-injectivity and measures in bits the amount of erroneous information that can be masked by this program. However, to the best of our knowledge, this metric is only theoretically presented and manually computed. Also, its empirical validation as quantitative measure of erroneous information that a program may mask, still required. Hence, we aim in this paper to empirically assess this metric. So, we ought to propose an automated support tool to automatically generate it and to identify its statistical relationship with two other semantic metrics which are initial and final state redundancy. The experimental study we perform consists of a set of java programs from which we generate the value of these metrics. This study is benefit since the empirical assessment of this metric will help developers to identify the amount of erroneous information that can be masked by their programs.

Keywords: Fault tolerance · Software redundancy ·
Semantic metrics · State redundancy · Functional redundancy ·
Error masking

1 Introduction

Software quality is an important concept mainly in delicate systems. Poor software quality could cause economical and financial loss and it may cause catastrophic situations. Thus, the major goal of any organization using software systems is to improve its quality.

© Springer Nature Switzerland AG 2019
R. Silhavy (Ed.): CSOC 2019, AISC 984, pp. 170–179, 2019.
https://doi.org/10.1007/978-3-030-19807-7_17

Software quality is generally described through different attributes which may be internal like complexity and cohesion or external like reliability and maintainability. These attributes are required by developers to make appropriate decisions about the quality of their programs and by users to reflect the degree of their satisfaction [1]. From the user perspective, software quality is described by a common concept that is dependability. This concept is caracterized by three basic dimensions; (1) quality attributes which are reliability, safety, integrity, availability and confidentiality, (2) dependability threats which are faults, errors and failures, and, (3) dependability means that involve fault prevention, fault removal and fault tolerance. These means are required to avoid dependability threats [2].

As noted above, dependability involves a set of quality attributes including reliability. Reliability is the probability that the software product performs its required functions for a given amount of time without violating its specification [3]. Assessing this attribute is not a simple task since there are no obvious definition of its related aspects [1]. One of the quantifiable attributes which directly influences it is fault tolerance [4]. Fault tolerance is the ability of a system to continue performing its intended functions in the presence of faults in order to avoid failures [5,6].

There are different techniques which are proposed to ensure fault tolerance in software systems. These techniques are based around a basic concept which is redundancy [3]. Redundancy consists on using extra elements like hardware components or software components i.e instructions, programs, functions, etc [7,8].

To quantitatively assess this redundancy, a set of software semantic metrics is suggested by Mili et al. [3]. The purpose of these metrics is to use the program redundancy in order to detect and mask faults [3,4].

The proposed suite consists of four basic metrics namely state redundancy, functional redundancy, error masking and error recovery. State and functional redundancy metrics are proposed for error detection. Error masking metric aims to measure the ability of a program to mask errors. Concerning error recovery, its objective is to determine the non-determinacy of a program [3,4]. The purpose of these metrics is very important since error detection and masking are the basic notions of fault tolerance which is one of the quantifiable attributes that help achieving dependable and reliable software systems. To the best of our knowledge, most of the proposed software metrics are succeeding in measuring internal quality attributes like cohesion and complexity but they are still limited for measuring external ones which is the purpose of the new suite.

Motivated by the importance of the proposed suite and the lack of empirical studies that focus on its assessment, we aim in this paper to empirically validate one of these metrics which is error masking as a quantitative measure of the amount of errors that a program may mask. To achieve this objective, we need to propose an automated support tool to automatically generate this metric for different java programs. Moreover, as the automatic computing of state redundancy metrics is performed in a previous work [9], we aim to identify the correlation between these two metrics and the error masking one. Two major contributions are driven from the performed study:

- To start with, error masking is only theoretically presented and manually computed for simple procedural programs, hence, we showed that it is possible to compute it for different object oriented programs including open source projects.
- Moreover, we proved the existence of a statistical relationship between this metric and the state redundancy ones. This means that it is possible to use one of these metrics to reflect the error masking one.

This paper is organized as follows. First, we present a literature review of software fault tolerance and redundancy concepts in Sect. 2. Then, we clarify in Sect. 3 the concept of error masking metric and we present its formulation as well as an illustrative example. Section 4 presents details of the empirical assessment of this metric, the data collection procedure, the statistical data analysis and the results' interpretations. Finally, conclusion and perspectives will be drown in Sect. 5.

2 Software Fault Tolerance

As mentioned earlier, to achieve dependable software systems, different means are defined which are fault prevention, fault removal and fault tolerance as shown in Fig. 1 [7,10,11]. Both fault prevention and fault removal are based on exhaustive testing and program correctness [11]. However, there are no reliable tools to guarantee that complicated software systems are fault-free since exhaustive testing is not practical in most cases [8,12]. Moreover, program correctness may never be practical for use with very complex software-based systems [13]. Consequently, fault tolerance is defined as an alternate technique defined as the ability of a system to continue its function by avoiding the service failure despite the presence of faults [11].

Fault tolerance is not a new concept. It is one of the important means of software dependability [8,10]. It is defined as the ability of a system to continue its function by avoiding the service failure despite the presence of faults [11].

Figure 1 shows also that to achieve fault tolerance systems, different techniques are required which are widely discussed in literature [5,7,8,10,11,15]. Literature shows that these techniques are based on a common concept that is redundancy (See Fig. 1).

Redundancy consists on using extra elements like hardware components or software components i.e instructions, programs, functions, etc [6,7]. In software systems, three major types of redundancy are defined [5,15,16]:

- Software redundancy: is the generation of different program' versions (different algorithms, few lines of code, implementation details) with the same functionality and having the same specification in order to reduce the probability of simultaneous failure of all versions [1,5,8].

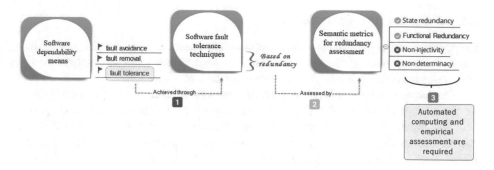

Fig. 1. Basics of software dependability and fault tolerance

- Information redundancy: is related to the information (coding) theory originally defined by Claude Shannon and Richard Hamming [17,18]. It is defined as a redundant data added to the actual data in order to detect and correct errors [1,7].
- Time redundancy: consists on repeating the execution of the failed process [5,7].

In this paper, we are interested by the information redundancy type since the semantic metric that we aim to empirically assess (error masking) is based on this type of redundancy.

As noted above, most of fault tolerance techniques are based on the redundancy concept. Different studies are focused on redundancy use to achieve fault tolerant systems. We cite for instance those proposed by Avizienis [19], Eckhardt et al. [20], Lyu [14], Jiang and Su [21], and Carzanigua et al. [16].

The cited studies show up the importance of redundancy concept to reflect the program ability to tolerate faults. However, to the authors' best of knowledge none of these studies are focusing on assessing the redundancy of programs in a quantitative way needed to make decisions and comparisons.

As software metrics are widely discussed in literature as quantitative measures of different quality characteristics [1,8], a semantic suite is proposed by Mili et al. [3] whose objective is to assess programs redundancy in order to reflect their ability to tolerate faults.

This suite consists of four semantic metrics. The purpose of these metrics is to ensure fault tolerance through error detection, error masking and error recovery. Hence, two of these metrics called state and functional redundancy are proposed as measures of error detection based on the information redundancy concept. An empirical study is performed to assess these two metrics [9]. The two other metrics are called error masking (error non-injectivity) and error recovery (non-determinacy). We are focusing in this study on the empirical assessment of the error masking metric described in the following section.

3   Error Masking Semantic Metric

This section describes the error maskability semantic metric. So, we present its purpose and formulation as well as an illustrative example.

3.1 Purpose and Formulation

Error masking is one of the fault tolerance phases defined as the systematic application of error compensation where enough redundancy is required in the erroneous state in order to deliver the error-free one [3].

The objective of this metric is to answer the following question: how much errors a program can mask? This metric is expressed in Shannon bits and reflects the non-injectivity of program functions which is their ability to map distinct states into a single image [3]. Hence, the more non-injective a program, the more damage it can mask [4]. Mathematically speaking, the Non-injectivity of a program g (error masking) on space S is defined by the conditional entropy (expressed in bits) as follows [3]:

$$\phi(g) = H(\sigma_1|\sigma_f)/H(\sigma_1) \tag{1}$$

– $H(\sigma_1)$ is the entropy of the initial state space σ_1 of the program g defined as the number of bits required to store the used variables before executing it.
– $H(\sigma_f)$: is the entropy of the final state space σ_f of the output produced by g defined as the number of bits required to store the result of the program execution,
– $\phi(g)$ is the conditional entropy that measures how much do we know about σ_1 if we observe σ_f defined above.

3.2 Illustrative Example

An example of an error masking inspired from Mili and Tchier [4] is illustrated in Fig. 2:

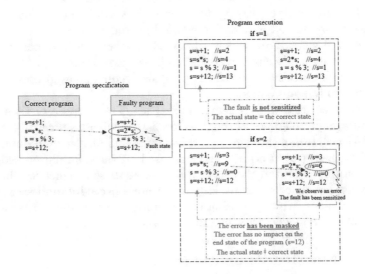

Fig. 2. Illustrative example of error maskability

Figure 2 shows that if the faulty state (See Fig. 2) has no effect on the program output, then the fault is not sensitized and there is no error to be observed $(s = 1)$. However, when the faulty state has an effect on the output, then, an error is detected $(s = 2)$. This error will be propagated and two cases will be possible:

- If the propagated error has an effect on the program' output, then it will cause a failure,
- If the propagated error has no effect on the program' output, then a failure is avoided and the error is masked.

4   Empirical Assessment of Error Masking Metric

We present in this section the data collection methodology. Moreover, we study the statistical relationship between this metric and the state redundancy ones studied in our previous work [9].

4.1   Data Collection Methodology

To collect our data set, we resort to Java language and the Eclipse development environment (version: Neon.3 Release (4.6.3)). The experiment consists of seven java programs that include basic and open source programs. The selected basic programs are the maximum value program (MaxValue), the Greatest Common Divisor (GCD), the linear search (Lsearch) and the Power program (Power). The open source programs we used are Account, jtCas and loseNotify. For collecting target open source projects, we used the Software-artifact Infrastructure Repository (SIR sir.unl.edu [24]).

We notice that these programs are chosen as first step to validate our metrics. Other added programs will be considered to improve our validation in future work. The process of computing the different metrics is illustrated in Fig. 3.

Figure 3 illustrates a part of the metric computing process for the GCD program using Eq. 1 as it is shown in line 72 in Fig. 3. We also generate the state redundancy metrics; initial state redundancy (ISR) and final state redundancy (FSR) [9] to study their correlation with the error masking (EM) one. Data are generated from 1.000 random input values (x and y) for each program (See lines 26 and 28 in Fig. 3).

4.2   Data Analysis

To study the existence of a statistical relationship between the cited metrics, we need to perform normality tests. The objective is to identify which correlation coefficient is the appropriate to use. Different tests may be used to check normality including Kolmogorov-Smirnov and Jarque-Bera (JB) tests [22].

We use in our study the Jarque-Bera (JB) test which indicates that data are normally distributed if p-value is less than 0.05; the larger the JB statistic, the

```
26 int x= rand.nextInt(1000) + 1;

28 int y= rand.nextInt(1000) + 1;

31 int values= Integer.MAX_VALUE;
32 int statespace=2*sizeOfBits(values);

36 int initialsizex= sizeOfBits(x);
37 int initialsizey= sizeOfBits(y);
38 int initialstatespace=initialsizex+
                          initialsizey;

53 int finalsizex= sizeOfBits(x);
54 int finalsizey= sizeOfBits(y);
55 int finalstatespace=finalsizex+finalsizey;

72 float errormasking= (float)
   (initialstatespace-finalstatespace)/finalstatespace;
```

Fig. 3. Part of the GCD java program for error masking computing

more the data deviates from the normal distribution [22, 23]. The test statistic JB is a function of the measures of skewness S and kurtosis K computed from the sample.

Under normality, the theoretical values of S and K are 0 and 3, respectively. If these values are different respectively from 0 and 3 then the data distribution is not normal [23]. Results of the normality tests are presented in Table 1.

Table 1. Normality tests for the generated metrics.

	Jarque-Bera test	Skewness test	Kurtosis test
Error Masking (EM)	0.834	0.597	−1.592
Initial State Redundancy (ISR)	0.834	1.619	0.795
Final State Redundancy (FSR)	0.834	1.619	0.795

Table 1 shows that the output variables distribution is not normal because of two things. To begin with, the values of the JB test are large (0.834) and greater than 0.05. Secondly both skewness and kurtosis values are different from the conventional ranges; 0 for skewness and 3 for kurtosis [19]. Consequently, the Spearman correlation is performed to study the statistical relationship between error masking and the three other metrics based on the following hypothesis:

- $H_0 : \rho = 0$ (null hypothesis) there is no significant correlation between these variables
- $H_1 : \rho \neq 0$ (alternative hypothesis) there is a significant correlation

Results of Spearman correlation are shown in Table 2.

Based on the Hopkin's correlation rating [25], Table 2 indicates that the correlation coefficient between the error masking metric and the two initial and

Table 2. Correlation between error masking with initial and final state redundancy metrics.

	Initial state redundancy	Final state redundancy
Erro Masking	0.864 p-value (0.011)	0.865 p-value (0.011)

final state redundancy ones are strong and positives; 0.864 and 0.865 respectively. Moreover, the p-values are good (less than 0.05). On this basis, the null hypothesis H_0 of no correlation between error masking and these metrics is rejected. Hence, it can be assumed, according to the alternative hypothesis H_1, that there is a strong positive correlation between error masking and state redundancy metrics for the different java programs. Consequently, we state that if the initial and final state redundancy metrics tend to increase (decrease), then the error masking metric tends to increase (decrease) too.

4.3   Overall Interpretation of the Results

The presented study leads us to draw different interpretations. To begin with, the error masking metric is positively correlated with state redundancy metrics (initial and final states [9]). One explanation of this correlation is that if the initial and final state redundancy increases, then, we have a large number of bits which are declared but not used by the program in both its initial and final states. Thus, these unused bits will increase the ability of a program to mask more errors. Consequently, redundancy bits generated by the initial and final state redundancy metrics can be used to reflect the number of erroneous states that the program can mask.

To sum up, the presented study is benefic and presents different contributions compared with the related work [3, 4] as shown in Table 3:

Table 3. Comparison between the performed study and the related work.

Related work [3]	Our work
– Error masking is manually computed for procedural programs	– Functional redundancy metric is computed for different oriented object java programs
– The paper do not perform the statistical relationship between the different metrics	– In this paper, a study of the statistical relationship between this metric and state redundancy one is performed

As seen in Table 3, two major contributions are driven from this study. First, we show that it is possible to automatically compute the error masking metric for different programs which leads as to quantitatively assess it.

Additionally, this metric is positively correlated with state redundancy metrics (initial and final states). This means that if the program' state redundancy increases, then the error masking will increase too and vice versa.

5 Conclusion

The presented paper details a way to empirically assess the error masking metric by proposing an automated way to automatically generate it. Moreover, authors are focused on the statistical relationship between this metric and the other ones. The use of different object-oriented java programs including open source projects shows that it is possible to automatically generate this metric.

Different benefits are driven from the presented empirical study. First, we succeed to automatically compute the error masking semantic metric for different java programs including open source projects. Second, we show that the error masking metric as measure of the amount of erroneous information that can be masked by a program, may be predicted by initial and final state redundancy metrics due to the strong positive correlation between them.

Although these benefits, we could enhance the quality of this work by expanding the obtained database to incorporate further programs. Furthermore, we focus on empirically assess the fourth semantic metric termed error recovery for basic programs and open source projects.

Acknowledgements. We would like to thank Ezzeddine Fatnassi for his valuable input to this work and his helpful comments that greatly improved this study.

References

1. Fenton, N., Bieman, J.: Software metrics: a rigorous and practical approach. CRC Press, Boca Raton (2014)
2. Laprie, J.C.: Dependability: basic concepts and terminology. In: Dependability: Basic Concepts and Terminology, pp. 3–245. Springer, Vienna (1992)
3. Mili, A., Tchier, F.: Software Testing: Concepts and Operations. Wiley, Hoboken (2015)
4. Mili, A., Jaoua, A., Frias, M., Helali, R.G.M.: Semantic metrics for software products. Innov. Syst. Software Eng. **10**(3), 203–217 (2014)
5. Dubrova, E.: Fault-Tolerant Design, pp. 55–65. Springer, New York (2013)
6. Lee, P.A., Anderson, T.: Fault tolerance. In: Fault Tolerance, pp. 51–77. Springer, Vienna (1990). https://doi.org/10.1007/978-3-7091-8990-0_3
7. Pullum, L.L.: Software Fault Tolerance Techniques and Implementation. Artech House, Norwood (2001)
8. Lyu, M.R.: Handbook of software reliability engineering (1996)
9. Amara, D., Fatnassi, E., Rabai, L.: An automated support tool to compute state redundancy semantic metric. In: International Conference on Intelligent Systems Design and Applications, pp. 262–272. Springer, Cham, December 2017
10. Randell, B.: System structure for software fault tolerance. IEEE Trans. Software Eng. **2**, 220–232 (1975)

11. Avizienis, A., Laprie, J.C., Randell, B., Landwehr, C.: Basic concepts and taxonomy of dependable and secure computing. IEEE Trans. Dependable Secure Comput. **1**(1), 11–33 (2004)
12. Rizwan, M., Nadeem, A., Khan, M.B.: An evaluation of software fault tolerance techniques for optimality. In: 2015 International Conference on Emerging Technologies (ICET), pp. 1–6. IEEE, December 2015
13. Scott, R.K., Gault, J.W., McAllister, D.F.: Fault-tolerant software reliability modeling. IEEE Trans. Software Eng. **5**, 582–592 (1987)
14. Lyu, M.R., Huang, Z., Sze, S.K., Cai, X.: An empirical study on testing and fault tolerance for software reliability engineering. In: 14th International Symposium on Software Reliability Engineering. ISSRE 2003, pp. 119–130. IEEE, November 2003
15. Asghari, S.A., Marvasti, M.B., Rahmani, A.M.: Enhancing transient fault tolerance in embedded systems through an OS task level redundancy approach. Future Gener. Comput. Syst. **87**, 58–65 (2018)
16. Carzaniga, A., Mattavelli, A., Pezz, M.: Measuring software redundancy. In: Proceedings of the 37th International Conference on Software Engineering, vol. 1, pp. 156–166. IEEE Press, May 2015
17. Hamming, R.W.: Error detecting and error correcting codes. Bell Syst. Tech. J. **29**(2), 147–160 (1950)
18. Shannon, C.E.: A mathematical theory of communication. ACM SIGMOBILE Mob. Comput. Commun. Rev. **5**(1), 3–55 (2001)
19. Avizienis, A.: The N-version approach to fault-tolerant software. IEEE Trans. Software Eng. **12**, 1491–1501 (1985)
20. Eckhardt, D.E., Caglayan, A.K., Knight, J.C., Lee, L.D., McAllister, D.F., Vouk, M.A., Kelly, J.P.J.: An experimental evaluation of software redundancy as a strategy for improving reliability. IEEE Trans. Software Eng. **17**(7), 692–702 (1991)
21. Jiang, L., Su, Z.: Automatic mining of functionally equivalent code fragments via random testing. In: Proceedings of the Eighteenth International Symposium on Software Testing and Analysis, pp. 81–92. ACM, July 2009. https://doi.org/10.1145/1572272.1572283
22. Yazici, B., Yolacan, S.: A comparison of various tests of normality. J. Stat. Comput. Simul. **77**(2), 175–183 (2007). https://doi.org/10.1080/10629360600678310
23. Thadewald, T., Bning, H.: JarqueBera test and its competitors for testing normalitya power comparison. J. Appl. Stat. **34**(1), 87–105 (2007). https://doi.org/10.1080/02664760600994539
24. Do, H., Elbaum, S., Rothermel, G.: Supporting controlled experimentation with testing techniques: an infrastructure and its potential impact. Empirical Software Eng. **10**(4), 405–435 (2005)
25. Hopkins, W.G.: A New View of Statistics. Will G Hopkins, Melbourne (1997)

Intelligent Software Agents for Managing Road Speed Offences

Maythem K. Abbas[1(✉)], Low Tan Jung[1], Ahmad Kamil Mahmood[1],
and Raed Abdulla[2]

[1] High Performance Cloud Computing Center (HPC3),
Computer and Information Sciences Department,
Universiti Teknologi PETRONAS, Seri Iskandar, Malaysia
{maythem.aladili,lowtanjung,kamilmh}@utp.edu.my
[2] Faculty of Engineering, Asia Pacific University of Technology and Innovation,
Kuala Lumpur, Malaysia
dr.raed@apu.edu.my

Abstract. The increasing fatal road accidents because of fast driving had urged police departments to set many cameras and speed traps at some parts of the roads in an attempt to reduce the problem. Many drivers know where those cameras and speed traps are, which makes it very easy to be avoided by the drivers who intend to drive faster than the road speed limit. This research proposes a system for smart cities and/or highways to control the problem via the placement of a set of Vehicular Ad-Hoc Network (VANET) devices on the roadside and within the vehicles to achieve side to side communication. The provided communication would be responsible for delivering real time information about the car, including its speed, which would help to detect the speeding offenders. The aims of this research are to identify the factors causing the accidents to happen, warn the speeding drivers at any time about their offence and to ensure a fine will be issued for every offender. In addition, the proposed system is expected to help traffic policemen in doing their job of identifying and proving that a driver has broken the traffic rules.

Keywords: Intelligent Transportation System (ITS) · Road traffic control · Smart cities technologies · Vehicular Ad-hoc Network (VANET) · Vehicle-to-Road side communication (V2R)

1 Introduction

Malaysia, for instance, is considered one of the highest countries in terms of having fatal road accidents as it was ranked as the 20th out of almost 200 countries, and was number 17 out of 200 countries in terms of the most dangerous roads [1–3]. The fatal accident count is still increasing because of the continuous increment in the number of vehicles on the road and the existence of fast driving. Police departments have set many speed traps and cameras at various parts of the roads trying to reduce the problem. However, many drivers know where those cameras and speed traps are, so they reduce their speed when they are within the coverage of those cameras, then return to their previous speed again [2]. This research proposes a system for smart cities or highways

© Springer Nature Switzerland AG 2019
R. Silhavy (Ed.): CSOC 2019, AISC 984, pp. 180–191, 2019.
https://doi.org/10.1007/978-3-030-19807-7_18

to control the problem via the Vehicular Ad-Hoc Network (VANET) Technology. A set of devices would be placed on the roadside and another set of devices would be placed within the vehicles. Thus, the aim of this research is to develop a model that can be used in cities and highways to reduce accidents caused by irresponsible high speed driving. It is expected that the outcome of this research will help identify the factors causing the accidents to happen, warn the speeding drivers at any time about their offence, and ensure that a fine will be issued for every offender. In addition, the proposed system is expected to help traffic policemen in doing their job of identifying and proving that a driver has broken traffic rules.

2 Problem Statement

This research focuses on the problem of the inability of the currently used on-road technology to detect those violating traffic rules and those causing the occurrence of fatal road accidents. In other words, the currently used technology to detect traffic rule offenders is insufficient to solve road problems.

3 Literature Review

Fatal road accidents have become a major problem in most countries. According to the statistical report of the Association of Safe International Road Travel (ASIRT) in the United States of America that was released on January 2016 [3], for the current decade, about 1.3 million people have died or are estimated to die annually because of fatal road accidents, averaging 3287 people per day. Additionally, around 20–50 million others have been disabled or injured. Similar numbers were confirmed by many different worldwide specialized organization reports, such as [1, 4–7].

Malaysia has the second highest number of fatal car accidents among South-east Asian countries [1]. The Malaysian Institute of Road Safety Research (MIROS) and the Road Safety Department have announced that the average annual death during the last 5 years was 6889, and it is estimated to hit 10716 deaths in 2020. Those numbers were estimated by the Auto-regressive Integrated Moving Average (ARIMA) Model [7, 8].

The report of ASIRT in 2016 [3] states that if no action is taken in the near future to fix the current situation of the road traffic control, then the ratio of fatal road accidents will increase to become the fifth leading cause of death by the next decade. As for Malaysia, fatal accidents are expected to increase, leading the death rate to increase by 20% more than the current rate.

Road accidents occur due to many different factors. According to the National Highway Traffic Safety Administration (NHTSA), those factors can be categorized into three main categories; human factors, road design, and vehicle design and maintenance. According to [9], human error causes around 93% of the total number of the worldwide accidents. One of those human factors is driving at high speed which is a huge contributor for vehicles crashes [10]. In reality, many of the specialized organizations support the fact that reducing or controlling vehicle's driving speed would enormously reduce traffic crashes [1].

Many researchers have looked at this problem, to derive with solutions, however, most of them could not solve the problem. For instance, the researchers of [11] have developed an application level protocol to reconstruct accidents using VANET. This system looks at the cause of the accident after it has happened.

4 Objectives

The main objective of this research is to design a conceptual model for a system that would overcome the downsides of the currently used technology to detect traffic rule offenders. Another objective is to design the system entities' hardware setup and the auto-fining protocol. Finally, it aims to evaluate the developed system and the protocol, a simulation was carried out using MATLAB and a physical test-bed was developed as well.

5 The Conceptual Model

This research has proposed a networking system mainly designed to be deployed in smart cities or highways to add both warning and controlling elements for the solution of the road speed offence problem via the deployment of the Vehicular Ad-Hoc Network (VANET) Technology. The main concept is setting up a set of roadside devices near the smart city's roads or highways to provide a fixed network infrastructure, with another set of devices being implanted within the vehicles to act as the mobile ad-hoc nodes of the network, similar to that introduced in [12] as shown in Fig. 1.

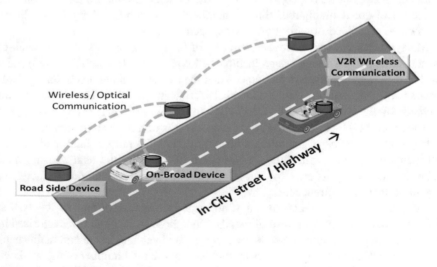

Fig. 1. Visual description for the communication between the roadside devices and the mobile devices

The main aim of the concept is to capture the vehicle's speed in a real-time manner and check if it exceeds the road's permitted speed. That can be achieved via a continuous reporting about the vehicle's instant speed by the mobile on-board devices in the moving vehicles. After getting a positive detection for a speeding offence, a warning will be issued by the nearest roadside device from the vehicle and it will be unicast to that vehicle. The warning message would be instantiated with the hope that the driver would reduce the driving speed as soon as the message is received, as psychologically expected [13]. In addition, a countdown timer would be set for that vehicle to respond to the warning. In case the offending vehicle's driver does not respond to the warning within the specified time, a fine would be issued for the vehicle's owner based on the registration number for that vehicle.

The proposed system is meant to reduce the fatal accidents from happening. That is why it would be important to handle the case when the driver had already been notified about the fine issuance and he/she kept on with the high speed driving. The proposed system would duplicate the fine amount and send a unicast message to the vehicle to either terminate the use of the car and prevent it from being turned on; or to set a maximum speed limit for the car to be driven at until the fine has been settled at the traffic police department.

6 The Auto-Fining Protocol

With the assumption that the city is equipped with the roadside devices and all of the vehicles have the embedded on-board VANET devices, as suggested by several studies such as [14, 15], and [16], the auto-fining protocol could be set up and deployed over the smart city network devices. This section illustrates the details of the protocol.

6.1 Protocol Messages

According to the conceptual concept explained in Sect. 5, the auto-fining protocol makes the use of 6 types of messages that will be exchanged between the fixed infrastructure devices themselves and with the mobile nodes (vehicles). Figure 2 shows the messages exchanged between the different devices of the developed system (Table 1).

6.2 Session Management

The proposed protocol pre-initiates its session by broadcasting an MtCA, waiting for a vehicle to respond with an MtCB message to initiate the session. Ideally, the session will stay ON till the vehicle handshakes with another RSE.

The proposed protocol suggests three different endings for the session:

1. The first ending would happen when the vehicle handshakes with an RSE then handshakes with the next RSE without exceeding the speed limit. This scenario would be called as 'No Offence and No Summon' or NONS.

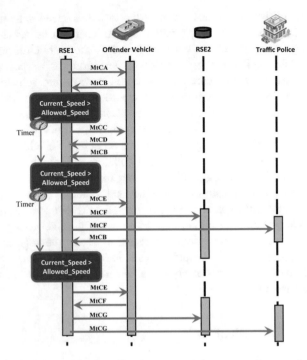

Fig. 2. Sequence diagram of the auto-fining protocol messaging

Table 1. The details of the proposed system's protocol messages

Msg Name	Src	Des	Message fields			Description
MtCA	RSE	Veh	0001	RSE_ID	Allowed_Spd	Broadcast – Hello Message
MtCB	Veh	RSE	0010	Veh_ID	Current_Spd	Unicast Reply – Vehicle's Report
MtCC	RSE	Veh	0011	Allowed_Spd	Offence_Spd	Unicast – Offence Warning (OW)
MtCD	Veh	RSE	0100	Veh_ID	Current_Spd	Unicast – OW Acknowledgment
MtCE	RSE	Veh	0101	Veh_ID	Offence_Spd	Unicast – Summon Issued (SI)
MtCF	RSE	RSE	0110	Veh_ID	Offence_Spd	Unicast – SI Acknowledgment
MtCG	RSE	RSE	0111	Veh_ID	Offence_Spd	Unicast – SI Redundancy

2. The second suggested ending would happen when the driver exceeds the speed limit, getting an Offence Warning (OW) and, finally, responds by reducing the current vehicle speed to be within the speed limit of the road. This scenario will be named as 'Offence, No Summon' or ONS.
3. The third scenario would be when the driver does not reduce the driving speed after receiving the OW, then the summon will be issued. In this case, it will be named as 'Offence with Summon' or OWS.

6.3   The Protocol's Tables

The proposed protocol consists of four tables for different purposes.

6.3.1   RSE's Table - Passed_by_Vehicles
This table lists down all the information of the vehicles handshake with the RSE. The columns of the table are the Vehicle's Fixed information (FI), Vehicle's Instant Status (IS), Street_Name, Position and Time_Stamp.

6.3.2   Vehicle's Table - Passed_by_RSEs
This table lists down all the information of the RSE devices the vehicle had passed by. This table keeps information about the RSE_ID, street's name, vehicle's position, and the time stamp of the last handshake with the RSE.

6.3.3   Vehicle's Summons/Warnings Table
This table keeps all the information about all the summons or warnings issued for the vehicle. Named as Warning, the warning information will be erased from the table after the driver has responded to the warning and reduced the speed of the vehicle to the limit. In the case that the driver has not responded to the warning and insisted on speeding, the warning record will be replaced by a summon record.

6.3.4   RSE's Table – Issued Summons/Warnings
This table keeps information about all the summons or warnings issued because of vehicles' offences. The record will stay in the table permanently until the authorities deal with it.

7   Experimental Work

To test the developed protocol, it was implemented on a raspberry pi-based platform after being simulated on a customized simulation for the vehicular ad hoc network (VANET) in MATLAB that was created to prove the introduced messaging concept using three experimental scenarios.

7.1 Simulation Work

The presented protocol was simulated on MATLAB Simulink as shown in Fig. 3, with the help of M-files. Both the algorithms of the vehicle and the RSEs were programmed in M-Files and embedded into the corresponding devices. The vehicle's device was programmed to respond based on the incoming messages. On the other hand, the RSE's device was programmed to perform timed actions in addition to responding to incoming messages.

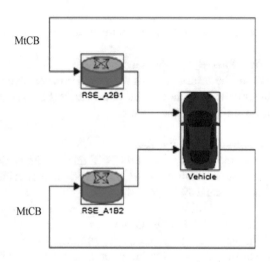

Fig. 3. Simulated VANET environment

7.2 Implemented Test-Bed

VANET uses Dedicated Short Range Communication (DSRC), a wireless technology which provides the V2I and V2V communications. However, for conceptual proof, Bluetooth wireless technology (version 4.0) was used instead of the DSRC. It can be seen in Fig. 4 that the RSEs were stationed and separated by 25 m while the mobile VANET device had a horizontal space of less than 10 m from the RSEs.

7.3 Experimental Scenario

Both the simulation and the implemented test-bed were set up to run and prove the introduced concepts of the three types of sessions: NONS, ONS and OWS. The Formal Specification Language for VANET (FSL-VANET) [17] is presented in Fig. 5 to describe the scenario used in the simulation and testing. During the simulation and testing, a set of five (5) outcomes was checked. Figure 5 shows the validation of the developed protocol. The results are presented in the experimental results section.

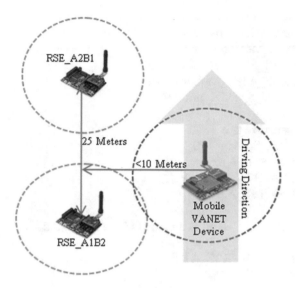

Fig. 4. Developed VANET test-bed

8 Experimental Results

In this section, the five results were collected from the experiments on MATLAB and on the test-bed. The results are presented in five tables along with the scenario, as described in Fig. 5.

The scenario begins with the initialization stage of its environment. The vehicle mobility feature was set to ON, while the same feature for the RSEs was set to OFF. The road's speed limit was set to be at 90 km/h, while the vehicle was driven at the speed of 110 km/h.

The next step was to run the test and monitor the interaction between the devices. When the vehicle entered the coverage of the first RSE, it received the broadcasted message (MtCA) that announced the existence of the RSE. At this point, the vehicle had to add a record of the RSE to its Passed_By_RSEs Table, which was the first check point (CHECK-1), of which the results are shown in Table 2.

In response, the vehicle replied with an MtCB message achieved a handshaking session with the RSE. In addition, the RSE was supposed to add a record of the handshake to its Passed_By_Vehicles. This point is named CHECK-2, which is shown in Table 3.

The vehicle sent its current speed with the MtCB. The RSE compared the vehicle's speed with the road's speed limit to check whether it was exceeding it or not. If it was exceeded, then an Offence Warning (OW) would be sent to the Vehicle. The RSE would have kept a record of that OW in its Issued Summons/Warnings Table. See Table 4.

// Devices mobility setup
Let veh_Mobility: On
Let RSE_A1B2_Mobility: Off
Let RSE_A2B1_Mobility: Off
Let Road_Speed_Limit = 90

// Scenario Starts
Let Veh_Speed = 110
Veh ▶ RSE_A1B2
RSE_A1B2 ~**MtCA**○ Veh
Veh ➔ RSE_A1B2

CHECK-1: Veh adds a record of the RSE_A1B2

Veh ~**MtCB>** RSE_A1B2
Veh ⟺ RSE_A1B2

CHECK-2: RSE_A1B2 adds a record of Veh

If(Veh_Speed > Road_Speed_Limit)
     RSE_A1B2 ~**MtCC>** Veh
     RSE_A1B2 Adds OW record

CHECK-3: RSE_A1B2 adds OW record of Veh

Veh <**MtCC**~ RSE_A1B2
Veh ~**MtCD>** RSE_A1B2

CHECK-4: Vehicle adds OW record

Veh ~**MtCB>** RSE_A1B2
If(Veh_Speed > Road_Speed_Limit)
     RSE_A1B2 ~**MtCE>** Veh
     RSE_A1B2 Adds SI record
Veh <**MtCE**~ RSE_A1B2
Veh ~**MtCF>** RSE_A1B2

CHECK-5: RSE_A2B1 adds SI record
CHECK-6: Veh adds SI record
// Scenario Ends

Fig. 5. Experimental scenario specification

Table 2. CHECK-1 result from the vehicle's table: Passed_By_RSEs

RSE_ID	Street name	Position	Time stamp
A1B2	A	A1/B2	05:12:39-2018.10.06

Table 3. CHECK-2 results from the RSE's table: Passed_By_Vehicles

Fixed info. of the vehicle	Record type	Current speed	Street name	Position	Time stamp
Veh_XYZ	R	110	A	A1/B2	05:12:44-2018.10.06

Table 4. CHECK-3 results from the RSE's table: issued Summons/Warnings

Fixed info. of the vehicle	Record type	Speed limit	Vehicle speed	Street name	Position	Time stamp
Veh_XYZ	OW	90	110	A	A1/B2	05:12:45-2018.10.06

At this moment, the vehicle would have received the offence warning that was issued and sent by the nearest RSE. The OW message would have contained the details of the offence, such as the current speed of the vehicle, and the approximate position of the vehicle on the road where the offence happened. The fourth check point in this test is shown in Table 5.

Table 5. CHECK-4 results from the vehicle's table: vehicle's Summons/Warnings

Fixed info. of the vehicle	Record type	Speed limit	Vehicle speed	Street name	Position	Time stamp
Veh_XYZ	OW	90	110	A	A1/B2	05:12:47-2018.10.06

Finally, the driver would have received the warning on the vehicle information screen, giving him/her two choices to be made. The driver might respond to the warning and reduce the driving speed, which would have the OW record erased from the RSE and the vehicle table itself. Alternatively, the driver might ignore the OW and insist on breaking the speed limit assigned for that road. In this case, the system would change the Offence Warning record types, in the vehicle Summons/Warnings Table in both of the RSE and the vehicle, into a Summon Issued record (SI), as shown in Tables 6 and 7.

Table 6. CHECK-5 results from the RSE's table: issued Summons/Warnings

Fixed info. of the vehicle	Record type	Speed limit	Vehicle speed	Street name	Position	Time stamp
Veh_XYZ	SI	90	110	A	A1/B2	05:12:55-2018.10.06

Table 7. CHECK-6 results from the vehicle's table: vehicle's Summons/Warnings

Fixed info. of the vehicle	Record type	Speed limit	Vehicle speed	Street name	Position	Time stamp
Veh_XYZ	OW	90	110	A	A1/B2	05:12:57-2018.10.06

9 Conclusion

Although police departments have set many speed traps and cameras along the roads aiming to reduce road accident rates, fatal accidents are increasingly occurring because of high speed driving. The current practice has no control on vehicles' speeding at all. The authors of this paper have suggested a system that adds some control factors to limit the offending vehicle's speed until the owner pays the summon issued because of the offence. The core concept is to install devices on road sides to achieve communication with other devices which are attached to the cars, revealing the vehicle's driving speed. The concept of detecting the offending vehicles has been proven via MATLAB and the developed test-bed, the results shown in the paper have proved that all of the six check points of the concept's protocol were successfully fulfilled, which support the proposed concept's validity and functionality.

Acknowledgments. Our appreciation goes to Universiti Teknologi PETRONAS for funding this project and for High Performance Cloud Computing Center (HPC3) for providing the required facilities.

References

1. World Health Organization (WHO). 10 facts on global road safety (2015). http://www.who.int/features/factfiles/roadsafety/facts/en/index.html
2. Transport, Local Government and the Regions - Ninth Report. House of Commons, United Kingdom, parliament Session 2001–02, Enforcement section. https://publications.parliament.uk/pa/cm200102/cmselect/cmtlgr/557/55708.htm
3. Association of Safe International Road Travel (ASIRT). Annual Global Road Crash Statistics, January 2016. http://asirt.org/initiatives/informing-road-users/road-safety-facts/road-crash-statistics

4. Insurance Institute for Highway Safety, Highway Loss Data Institute, "General Statistics from the U.S. Department of Transportation's Fatality Analysis Reporting System (FARS): Status Report", February 2016. http://www.iihs.org/iihs/topics/t/general-statistics/fatalityfacts/state-by-state-overview
5. Global Health Observatory (GHO). Number of Road Traffic Deaths (2010). http://www.who.int/gho/road_safety/mortality/traffic_deaths_number/en/
6. World Health Organization (WHO). Global status report on road safety 2013 (2013). http://www.who.int/violence_injury_prevention/road_safety_status/2013/en/
7. International Traffic Safety Data and Analysis Group (IRTAD), "Road Safety Annual Report 2015", OECD Publishing, Paris. http://dx.doi.org/10.1787/irtad-2015-en
8. Sarani, R., Mohamed Rahim, S.A.S, Marjan, J.M., Voon, W.S.: Research Report "Predicting Malaysian Road Fatalities for Year 2020", Malaysian Institution of Road Safety Research (2012)
9. Lum, H., Reagan, J.A.: Interactive Highway Safety Design Model: Accident Predictive Module. Public Roads Magazine, Winter 1995
10. Lee, Y.M., Chong, S.Y., Goonting, K., Sheppard, E.: The effect of speed limit credibility on drivers' speed choice. Transp. Res. Part F: Traffic Psychol. Behav. **45**, 43–53 (2017). ISSN 1369-8478, http://dx.doi.org/10.1016/j.trf.2016.11.011
11. Farkas, C., Kopylova, Y.: Application Level Protocol for Accident Reconstruction in VANETs", Master's Thesis, University of South Carolina, September 2007
12. Brik, B., Lagraa, N., Lakas, A., Cheddad, A.: DDGP: distributed data gathering protocol for vehicular networks. Veh. Commun. 4 February 2016. ISSN 2214-2096, http://dx.doi.org/10.1016/j.vehcom.2016.01.001
13. Steinbakk, R.T., Ulleberg, P., Sagberg, F., Fostervold, K.I.: Analysing the influence of visible roadwork activity on drivers' speed choice at work zones using a video-based experiment. Transp. Res. Part F: Traffic Psychol. Behav. **44**, 53–62 (2017). ISSN 1369-8478. http://dx.doi.org/10.1016/j.trf.2016.10.003
14. Wietfeld, C., Ide, C.: 1 - Vehicle-to-infrastructure communications. In: Chen, W. (ed.) Woodhead Publishing Series in Electronic and Optical Materials. Woodhead Publishing, pp. 3–28 (2015). Vehicular Communications and Networks, ISBN 9781782422112. https://doi.org/10.1016/B978-1-78242-211-2.00001-5
15. Manvi, S.S., Tangade, S.: A survey on authentication schemes in VANETs for secured communication. Veh. Commun. **9**, 19–30 (2017). ISSN 2214-2096, http://dx.doi.org/10.1016/j.vehcom.2017.02.001
16. Karanki, S.S., Khan, M.S.: SMMV: secure multimedia delivery in vehicles using roadside infrastructure Veh. Commun. **7**, 40–50 (2017). ISSN 2214-2096. http://dx.doi.org/10.1016/j.vehcom.2016.12.002
17. Abbas, M.K.: Formal Specification Language for Vehicular Ad Hoc Network, Master's thesis, Universiti Teknologi PETRONAS, Department of Computer and Information Sciences (2009)

Hybrid Artificial Bee Colony Algorithm for *t*-Way Interaction Test Suite Generation

Ammar K. Alazzawi$^{(\boxtimes)}$, Helmi Md Rais, and Shuib Basri

Department of Computer and Information Sciences,
Universiti Teknologi PETRONAS, 32610 Bandar Seri Iskandar, Perak, Malaysia
{ammar_16000020,helmim,shuib_basri}@utp.edu.my

Abstract. The very large number of test cases and time consumption for a test, it is becoming hard to perform exhaustive testing for any software fault detection. For this reason, combinatorial testing (CT) also known as *t*-way testing, is one of the well-known methods that are used for fault detections to many software systems. Various existing research works are available in the literature to minimize the number of test cases, and the time to obtain an optimal test suite or competitive test suite. However, the interaction strength of the existing research works are supports up to $t = 2$ or $t = 3$, where t is the strength of parameter's interaction. The major purpose of this research is to suggest a new *t*-way strategy to minimize the test cases. This is called hybrid artificial bee colony (HABC) strategy, which is based on hybridize of an artificial bee colony (ABC) algorithm with a particle swarm optimization (PSO) algorithm. This is to provide a high-interaction strength combinatorial test suite up to $t = 6$. From experimental results, HABC strategy performed best when compared with existing methods in terms of generating the optimum test case.

Keywords: Meta-heuristics · Hybrid Artificial Bee Colony ·
Optimization algorithms · *t*-way testing · Software testing

1 Introduction

To meet the software specifications before producing it to the market, it is important to validate and verify that the software is free from any faults and defects by carrying out software testing [1]. For this reason, it is vital element in any software development system. Numerous testing techniques exist such as Black Box testing, White Box testing, Regression testing, *t*-way testing etc. Some of these techniques are represented as an optimization problem, where combinatorial testing is one of these problems [2]. In order to be able to detect the errors and compare them with specifications, there is a need for software testing to apply test cases on the developed software. However, a very large number of the generated test cases, it is very difficult to minimize. Therefore, combinatorial testing (CT) strategy is used to minimize the number of test cases and generate the best test suite that can cover all possible combinations within the required strength of interaction [3–6].

In addition, CT can produce a suite with fewer numbers of test cases similar to the original test set to meet all the specifications in the original test suite. The interactions

© Springer Nature Switzerland AG 2019
R. Silhavy (Ed.): CSOC 2019, AISC 984, pp. 192–199, 2019.
https://doi.org/10.1007/978-3-030-19807-7_19

between parameters of the system under test (SUT) are covered by using combinatorial testing. To apply CT, it is necessary to figure out the number of test inputs that would cover all the possibilities of *t*-way combination values of each parameter, and it compares the inputs set to the output set. Many test cases can be interloped with other test cases with the same requirements. These test cases can be considered as redundant cases that can be removed from the final test suite. Consequently, reducing the number of test cases lead to reduction of the cost of the software.

There are many existing strategies available in the literature (e.g. *t*-way strategies). These strategies are designed to minimize the number of test cases based on the AI-algorithms. The recent designed strategies include Test Generation Flower Pollination (TGFP) [7], Ant Colony Optimization Algorithm (ACA) [4, 8], Simulated Annealing Algorithm (SA) [9], Genetic Algorithm (GA) [10], Particle Swarm Optimization Algorithm (PSO) [11], Harmony Search Algorithm (HSS) [12], bat-inspired Strategy [13–19] and others [5, 20–24]. However, these strategies might not produce the optimum test suite for every single configuration system due to its complexity. As such, this research is a continuation to the previous work [25–27], where by a new optimization algorithm proposed and named as hybrid artificial bee colony (HABC) algorithm. HABC is based on the hybrid of an artificial bee colony (ABC) algorithm with a particle swarm optimization (PSO) algorithm for *t*-way test generation. HABC algorithm is mimicking the behavior of honeybee inside the hive; it merges the advantages of ABC algorithm with the advantages of PSO algorithm.

The rest of the paper is organized as follows: Sect. 2 describes the proposed test case generation algorithm. Section 3 evaluates the HABC through different benchmarking experiments in terms of efficiency and performance. Section 4 concludes the research and suggest possible future work improvements.

2 Test Suite Generation by Hybrid Artificial Bee Colony Algorithm

In 2005, Karaboge have suggested one of the most important optimization algorithms nowadays named as Artificial Bee Colony (ABC) algorithm. Designed based on the mimicking of foraging behavior by honey-bee within the hive [28]. The ABC execute certain tasks by the bees. These bees are divided into three different types, where each type has a specific duty inside the hive to increase the amount of nectar. These bees are Employed bees, Onlooker bees and Scout bees. The first type of bee is employed bee avails the advantage of the explored food sources having a high nectar amount and exchange the vital information of food source such as direction, distance and profitability with other types of bee inside the hive (e.g. the number of the food source represents the number of test case). The second type is Onlooker bees are working to select the food source with high nectar based on the information shared by employed bee. The third type is the Scout bees are randomly search the environment to detect a new or better food source. Employed bee represents half of the colony and the rest representing the Onlooker bee. In essence, the number of food source equals to the number of Employed bees.

However, ABC algorithm is similar to the other meta-heuristic algorithms that have advantages and disadvantages. Owing to the fact of meta-heuristic algorithms randomization, is causing it hard to be an optimization algorithm to obtain a global optimum for optimization problems. The weaknesses of ABC, there are insufficiencies of solution development process due to the simple operation process. In addition, the convergence speed of the algorithm is increased for some complex problems [29]. For this reason, the algorithm may be driven to stuck in the local optimum for complex problems due to the occurrence of fast convergence. The ABC information sharing process has demonstrated unworthy performance during the experiments execution, (note that the information-sharing activity is defined by using Eq. (1)) [30]. Several modifications made by researchers for ABC algorithm in order to overcome it is disadvantages such as change on the original ABC or hybridizing with other algorithms [29, 31].

$$V_{ij} = X_{i,j} + \text{rand} [-1, 1] (X_{ij} - X_{kj}) \tag{1}$$

In this research, the authors propose a new hybrid artificial bee colony (HABC) algorithm by leveraging PSO algorithm to overcome the disadvantage in ABC algorithm. The inspiration is on the particle movement operation of PSO algorithm. The PSO solution improvement mechanism and information sharing processes are totally different and unique unlike the information sharing activity in ABC algorithm. PSO has distinguish and unique parameter named as Weight Factor (w). The improvement solution is made by Velocity parameter by depend on the previous solution. Moreover, to determine the relative impact of cognitive and social components there is a learning factors parameters (C1 and C2), respectively using Eq. (2).

$$V_{i,d}^{t+1} = W^t * V_{i,d}^t + C_1^t * r_1 * (pbest_{i,d}^t - X_{i,d}^t) + C_2^t * r_2 * (gbest_{i,d}^t - X_{i,d}^t) \tag{2}$$

The motion process of PSO particles depends on the variable velocity, which is not randomly or arbitrary like the solution improvement in ABC algorithm. The best solution variable of local optimum, it is provide the local information by interacting with selected particles of value to the next move. The particle's next move also affected by the best solution variable of global optimum. These aforementioned advantages of PSO not available in ABC algorithm. This research combining the advantages of ABC and PSO together to overcome the optimization problem. The smart behavior of the proposed hybrid artificial bee colony (HABC) can be clarify as follows:

1. Initial step: the initial process of HABC algorithm is same the original in ABC and PSO algorithm, it begins by randomly searching the environment looking for food sources. Producing the initial food sources relies on the range of boundaries for the algorithm's parameters that defined by using Eq. (3).

$$x_{ij} = x_{min,j} + \text{rand}(0, 1)(x_{max,j} - x_{min,j}) \tag{3}$$

2. Employed bee step: each employed bee exploit one food source (where the number of bees is similar to the number of food source as equally). Therefore, the employed bee starts to avail the detected food source and collects the information about the nectar amount. Then, the employed bee come back to the hive to communicate the details of the food source with other bees waiting at the hive in the dance area (Where the information sharing by the dance). After exhausted the nectar of food source, the employed become a scout's bee and start again to search randomly for a better or new source. Local search for a new food source is defined by using the local search of PSO in Eq. (2) instead of Eq. (1).

After detecting the food source, the probability of selecting food source is defined by using (4).

$$
fitness_i = \begin{cases} \frac{1}{1+f_i}, & if\ f_i \geq 0 \\ 1 + |f_i|, & if\ f_i < 0 \end{cases} \tag{4}
$$

3. Onlooker bee step: the selection of the food source by the onlooker bee criteria relies on the nectar amounts, where the nectar amounts evaluated based on the information inside the dance area by the employed bee. The probability selection of the food source is defined by using Eq. (5).

$$
Pi = \frac{fit_i}{\sum_{n=1}^{sn} fit_n} \tag{5}
$$

4. Scout bee and Limit step: the global search mechanism of scout bee, able to provides the capability to reduce the convergence problem of early premature. This feature of mechanism is not available in PSO. In addition, during the environmental search the algorithm may fall down in the local minima. Thence, the "limit parameter" of the ABC utilized in the suggested HABC. The "limit" prevents the HABC from falling down in the local minima search, where from time to time the "limit parameter" insert selected solution randomly to search space. After both of employed and onlooker bee's tasks complete, searches the environment by the algorithm again to be deserted of the exhausted source. The abandoned food source decision relies on counters called limit, which is defined by Eq. (6).

$$
limit = c.ne.D \tag{6}
$$

During the searching process, the algorithm will update the counter value, if the value higher than limit value then the related food source with counter value will be abandoned. The new source that discovered by scout bee will be instead of the abandoned food source. The main steps of HABC algorithm are shown as follows in Fig. 1.

1: *Initialization step: The same process as the original ABC and PSO algorithms.*
2: *REPEAT*
3: *Move the employed bees onto their food sources and determine their nectar amounts.*
4: *Calculate the probability value of the sources with which they are preferred by the onlooker's bee.*
5: *Move the Onlookers onto the food sources and determine their nectar amounts.*
6: *Move the scouts to search for new food sources replacing the abandoned ones.*
7: *Memorize the best food source found so far.*
8: *UNTIL (requirements are met).*

Fig. 1. HABC algorithm pseudocode.

3   Result

This section explains the benchmark the performance of the proposed HABC strategy the existing *t*-way strategies by referencing to the found experiments in [7, 12, 32] such as HSS, TGFP, PSTG, Jenny, WHICTH, IPOG and TConfig. This research have adopted two experiments to evaluate HABC strategy. The first experiment adopted CA $(N, t, 3^7)$, where t is variable interaction strength from 2 to 6. The second experiment have adopted CA $(N, t, 2^{10})$, where t is variable interaction strength from 2 to 6.

The HABC strategy parameters are set at Nbees = 5, maxCycle = 1000, limit = 100, C1 & C2 = 2.0 and W = 0.9. The experiments were implemented twenty independent runs to obtain the best result due to the randomization characteristic. The experiments were conducted on a Windows 7 (OS) desktop computer with 3.40 GHz Xeon (R) CPU E3 and 8 GB RAM. The Java language JDK 1.8. it was used to code and implement the HABC. Tables 1 and 2, presents the experimental result, and each table presented the optimal test suite size for each configuration. The dark cell with (*) represents the optimal test suite size, the dark cell without (*) represents the best test suite size that shared with other strategy and the cell with NA represents (not available).

As shown in Table 1, HABC strategy have produced the most optimal test suite size for CA $(N, 4, 3^7)$ is 149 test cases and for CA $(N, 6, 3^7)$ are 810 test cases comparing to all other strategies. For CA $(N, 2, 3^7)$, HABC matches with HSS and PSTG by producing 14 test case. Whereas WHICTH and TGFP produce the most optimal test cases for both CA $(N, 3, 3^7)$ and CA $(N, 5, 3^7)$ with 45 test cases by WHICTH, and 435 test cases by TGFP respectively. However, HABC strategy produced competitive and very close results to the optimal result as shown in CA $(N, 3, 3^7)$ and CA $(N, 5, 3^7)$.

Table 1. CA (N, *t*, 3^7) with variable interaction strength from 2 to 6.

t	HSS Best	IPOG Best	WHITCH Best	Jenny Best	TConfig Best	PSTG Best	TGFP Best	HABC Best	HABC Average
2	**14**	17	15	16	15	**14**	15	**14**	15.300
3	50	57	45*	51	55	50	52	46	51.200
4	157	185	216	169	166	160	157	149*	158.25
5	437	561	NA	458	477	444	435*	438	445.15
6	916	1281	NA	1087	921	955	965	810*	932.35

Table 2. CA (N, *t*, 2^{10}) with variable interaction strength from 2 to 6.

t	HSS Best	IPOG Best	WHITCH Best	Jenny Best	TConfig Best	PSTG Best	TGFP Best	HABC Best	HABC Average
2	8	10	6*	10	9	8	8	8	8.3
3	**16**	19	18	18	20	17	17	**16**	16.7
4	**37**	49	58	39	45	**37**	39	38	41.55
5	**81**	128	NA	87	95	82	82	**81**	84.55
6	158	352	NA	169	183	158	160	**156***	165.1

In Table 2, HABC strategy produced 8 test cases similarly to TGFP, HSS and PSTG, while WHITCH produced 6 test cases for CA (N, 2, 2^{10}). HABC has produced the same results shared with HSS for both CA (N, 3, 2^{10}) 16 test cases and the CA (N, 5, 2^{10}) 81 test cases. However, HABC strategy produced competitive and similar results to the optimal result as shown in CA (N, 4, 2^{10}). Whereas HABC strategy obtains the most optimal test suite size comparing to other strategies for CA (N, 6, 2^{10}) by producing 156 test cases.

4 Conclusions

This research has proposed HABC strategy for *t*-way approach for test suite size generation by relying on the hybrid artificial bee colony algorithm. The experiments results have shown HABC strategy produced a good performance in able to generate better optimal test suite size, as compared to other *t*-way strategies. HABC has demonstrated a very similar result to the optimal test suite. As part of the future work, we are looking to extend the HABC strategy to support high interaction parameters to use in future for software product line testing (SPL) as well as to support the variable strength interaction.

References

1. Cohen, M.B.: Designing Test Suites for Software Interactions Testing. University of Auckland (New Zealand) (2004)
2. Nie, C., Leung, H.: A survey of combinatorial testing. ACM Comput. Surv. (CSUR) **43**, 11 (2011)
3. Cohen, D.M., Dalal, S.R., Parelius, J., Patton, G.C.: The combinatorial design approach to automatic test generation. IEEE Softw. **13**, 83–88 (1996)
4. Zamli, K.Z., Alsewari, A.R., Al-Kazemi, B.: Comparative benchmarking of constraints t-way test generation strategy based on late acceptance hill climbing algorithm. Int. J. Softw. Eng. Comput. Sci. (IJSECS) **1**, 14–26 (2015)
5. Zamli, K.Z., Din, F., Kendall, G., Ahmed, B.S.: An experimental study of hyper-heuristic selection and acceptance mechanism for combinatorial t-way test suite generation. Inf. Sci. **399**, 121–153 (2017)
6. Rabbi, K., Mamun, Q., Islam, M.R.: A novel swarm intelligence based strategy to generate optimum test data in t-way testing. In: International Conference on Applications and Techniques in Cyber Security and Intelligence, pp. 247–255. Springer (2017)
7. Alsewari, A.A., Har, H.C., Homaid, A.A.B., Nasser, A.B., Zamli, K.Z., Tairan, N.M.: Test cases minimization strategy based on flower pollination algorithm. In: International Conference of Reliable Information and Communication Technology, pp. 505–512. Springer (2017)
8. Chen, X., Gu, Q., Li, A., Chen, D.: Variable strength interaction testing with an ant colony system approach. In: Asia-Pacific Software Engineering Conference, APSEC 2009, pp. 160–167. IEEE (2009)
9. Cohen, M.B., Colbourn, C.J., Ling, A.C.: Constructing strength three covering arrays with augmented annealing. Discrete Math. **308**, 2709–2722 (2008)
10. Esfandyari, S., Rafe, V.: A tuned version of genetic algorithm for efficient test suite generation in interactive t-way testing strategy. Inf. Softw. Technol. **94**, 165–185 (2018)
11. Ahmed, B.S., Zamli, K.Z., Lim, C.P.: Application of particle swarm optimization to uniform and variable strength covering array construction. Appl. Soft Comput. **12**, 1330–1347 (2012)
12. Alsewari, A.R.A., Zamli, K.Z.: Design and implementation of a harmony-search-based variable-strength t-way testing strategy with constraints support. Inf. Softw. Technol. **54**, 553–568 (2012)
13. Alsariera, Y.A., Zamli, K.Z.: A real-world test suite generation using the bat-inspired t-way strategy. In: The 10th Asia Software Testing Conference (SOFTEC2017), vol. 10 (2017)
14. Alsariera, Y.A., Alamri, H.S., Zamli, K.Z.: A bat-inspired testing strategy for generating constraints pairwise test suite. In: The 5th International Conference on Software Engineering & Computer Systems (ICSECS), vol. 5 (2017)
15. Alsariera, Y.A., Nasser, A., Zamli, K.Z.: Benchmarking of bat-inspired interaction testing strategy. Int. J. Comput. Sci. Inf. Eng. (IJCSIE) **7**, 71–79 (2016)
16. Alsariera, Y.A., Zamli, K.Z.: A bat-inspired strategy for t-way interaction testing. Adv. Sci. Lett. **21**, 2281–2284 (2015)
17. Alsariera, Y.A., Majid, M.A., Zamli, K.Z.: Adopting the bat-inspired algorithm for interaction testing. In: The 8th Edition of Annual Conference for Software Testing, p. 14 (2015)
18. Alsariera, Y.A., Majid, M.A., Zamli, K.Z.: SPLBA: an interaction strategy for testing software product lines using the bat-inspired algorithm. In: 2015 4th International Conference on Software Engineering and Computer Systems (ICSECS), pp. 148–153. IEEE (2015)

19. Alsariera, Y.A., Majid, M.A., Zamli, K.Z.: A bat-inspired strategy for pairwise testing. ARPN J. Eng. Appl. Sci. **10**, 8500–8506 (2015)
20. Homaid, A.B., Alsweari, A., Zamli, K., Alsariera, Y.: Adapting the elitism on the greedy algorithm for variable strength combinatorial test cases generation. IET Softw. (2018)
21. Cai, L., Zhang, Y., Ji, W.: Variable strength combinatorial test data generation using enhanced bird swarm algorithm. In: 2018 19th IEEE/ACIS International Conference on Software Engineering, Artificial Intelligence, Networking and Parallel/Distributed Computing (SNPD), pp. 391–398. IEEE (2018)
22. Zamli, K.Z., Din, F., Baharom, S., Ahmed, B.S.: Fuzzy adaptive teaching learning-based optimization strategy for the problem of generating mixed strength t-way test suites. Eng. Appl. Artif. Intell. **59**, 35–50 (2017)
23. Zakaria, H.L., Zamli, K.Z.: Elitism based migrating birds optimization algorithm for combinatorial interaction testing. Int. J. Softw. Eng. Technol. **3** (2017)
24. Sheng, Y., Wei, C., Jiang, S.: Constraint test cases generation based on particle swarm optimization. Int. J. Reliab. Qual. Saf. Eng. **24**, 1750021 (2017)
25. Alazzawi, A.K., Rais, H.M., Basri, S.: Artificial bee colony algorithm for t-way test suite generation. In: 2018 4th International Conference on Computer and Information Sciences (ICCOINS), pp. 1–6. IEEE (2018)
26. Alsewari, A.A., Alazzawi, A.K., Rassem, T.H., Kabir, M.N., Homaid, A.A.B., Alsariera, Y.A., Tairan, N.M., Zamli, K.Z.: ABC algorithm for combinatorial testing problem. J. Telecommun. Electron. Comput. Eng. (JTEC) **9**, 85–88 (2017)
27. Alazzawi, A.K., Homaid, A.A.B., Alomoush, A.A., Alsewari, A.A.: Artificial bee colony algorithm for pairwise test generation. J. Telecommun. Electron. Comput. Eng. (JTEC) **9**, 103–108 (2017)
28. Karaboga, D.: An idea based on honey bee swarm for numerical optimization. Technical report-tr06, Erciyes University, Engineering Faculty, Computer Engineering Department (2005)
29. Karaboga, D., Akay, B.: A survey: algorithms simulating bee swarm intelligence. Artif. Intell. Rev. **31**, 61–85 (2009)
30. Kıran, M.S., Gündüz, M.: A novel artificial bee colony-based algorithm for solving the numerical optimization problems. Int. J. Innov. Comput. Inf. Control **8**, 6107–6121 (2012)
31. Yan, X., Zhu, Y., Zou, W.: A hybrid artificial bee colony algorithm for numerical function optimization. In: Hybrid Intelligent Systems (HIS), pp. 127–132. IEEE (2011)
32. Ahmed, B.S., Zamli, K.Z.: PSTG: a t-way strategy adopting particle swarm optimization. In: The Fourth Asia International on Mathematical/Analytical Modelling and Computer Simulation (AMS), pp. 1–5. IEEE (2010)

Identification, Assessment and Automated Classification of Requirements Engineering Techniques

Aleksander Jarzębowicz[(✉)] [iD] and Kacper Sztramski

Department of Software Engineering, Faculty of Electronics,
Telecommunications and Informatics, Gdańsk University of Technology,
Narutowicza 11/12, 80-233 Gdańsk, Poland
olek@eti.pg.edu.pl

Abstract. Selection of suitable techniques to be used in requirements engineering or business analysis activities is not easy, especially considering the large number of new proposals that emerged in recent years. This paper provides a summary of techniques recommended by major sources recognized by the industry. A universal attribute structure for the description of techniques is proposed and used to describe 33 techniques most frequently quoted by reviewed sources. A pilot study of automated classification of techniques based on attribute values is also reported. The study used fuzzy c-means clustering algorithm and produced pairings of complementary techniques, most of which successfully passed validation conducted by business analysis practitioners.

Keywords: Requirements Engineering · Business Analysis · Techniques · Clustering · Industrial standards

1 Introduction

Requirements Engineering (RE) aims at establishing stakeholder's viewpoints and determining requirements that reflect the purpose of software system to be developed. It is considered a crucial factor of software development project, as it strongly influences software quality and final project outcome [1, 2]. RE is recognized as an important topic in research and industrial practice since many years [3, 4]. Several books (e.g. [5, 6]) and international standards (e.g. [7, 8]) describing recommended processes and practices are available.

In recent years, however, two new observations can be made. One is the emergence of Business Analysis (BA), which, while being closely related to RE, has a broader scope and puts more emphasis on facilitating business change [9]. Another observation is an increased interest of industrial community, which resulted in founding professional associations e.g. International Institute of Business Analysis (IIBA) or International Requirements Engineering Board (IREB) and their subsequent activity in the fields of Requirements Engineering and Business Analysis (RE/BA). Such activities include education, certification and publishing standards and other guidelines. As result, several new sources of knowledge became available, including industrial

© Springer Nature Switzerland AG 2019
R. Silhavy (Ed.): CSOC 2019, AISC 984, pp. 200–212, 2019.
https://doi.org/10.1007/978-3-030-19807-7_20

standards (BABOK [9], PMI Guide [10]) and training materials associated with certification schemes (IREB [11], REQB [12]).

Such sources usually cover various RE/BA aspects e.g. definitions of processes and activities, good practices, competencies expected from an analyst and RE/BA techniques. There are differences between particular sources with respect to most of mentioned aspects. In this paper, however, we will focus solely on RE/BA techniques. Adopting the corresponding definition from BABOK glossary [9], we define RE/BA technique as a manner or method for conducting a particular RE/BA task or for shaping its output. For example, interview, observation and document analysis are techniques used for requirements elicitation, while use cases, business process modeling or user stories belong to requirements specification (documentation) techniques.

As mentioned, particular sources propose different sets of techniques. Consequently, a large number of techniques is included in the collective body of knowledge available. An analyst looking for tools to do his/her work has a potentially wide choice of techniques, but this can also become a problem. It may be difficult to pick a technique (or a combination of techniques to be used together) that is appropriate for a given task in a given context of software project from so many candidates. Moreover, differences in techniques' names and levels of abstraction can increase confusion. As result, it is likely, that such selection would be based on personal preferences and (limited) knowledge about available techniques. Unfortunately, there are no clear and comprehensive summaries of techniques recommended by state of the art sources nor methods guiding analysts in selection of those techniques.

In this paper we aim to provide a solution. We made a thorough review of four sources published by industrial professional associations and one additional source being an international standard. We extracted RE/BA techniques from them and created a unified summary, matching corresponding techniques appearing under different names and/or abstraction levels. We selected 33 techniques recommended by at least 2 sources for further consideration. Next, we proposed a set of attributes describing techniques and their applicability. For each of 33 techniques we assessed them by assigning values to attributes.

Apart from providing summary of techniques together with their structured descriptions, we made an initial attempt to use it as a dataset for automated analysis (clustering algorithm) to group similar techniques and to identify complementary techniques i.e. those recommended to be used jointly. Recommendations given by the algorithm were validated by two experienced business analysts and results can be considered promising.

Hence, the main contributions provided by this paper are:

- A summary of RE/BA techniques recommended by present industrial standards;
- A set of attributes for technique's description (which can be used both by humans and as input of automated analysis) and attribute values for 33 techniques;
- A pilot study applying automated analysis aiming at identifying complementary RE/BA techniques.

The remainder of this paper is structured as follows. In Sect. 2 we describe related work. In Sect. 3 we present the processes of: techniques identification (through reviewing the sources) and techniques assessment (using a pre-defined attribute structure), as well as outcomes of both those processes. Section 4 outlines the study of automated classification of techniques and the validation conducted afterwards. We conclude the paper in Sect. 5, by summarizing contributions, discussing limitations of our work and sharing ideas on future research directions.

2   Related Work

There are several studies on defining attributes or criteria of particular RE/BA techniques, in order to compare them and/or provide guidelines about the context a given technique should be used in.

Hickey and Davis [13] provided a formalized process model for selection of requirements elicitation techniques. Their proposal lacks particular selection criteria but the process described by them sets a foundation for further works. Escalona and Koch [14] assessed several techniques (for elicitation, specification and validation of requirements) with respect to their ability to be used in particular methods of web application development and to be applied to various categories of requirements.

Jiang et al. [15] developed a framework for selection of requirements engineering techniques. They compiled an extensive list of 46 techniques and described each one using a set of attributes reflecting technique's abilities. They also used fuzzy clustering algorithms to group similar techniques. On the basis of work by Jiang et al., other research studies were conducted [16, 17]. Kheirkhah and Deraman [16] slightly modified the set of attributes and extended it by organizational viewpoint. Tiwari and Rathore [17] developed a framework based on characteristics of 3 aspects (project, people and process) to select requirements elicitation techniques most suitable in a given context defined by those 3 aspects.

des Santos Soares et al. [18] defined an attribute structure for requirements documentation techniques expressing mainly technique's abilities, but also e.g. maturity or popularity and assessed 8 techniques with respect to such criteria. Besrour et al. [19] conducted an experiment in academic setting to assess 3 popular requirements elicitation techniques in several dimensions e.g. usability or communicating ability. Darwish et al. [20] used a set of 42 attributes grouped 8 categories to characterize 14 techniques and applied an artificial neural network as a tool for techniques selection.

For other related research studies (limited to requirements elicitation techniques) a reader can also be referred to a systematic mapping study by Carrizo et al. [21]. The research gap that can be identified is based on the following observations:

- No research study is based on the comprehensive review of state of the art techniques recommended by current industrial standards. The only exception is the previous work of one of us [22], which however was based on smaller number of standards and in some cases their older versions were used.
- Most studies are limited to a relatively small subset of techniques (e.g. dedicated to requirements elicitation or documentation only).
- Only two studies ([15, 20]) use automated methods of selecting most effective techniques for a given context (despite the fact that a large number of attributes and their values is difficult to comprehend for humans).

3   Selection and Description of RE/BA Techniques

The following sources were selected and reviewed to extract RE/BA techniques from their contents:

1. A Guide to the Business Analysis Body of Knowledge (BABOK Guide v3, 2015) [1] – It is a widely recognized industrial standard published by International Institute of Business Analysis. Its purpose is to define the profession of business analysis and provide a set of commonly accepted practices. It aims to help practitioners discuss and define the skills necessary to effectively perform BA.
2. PMI Business Analysis for Practitioners: A Practice Guide (2015) [10] – Another recognized standard issued by Project Management Institute (PMI), known from other standards and methodologies e.g. PMBOK. The intent of this publication is to provide o comprehensive guidance on how to apply BA practices to projects.
3. IREB Certified Professional for Requirements Engineering Syllabi – International Requirements Engineering Board (IREB) is a non-profit organization focusing on certification of RE practitioners. A 3-level certification scheme is available. IREB published several syllabi summarizing the scope of knowledge required on certification exams. In our study we used the following syllabi from Foundation and Advanced levels: [3, 23, 24].
4. REQB Certified Professional for Requirements Engineering Syllabi - Requirements Engineering Qualification Board (REQB) was another organization, which developed a certification scheme for requirement engineers. In January 2017 it was merged with IREB. However, as REQB certificates were recognizable for many years and associated syllabi were available, we decided to use the following documents: [4, 25].
5. IEEE Guide to the Software Engineering Body of Knowledge (SWEBOK Guide v. 3.0, 2014) [8] – SWEBOK is a guide developed by IEEE and later adopted as an international standard (ISO/IEC TR 19759:2015). It defines Software Engineering discipline and 15 knowledge areas summarizing the expected knowledge of a qualified software engineer. One of these areas is Software Requirements and its description by SWEBOK provided input to our study.

It should be stressed that all of these sources were most up to date versions at the time of conducting our review and search for RE/BA techniques (and no updated versions have been published till present day). The only exception was REQB – as already mentioned, despite the fact that in January 2017 REQB and IREB merged under unified IREB brand, we decided to include it, because REQB certificates were recognizable for many years. It is also worth mentioning that at the time of writing this paper still no new source incorporating IREB's and REQB's approaches and ideas was available.

We reviewed all sources to identify RE/BA techniques described or even mentioned by them. In some cases (BABOK) such task is rather easy, as techniques are indexed and presented in a separate section. In other cases the task required reading all con-tents of sources. The next step was to unify different sets originating from various sources. This task included the following actions:

- Resolving simple name differences e.g. Backlog Management (BABOK) vs. Maintaining Product Backlog (PMI);
- Unifying different but closely related techniques (including cases when one technique extended other with additional tasks/tools) e.g. Persona (IREB) and Stakeholder List, Map, or Personas (BABOK);
- Providing a common level of abstraction as some sources distinguish more specific variants of a given technique e.g. there is a number of various collaborative games (IREB) that can be treated as more specific forms of Brainstorming.

As result, we obtained a list of 82 techniques, which we considered too extensive to be processed in next steps of this study. We decided that only techniques mentioned by at least two of the reviewed sources would be considered further and consequently we were left with 33 techniques. The final set of RE/BA techniques is given in Table 1. The table also shows the sources in which a given technique is found. For detailed description of techniques, readers are referred directly to the sources, due to space limitations of this paper.

Table 1. A list of sources used to extract RE/BA techniques from their contents.

#	Technique	Sources
1	Activity Diagram	IREB; REQB
2	Approval Levels	PMI; REQB
3	Benchmarking	BABOK; PMI
4	Brainstorming	BABOK; PMI; IREB; REQB
5	Business Rules Catalog and Analysis	BABOK; PMI
6	Class Diagram	BABOK; IREB; REQB
7	Communication Diagram	IREB; REQB
8	Cost-Benefit Analysis	PMI; IREB
9	Data Dictionary	BABOK; PMI
10	Data Flow Diagram	BABOK; PMI; IREB; REQB
11	Dictionary (Glossary)	BABOK; IREB
12	Document Analysis	BABOK; PMI; IREB; REQB
13	Entity Relationship Diagram	BABOK; PMI; REQB
14	Facilitated Workshop	BABOK; PMI; IREB; REQB; SWEBOK
15	Focus Groups	BABOK; PMI
16	Interviews	BABOK; PMI; IREB; REQB; SWEBOK
17	Lessons Learned	BABOK; PMI
18	Maintaining Product Backlog	BABOK; PMI
19	Observation	BABOK; PMI; IREB; REQB; SWEBOK
20	Organizational Charts	BABOK; PMI
21	Peer Review	BABOK; PMI; IREB; REQB; SWEBOK

(continued)

Table 1. (*continued*)

#	Technique	Sources
22	Prioritization	BABOK; PMI; REQB; SWEBOK
23	Process Modeling	BABOK; PMI; IREB; REQB
24	Prototyping	BABOK; PMI; IREB; REQB; SWEBOK
25	Questionnaires	BABOK; PMI; IREB; REQB
26	Scope Modeling	BABOK; PMI
27	Sequence Diagram	BABOK; IREB; REQB
28	Stakeholders List, Map or Personas	BABOK; PMI; IREB; REQB
29	State Table/Diagram	BABOK; PMI; IREB; REQB
30	SWOT Analysis	BABOK; PMI
31	Traceability Matrix	PMI; REQB
32	Use Cases	BABOK; PMI; IREB; REQB; SWEBOK
33	User Stories	BABOK; PMI; IREB; REQB; SWEBOK

Next, we proposed a structure of attributes to describe each technique. The attributes and their value scales are shown in Table 2. The values are given in two forms: descriptive and numerical (the latter is for the purpose of automated analysis). Attributed are also divided into 3 groups, but it is for clarity sake only, it does not influence automated classification described in later sections.

Table 2. Attributes used to describe RE/BA techniques and their possible values.

Group	ID	Attribute	Values
Resources	A1	Required skill level	1 – low; 2 – medium
	A2	Required effort	3 – high
	A3	Required involvement of stakeholders	0 – none; 1 – low; 2 – medium; 3 – high
Abilities	A4	Ability to identify functional requirements	0 – lack of ability 1 – to a small extent
	A5	Ability to identify non-functional requirements	2 – to a moderate extent 3 – to a large extent
	A6	Ability to identify stakeholders, their roles and relationships between them	
	A7	Ability to support verification and validation	
	A8	Ability to support communication with stakeholders	
	A9	Ability to support requirements management, traceability and monitoring	

(*continued*)

Table 2. (*continued*)

Group	ID	Attribute	Values
Inherent characteristics	A10	Availability of graphical representation	0 – no representation 1 – limited representation 2 – complex graphical representation
	A11	Availability of precise guidelines/procedure of use	1 – requires analyst's interpretation 2 – generic/partial procedure defined 3 – detailed procedure defined
	A12	Degree of creativity enabled	1 – low; 2 – medium 3 – high

The set of attributes was kept relatively small for two reasons. The first is the intent to have a common set of attributes to describe all RE/BA techniques, regardless of the tasks they are used for (elicitation, analysis, validation etc.). The second reason was a practical one – to reduce the effort necessary to assess all 33 techniques. We treated it as a first trial and intended to refine this set of attributes, in case results of next steps (automatic classification and validation) suggest it.

Using the set of attributes given in Table 2, 33 techniques from Table 1 were assessed by assigning attributes with values. Two persons completed this task independently and later met to discuss their assessments and the rationale behind them. In most cases the discussion allowed them to reach consensus. In the remaining cases some disagreement remained, however the difference between assigned attribute's values was never greater than 1. If differences between assessors could not be resolved, then a mean arithmetic value was assigned to the attribute e.g. the disagreement whether Maintaining Product Backlog requires low (1) or medium (2) skill level, led to assigning 1.5 value. Examples of attribute values for 3 techniques used in different RE/BA areas and for different purposes are given in Table 3.

Table 3. Attribute values for example techniques.

Technique	A1	A2	A3	A4	A5	A6	A7	A8	A9	A10	A11	A12
Interviews	2	2.5	2.5	3	2.5	3	2.5	2.5	0.5	0	2	2
User stories	1	1	1.5	2.5	1.5	1.5	1	3	0.5	0.5	3	2
Maintaining product backlog	1.5	1.5	0.5	2	0.5	0	1.5	2	3	0.5	2.5	1

4 Automated Classification

We decided to apply data clustering approach for automated classification of RE/BA techniques. Clustering is a generic concept of unsupervised classification intended to identify natural groupings of data from a larger data set. Several methods based on this concept were proposed e.g. hierarchical, c-means or fuzzy clustering [26]. We acknowledge that our approach is similar to the one used in [15] and was inspired by that work.

We used Fuzzy C-Means Clustering algorithm implemented in Fuzzy Logic Toolbox library [27] for Matlab 2012. This algorithm was chosen because it provided additional information as its output – not only RE/BA technique's final classification into a given cluster, but also the degree the technique belongs to each cluster specified by a membership grade. As it was our first attempt of classifying RE/BA techniques, we intended to experiment with e.g. various numbers of clusters and such additional output information proved valuable. The *fcm* function available in Matlab's library requests as input: the data set, number of clusters and optionally parameters like maximum number of iterations or improvement of objective function. Its output includes: cluster centers, fuzzy partition matrix (indicating the degree of membership of each data point to each cluster) and objective function values. The objective function is defined as in Eq. (1) [27].

$$J_m = \sum_{i=1}^{D} \sum_{j=1}^{N} f_{ij}^m \left\| x_i - c_j \right\|^2 \tag{1}$$

The symbols used in Eq. (1):

- D is the number of data points.
- N is the number of clusters.
- m is fuzzy partition matrix exponent for controlling the degree of fuzzy overlap, with $m > 1$. Fuzzy overlap refers to how fuzzy the boundaries between clusters are, that is the number of data points that have significant membership in more than one cluster.
- x_i is the i-th data point.
- c_j is the center of the j-th cluster.
- f_{ij} is the degree of membership of x_i in the j-th cluster. For a given data point, x_i, the sum of the membership values for all clusters is one.

We run *fcm* function several times, using different input values for the number of clusters (3–10) and finally we decided to use the classification based on 8 clusters. It was a subjective decision based on our experience in RE/BA as we rejected classifications that grouped very different RE/BA techniques in one cluster. The resulting classification for 8 clusters is shown in Table 4 and indicates sets of "similar" RE/BA techniques as determined by the algorithm.

After assigning techniques to clusters, we proceeded to find out which pairs of techniques could be considered as complementary i.e. recommended to be used

Table 4. Classification of RE/BA techniques into clusters.

Cluster	Techniques
A	Benchmarking; Document Analysis; Lessons Learned; Questionnaires
B	Cost-Benefit Analysis; Data Dictionary; Maintaining Product Backlog; Prioritization
C	Focus Groups; Prototyping
D	Business Rules Catalog and Analysis; Process Modeling; Use Cases; User Stories
E	Approval Levels; Peer Review; Traceability Matrix
F	Brainstorming; Facilitated Workshop; Interviews; Observation
G	Activity Diagram; Class Diagram; Communication Diagram; Data Flow Diagram; Entity Relationship Diagram; Sequence Diagram; State Table/Diagram
H	Dictionary (Glossary); Organizational Charts; Scope Modeling; Stakeholders List, Map or Personas; SWOT Analysis

together, because they complement each other, as advantages of one technique counterbalance limitations of the other. We define complementary techniques M and N as:

1. Belonging to different clusters;
2. Satisfying the condition:

$$5 \leq \sum_{k=1}^{12} t_M(k) - t_N(k) \leq 5,5,$$ where $t_I(k)$ is value of k-th attribute of technique I

(where I belongs to {M, N});

The values 5 and 5,5 were determined considering the number of attributes and their value scales. It reflects the concept of complementary techniques which is a mix of similarities and differences (completely different techniques, having nothing in common would be difficult to use jointly). It is worth to mention that this complementarity relation between techniques is symmetric but not transitive. Based on this definition we identified 35 pairs of complementary techniques. They are listed in Table 5.

This result was a subject of validation through interviews with industry professionals working as business analysts. Two professionals participated in validation, none of them had been involved in earlier steps of research described here, thus they presented an independent viewpoint. One of them had 11 years of experience in RE/BA and was employed as Senior Business Analyst. The second person reported 5 years of experience in RE/BA and at that time held position of Business Analyst.

Each of two analysts was interviewed separately. Their main task was to review pairs of techniques (documented in Table 5) and for each one give a definite answer – to confirm or deny that such two techniques are really complementary and worth using in a joint manner. Additionally they provided some remarks about possible or preferred ways techniques could be used together, but we skip it here, focusing on main validation results, which are shown in "Validation" column of Table 5. The following possible outcomes are reported in this column: both analysts confirmed the pair of techniques is complementary (Y), they both rejected such proposal (N), or their

Table 5. Complementary RE/BA technique pairs and associated validation results.

Technique 1	Technique 2	Validation
Activity Diagram	Observation	Y
Activity Diagram	Use Cases	Y
Approval Levels	Process Modeling	Y
Benchmarking	Focus Groups	Y
Benchmarking	Organizational Charts	N
Brainstorming	Peer Review	N
Business Rules Catalog and Analysis	Peer Review	Y/N
Cost-Benefit Analysis	Document Analysis	Y/N
Cost-Benefit Analysis	Scope Modeling	Y
Cost-Benefit Analysis	Stakeholders List, Map or Personas	Y
Cost-Benefit Analysis	State Table/Diagram	Y/N
Data Dictionary	Peer Review	Y
Data Flow Diagram	Prioritization	N
Dictionary (Glossary)	Prioritization	N
Document Analysis	Observation	Y
Document Analysis	Process Modeling	Y
Document Analysis	Use Cases	Y
Facilitated Workshop	Process Modeling	Y
Focus Groups	User Stories	Y
Interviews	Prioritization	Y/N
Lessons Learned	Peer Review	Y
Maintaining Product Backlog	Process Modeling	Y/N
Observation	Scope Modeling	Y/N
Observation	Stakeholders List, Map or Personas	Y
Observation	State Table/Diagram	Y
Organizational Charts	User Stories	N
Peer Review	Prototyping	Y/N
Prioritization	SWOT Analysis	Y/N
Prioritization	Traceability Matrix	Y/N
Process Modeling	Questionnaires	Y/N
Process Modeling	Scope Modeling	Y
Process Modeling	Stakeholders List, Map or Personas	Y
Scope Modeling	Use Cases	Y
Stakeholders List, Map or Personas	Use Cases	Y
State Table/Diagram	Use Cases	Y

opinions differed (Y/N). In total, 20 pairs (out of 35) were confirmed by both analysts, 10 pair by one analysts only and 5 proposals were unanimously rejected.

5 Conclusions

This paper identified a set of RE/BA techniques recommended by present industrial standards and certification schemes. It also proposed a description structure to represent technique's abilities, required resources and other characteristics. Describing a technique by determining values of attributes can both support humans in decision-making (selecting a technique to be used in a given project) and provide input for automated classification. We conducted an initial exploration of the second possibility by applying fuzzy c-means clustering algorithm to group similar techniques and then we identified pairs of complementary techniques. That this initial attempt was promising – the majority of generated proposals were confirmed by at least one of experiences business analysts participating in validation.

Our research study had obviously several limitations that could influence its validity. The set of RE/BA techniques we collected does not necessarily have to reflect industrial practice, despite our effort to use sources recognized by industry practitioners. We also cannot exclude the possibility of omitting some techniques – our review was rather thorough, but there are cases that a technique is only briefly mentioned by a source, so overlooking, however unlikely, is still possible. Our attribute structure can be challenged, especially that we deliberately kept the number of attributes limited. Also, assessment of attribute values for particular techniques is based on experience and judgement of two people only. Finally, selection of parameters used in automated classification like number of clusters or the numbers used in the inequality in definition of complementarity was to some extent arbitrary. As for validation, we consider it as an initial step - for a more convincing confirmation a larger group of analysts should be involved.

As mentioned, it was the first trial of using automated classification. There are several opportunities of follow up work. It is possible to improve descriptions of techniques by expanding the attribute set and by "tuning" values of attributes e.g. by involving a larger group of assessors. More attributes reflecting the context a particular technique is best applied (software project constraints, developed product type, stakeholders' attitude etc.) can enable new possibilities. Moreover, knowledge and experience of analysts who are supposed to use RE/BA techniques can be considered more thoroughly instead of using a simplified "Required skill level" attribute. Finally, various relations between techniques (complementary, equivalent in general or with respect to a given characteristics etc.) can be defined and explored using automated classification.

References

1. Broy, M.: Requirements engineering as a key to holistic software quality. In: Proceedings 21st International Symposium on Computer and Information Sciences (ISCIS 2006), LNCS, vol. 4263, Springer, pp. 24–34 (2006)
2. Ellis, K., Berry, D.: Quantifying the impact of requirements definition and management process maturity on project outcome in large business application development. Requirements Eng. 18(3), 223–249 (2013)

3. Cheng, B., Atlee, J.: Research directions in requirements engineering. In: International Conference on Software Engineering (ICSE 2007), IEEE Computer Society, pp. 285–303. IEEE Computer Society, Washington DC (2007)
4. Bano, M., Zowghi, D., Ikram, N.: Systematic reviews in requirements engineering: a tertiary study. In: IEEE 4th International Workshop on Empirical Requirements Engineering (EmpiRE), pp. 9–16, IEEE (2014)
5. Pohl, K.: Requirements Engineering: Fundamentals, Principles, and Techniques. Springer Publishing Company, Heidelberg (2010)
6. Wiegers, K., Beatty, J.: Software Requirements, 3rd edn. Microsoft Press (2013)
7. ISO/IEC/IEEE Standard 29148-2011. Systems and Software Engineering - Life Cycle Processes - Requirements Engineering (2011)
8. IEEE: A Guide to Software Engineering Body of Knowledge SWEBOK 3.0 (2014)
9. International Institute of Business Analysis: A Guide to the Business Analysis Body of Knowledge (BABOK Guide) ver. 3 (2015)
10. Project Management Institute: Business Analysis for Practitioners A Practice Guide (2015)
11. International Requirements Engineering Board: IREB CPRE Foundation Level Syllabus ver. 2.2.2 (2017)
12. Requirements Engineering Qualifications Board: REQB CPRE Foundation Level Syllabus ver. 2.1 (2014)
13. Hickey, A., Davis, A.: Requirements elicitation and elicitation technique selection: model for two knowledge-intensive software development processes. In: Proceedings of the 36th Annual Hawaii International Conf. on System Sciences, pp. 96–105. IEEE (2003)
14. Escalona, M.J., Koch, N.: Requirements engineering for web applications - a comparative study. J. Web Eng. **2**(3), 193–212 (2004)
15. Jiang, L., Eberlein, A., Far, B., Mousavi, M.: A methodology for the selection of requirements engineering techniques. Softw. Syst. Model. **7**(3), 303–328 (2008)
16. Kheirkhah, E., Deraman, A.: Important factors in selecting requirements engineering techniques. In: Proceedings of International Symposium on Information Technology (ITSim 2008), pp. 1–5 (2008)
17. Tiwari, S., Rathore, S.: A methodology for the selection of requirement elicitation techniques (2017). https://arxiv.org/pdf/1709.08481.pdf
18. dos Santos Soares, M., Cioquetta, D.: Analysis of techniques for documenting user requirements. In: Proceedings of Computational Science and Its Applications, pp. 16–28 (2012)
19. Besrour, S., Bin Ab Rahim, L., Dominic, P.: Assessment and evaluation of requirements elicitation techniques using analysis determination requirements framework. In: 2014 International Conference on Computer and Information Sciences, pp. 1–6 (2014)
20. Darwish, N., Mohamed, A., Abdelghany, A.: A hybrid machine learning model for selecting suitable requirements elicitation techniques. Int. J. Comput. Sci. Inf. Secur. **14**(6), 1–12 (2016)
21. Carrizo, D., Dieste, O., Juristo, N.: Contextual attributes impacting the effectiveness of requirements elicitation techniques: mapping theoretical and empirical research. Inf. Softw. Technol. **92**, 194–221 (2017)
22. Jarzębowicz, A., Marciniak, P.: A survey on identifying and addressing business analysis problems. Found. Comput. Decis. Sci. **42**(4), 315–337 (2017)
23. International Requirements Engineering Board: IREB CPRE: Elicitation and Consolidation, Advanced Level Syllabus ver. 1.0 (2012)
24. International Requirements Engineering Board: IREB CPRE: Requirements Modeling, Advanced Level Syllabus ver. 2.2 (2016)

25. Requirements Engineering Qualifications Board: REQB CPRE Advanced Level Requirements Manager ver. 2.0 (2015)
26. Jain, A.K., Murty, M.N., Flynn, P.J.: Data clustering: a review. ACM Comput. Surv. (CSUR) **31**(3), 264–323 (1999)
27. MathWorks: MATLAB Documentation, Fuzzy Logic Toolbox. https://www.mathworks.com/help/fuzzy/fuzzy-clustering.html

RF-IoT: A Robust Framework to Optimize Internet of Things (IoT) Security Performance

G. N. Anil[✉]

Department of Computer Science and Engineering,
BMSIT&M, Bengaluru, India
gramaanil@gmail.com

Abstract. The current development in data-driven technology has significantly standardized the software defined networking aspects for various cloud enabled streamlined applications. The enormous growth in the statistics associated with the rate of connected devices over computer networks has led to the origins of ides and conceptual notions behind *Internet-of-things*. The evolution of IoT applications and their potential aspects is witnessed since last 5–6 years, which has made it an active area of research. The diverse and dynamic user requirement instances through on-line services supported with software defined networks (SDN) also leverages the operational cost of IoT. However, the collaboration of different wireless networking components has made IoT vulnerable to different types of security attacks while operated with SDN. The study has introduced a robust security framework namely (Framework for Internet of Things Security) abbreviated as RF-IoT to address security loopholes in IoT networking environment. The secure mechanism in the context of SDN targeted to resist maximum possible attacks in IoT. A numerical simulation environment in created to validate the performance of the RF-IoT in contrast with existing baseline.

Keywords: Internet of Things (IoT) · Software defined networking (SDN) · Data scheduling and flow control · Security

1 Introduction

The current era in the field of information technology and communication has made data very much expensive in terms of different means. The increasing rate of exchange of data between different hosts in the current networking scenario has introduced IoT environment which is a collaborative network comprises different types of networking attributes working with a common goal of strengthening the data driven various application performances [1–4]. The exponential growth of inter connected machine-to-machine (M2M) communication ensured faster way of data communication in the current era of IoT where SDN has simplified the computing and networking structure to achieve higher degree of quality levels from performance aspect in wireless networking scenarios [5, 6]. However, SDN operates with different open interfaces in the network layer where few of the components also get integrated to the transport layer to enhance the performance of data flow-control which makes it vulnerable to various

© Springer Nature Switzerland AG 2019
R. Silhavy (Ed.): CSOC 2019, AISC 984, pp. 213–222, 2019.
https://doi.org/10.1007/978-3-030-19807-7_21

network-layer attacks. Thereby, it consequences the malfunction in the SDN system integrated with the IoT environment [7–10]. To address this limitation associated with SDN driven IoT routing layer the proposed study introduces a novel security paradigm namely RF-IoT to safeguard the data transmission and routing along with a cost-effective data flow control using intermediary device performing functions (IDPF). The prime aim lies into securing the data transfer modeling through SDN deployment and managing data flows which in long term defines higher resistivity against different forms of routing attacks in IoT. The conceptualization of RF-IoT is purely intended for network layer protocols but to some extent its scope of optimized performance is extended to the transport layer also where the information flow control play a very crucial role.

The design principle adopted to enhance the security features associated with the RF-IoT realized the fact that the proper information flow management have a significant impact on the network latency and also influences intrusion detection with faster and reliable aspects. RF-IoT network modeling also introduces a novel route selection mechanism by means of handling different attacks at dynamic traffic flow and also applies an efficient load balancing paradigm to utilize computational resources cost-effectively.

The performance of RF-IoT has been validated by introducing an experimental simulation paradigm where the comparative analysis ensured that the proposed RF-IoT achieves higher level security with minimized latency as compared to the conventional secure SDN-based IoT system.

The manuscript structured pattern is planned as follows, Sect. 2 discusses about the existing research track which has focused on the similar problem followed by problem identification in Sect. 3. Section 4 discusses about the design methodology associated with RF-IoT where Sect. 5 introduces the design analysis. Finally, comparative analysis of accomplished result is discussed under Sect. 6 followed by conclusion in Sect. 7.

2 Related Work

This section discusses the prior research study towards security on IoT with different security related techniques and algorithms to enhance the performance of internet of things. He and Zeadally [11] have discussed the security of IoT using RFID authentication system and ECC algorithm for healthcare devices. It provides satisfied result in security in similar kind of cryptographic base work for IoT. Ning et al. [12] have introduced the hierarchical authentication system for various kind of application of IoT. It shows the improvement in security by reaching different application. In Liu et al. [13], the software defined networking scheme with secure mechanism is presented for IoT security in data transfer among the medium. The results found that the security enhancement and it effectively sharing the data one medium to another. Similarly, Pinto et al. [14] have presented the software based approach with security algorithm for edge IoT industrial based devices. It results as various applications in industrial field. Raza et al. [15] have discussed the developed datagram TLS (Transport Layer Security) using symmetric key for secure IoT application. It resulted that the techniques is performed well in constrained medium and scalable in large number of devices. Siegel [16] have discussed the IoT for future to reach security, efficiency and feasibility. Similarly, Shahzad and Singh [17] discussed the data authentication and authorization using various techniques for IoT with theoretical results in different case studies. Sarigiannidis [18] have introduced the security of IoT under various kinds of attacks using G-network scheme. It provides good results in light as well as heavy attacks in different media. Wolf [19] has discussed the security model for IoT and Cyber Physical System with related emerging techniques. That gives a research direction for future study to enhance the performance. Wu and Wang [20] have addressed the game theory related security identification method for various security threats in IoT scheme. The result is tested theoretically as well as practically under DoS attack and it performs well. Similar kind of security related study with different efforts for IoT are found in Zhang et al. [21], Zhang et al. [22], Yang et al. [23], Zhang [24], Xu et al. [25], Ulz [26], Wazid [27], Yang et al. [28], Roy et al. [29], Salameh [30], Manimuthu and Ramesh [31], Mughal et al. [32], Sajid et al. [33], Xu et al. [34], and Yin et al. [35].

Study in similar direction also witnessed in [36–42] where security is concerned in the context of mobile adhoc networks and cloud driven smart agents which infers an integral operations in the context of IoT communication. The following section infers the extracted open research gap which impactful to define the proposed study and its research goal.

3 Problem Description

The information circulation and content sharing has tremendously increased in the current scenario which has realized the need of replacement of conventional networking architectures with high-level optimized networking policies supported with interoperable configuration of protocols in IoT. The essentiality of SDN is realized in this aspect where the limitations of scalability when the data traffic increases to great extent is addressed. SDN is also capable of enabling logical structure which enhances the operational and design aspect of networking principles to attain more flexible, flow-centric, and adaptable way of communication which in turn improve the networking performance with a dramatic state. Although the implementation of SDN on the top of complex IoT has simplified the security architectural paradigm but at the same time poses susceptibility to various new forms of attacks. Thereby the data scheduling along with proper information flow control in each IoT devices is important. Thereby few of the significant networking challenges in complex IoT architectures are observed as follows:

- The 1st operational challenge is - as IoT involves dynamic data flows in between Machine-to-Machine (M2M), thereby, overcoming security loopholes along with higher efficient data scheduling is crucial. It can attain highly stable networking environment with scalable performance.
- On the other hand the 2nd research challenge which has been explored in the previous section where various existing archives are studied on the IoT network layer security but few concerned about the optimized security architecture with proper data flow control.

- A set of recent literatures are found to restrict the implementation of intermediary device performing functions (IDPF) under the collaborative control plane of SDN which is an integral part of network layer where the routing, switching technologies get enabled by means of virtual circuits. It basically introduces higher degree of network latency problems.

Thereby, it is essential and also challenging to deploy IDPF in a way where the performance of the IoT environment get optimized with respect to both data flow control, latency along with security. Thereby, it becomes an open research problem and break point of IDPF deployment to enhance network layer operations.

4 Proposed Methodology

The prime aim of the proposed RF-IoT is to implement a robust security paradigm which involves flow-centric SDN to perform secure data routing in the network layer associated with IoT. It applies distributed IDPF to speed up the routing between source hosts to destination host with information flow control mechanism. The IDPF architecture here deployed in SDN to attain higher degree of information security in terms of intrusion detection and prevention. The *stateful* art of communication along with complex and varied operations packets make it very much essential in the communication paradigm associated with IoT. It also helps tracking down the frequently updated data associated with every connection and every data packets. The conceptualization of RF-IoT is performed on the networking layer in the context of error handling, data congestion management and packet sequencing. To some extent the protocol is designed to be applied on transport layer also where IoT operations get better flow control of data which influences the better security performance optimization. The design aspects of RF-IoT configure IDPF in the SDN based IoT network layer in a way where the filtration of packets can happen in a cost effective way. It also considered flow of the information in a way where it should not exceed the capacity of networking components and switches. In order to formulate a stable networking operation various design and functional combinations are applied. A graph based mathematical formulation is applied by means of establishing a IoT network which is expressed with $f_{Graph}(n, l)$, here n refers to the number of sensor enabled machines or switches and l refers to the communication link in between them. The graph modeling is considered following the notion of undirected graph theory. The Fig. 1 shows a block based overview of the conceptual design of RF-IoT and further its design modeling is also discussed in this section and the consecutive section as well. The proposed RF-IoT also attempted to solve the network latency problems which is a significant measure of the network performance. Thereby proper data management and network resource scheduling are important.

RF-IoT imposes network efficient middle box placement strategy with cost-effective solution of optimizing the communication link selection which results in shortest path execution with imposing higher level security features. The study also investigated various hypothetical facts such as how to cost effectively place IDPF subsystem of middle-box through <External Web Source> where IP prefixes play a crucial role. RF-IoT basically implements a heuristic algorithm to select the optimized location of IDPF.

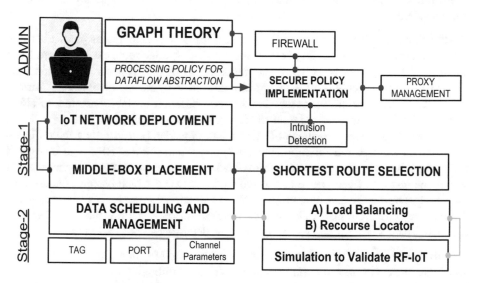

Fig. 1. Architectural stage-wise representation of RF-IoT

The proposed framework work optimizes the switch constraints by minimizing the estimation of occurrence of IDPF (κ) in physical sequence attributes (P_{Seq}). This has resulted in most appropriate route selection where cost optimized operations can be performed. The dataflow control implemented in a sequential manner where the load at each middle-box (L_ε) is balanced. Middle-box (M_b) here in the design process of RF-IoT also refers to the structured implementation of IDPF.

A linear programming notion of patterned execution is formulated to get the cumulative load of each L_ε and which is further minimized to perform load balancing in entire IoT environment with the following mathematical expression.

$$Min \sum_{i=1}^{M_b(i)} L_\varepsilon(\kappa) \text{ where } \kappa \leq P_{Seq} \tag{1}$$

The RF-IoT also implements a notion of data tracking to estimate the intrusion occurrence through relative tags and channels which helps in formulating secure data flow rules. The channel parameters basically indicates, the path status from that channel to the next switch element, whereas tags enables the information flow related status in corresponding switch entity. This notion of implementation has given RF-IoT a cost effective shape in terms of flexibility, scalability and effectiveness.

5 Design Analysis of RF-IoT

The design of the proposed technique is developed considering a numerical computing environment where different network topologies, data processing policy deployment are considered.

Design Stage-1: The middle-boxes and their IDPF also considered as per the specific user requirements. In the initial phase of the computing stage route establishment takes place as per the security policies and topology formation aspects. Simultaneously RF-IoT also implements IDPF enabled M_b placement by means of polychain formulation. The route selection process also assesses different switch constrains to effectively establish secure mode of data routing in the network layer.

The M_b deployment positions are taken into consideration in a way which influences the secure data transmission where, the conceptualization of RF-IoT also targeted to leverage the computing resources from cloud environment if IDS component want to assess the history of each node connected to the network which ensure reputation of each node. During the placement of M_b, the computational process takes the origin and destination node coordinates along with a pre-estimated departure time (d_t). Further, optimized path selection and extension process determines least expected time path.

Design Stage-2: In this phase different sequence of M_b along with different IoT topologies are concerned to ensure minimal traffic flow through each switch component. It limits the amount of traffic in each channel and communication link in a way where it should not surpass the switch and middle-box capacities. Finally the switch tunnel is performed to narrow down the network size so that network latency should minimize. Further it determine each P_{Seq} and add <state tag> to each packet header to perform detection and efficient flow control of the data. The SDN mainly imposes route selection and data flow management with the ease of computation to attain cost optimized security. This process basically controls two different constraints such as (1) cost optimization dynamic dataflow streams in IoT where efficient use of CPU, RAM and counter is essential. On the other hand, (2) It also look for volume restriction in SDN switches.

$$\sum In_{[TF]} \geq \mathfrak{E}_{L \text{ where}} In_{[TF]}[] \in \{0,1\} \tag{2}$$

The switch constraints are optimized using an integer linear programming feature which later got a significant impact on the cost-effective resource utilization. The route establishment in the network layer is performed in a way where the load balancing can be performed to a greater extent for this an indicator component is considered (In $_{[TF]}$). It basically ensure whether a sequence is selected or not with logical pattern of matrix attribute. The logical chain formulation is consider with respect to an exposure level $\mathfrak{E}_L$ which is defined as mentioned in the above Eq. 2. In the next phase RF-IoT basically divides the entire IoT network into small sequences to ensure the dynamic traffic flow in all the logical chains which is associated with P_{Seq}. Further, the load balancer minimizes the maximum load at each middle box component.

The security mechanisms are imposed to safeguard the data traversing through network layer. During secure routing phase the switch retrieve the efficient route which determines which tunnel to select. The switch basically sends the data associated with tunnel and its address to the base station controller. After applying the linear programming base path selection mechanism the controller decides the next hop where the data to be transmitted.

6   Results Discussion

This section shows the outcome obtained after simulating the RF-IoT in a numerical computing environment which requires minimum computing resources like 4 GB RAM and 1.2 GHz processing capacity.

The above Fig. 2 clearly shows that owing to the potential optimized operations RF-IoT attains significant performance improvement when intrusion identification accuracy (IDA) and Network Latency are considered as compared to the conventional approach [9].

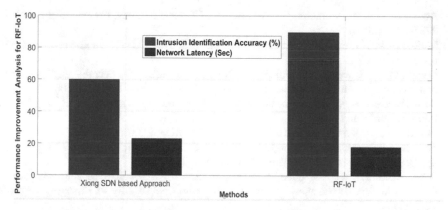

Fig. 2. RF-IoT performance assessment in terms of IDA and Network latency

Figure 3 also shows that RF-IoT attain better throughput performance owing to its potential data tracking policy and effective path selection mechanism. It also attains better resource utilization as compared to [9].

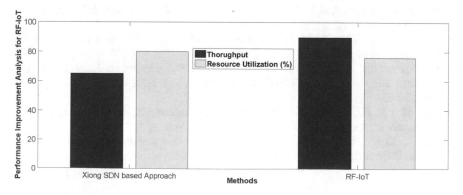

Fig. 3. RF-IoT performance assessment in terms of throughput and resource utilization

7   Conclusion

This paper introduces a novel security framework namely RF-IoT which ensue balancing the trade-off between latency and the security aspects to a significant extent. The conceptualization of RF-IoT is hypothetical and the design and development has been carried out in a numerical computing environment. The experimental simulation outcome further ensures its effectiveness in terms of different parameters such as throughput, intrusion detection accuracy, network latency and optimized resource allocation as compared to the existing system. The proposed study ensures its applicability into various futuristic IoT applications such as smart transportation, health monitoring, environmental monitoring etc.

References

1. Atzori, L., Iera, A., Morabito, G.: The Internet of Things: a survey. Comput. Netw. **54**(15), 2787–2805 (2010)
2. McKeown, N.: Software-defined networking. In: IEEE INFOCOM Keynote Talk, vol. 17(2), pp. 30–32, April 2009
3. Dacier, M.C., Konig, H., Cwalinski, R., Kargl, F., Dietrich, S.: Security challenges and opportunities of software-defined networking. IEEE Secur. Priv. **15**(2), 96–100 (2017)
4. Scott-Hayward, S., O'Callaghan, G., Sezer, S.: SDN security: a survey. In: IEEE SDN for future networks and services
5. Fayazbakhsh, S.K., Chiang, L., Sekar, V., Yu, M., Mogul, J.C.: Enforcing network-wide policies in the presence of dynamic middle-box actions using FlowTags. In: 11th USENIX Symposium on Networked Systems Design and Implement
6. Bates, A., Butler, K., Haeberlen, A., Sherr, M., Zhou, W.: Let SDN be your eyes: secure forensics in data center networks. In: The Workshop on Security of Emerging Network Technologies
7. Guo, L., Pang, J., Walid, A.: Dynamic service function chaining in SDN-enabled networks with middleboxes. In: 2016 IEEE 24th International Conference on Network Protocols
8. Kulkarni, S.G., et al.: Name enhanced SDN framework for service function chaining of elastic network functions. In: 2016 IEEE Conference on Computer Communications Workshops
9. Xiong, G., Sun, P., Hu, Y., Lan, J., Li, K.: An optimized deployment mechanism for virtual middleboxes in NFV-and SDN-enabling network. KSII Trans. Internet Inf. Syst. **10**, 3474–3497 (2016)
10. Jamshed, M.A., Kim, D., Moon, Y., Han, D., Park, K.: A case for a stateful middlebox networking stack. ACM SIGCOMM Comput.
11. He, D., Zeadally, S.: An analysis of RFID authentication schemes for Internet of Things in healthcare environment using elliptic curve cryptography. IEEE Internet of Things J. **2**(1), 72–83 (2015)
12. Ning, H., Liu, H., Yang, L.: Aggregated-proof based hierarchical authentication scheme for the Internet of Things. IEEE Trans. Parallel Distrib. Syst. **1**, 1 (2015)
13. Liu, Y., Kuang, Y., Xiao, Y., Xu, G.: SDN-based data transfer security for Internet of Things. IEEE Internet of Things J. **5**(1), 257–268 (2018)

14. Pinto, S., Gomes, T., Pereira, J., Cabral, J., Tavares, A.: IIoTEED: an enhanced, trusted execution environment for industrial IoT edge devices. IEEE Internet Comput. **21**(1), 40–47 (2017)
15. Raza, S., Seitz, L., Sitenkov, D., Selander, G.: S3K: scalable security with symmetric keys—DTLS key establishment for the Internet of things. IEEE Trans. Autom. Sci. Eng. **13**(3), 1270–1280 (2016)
16. Siegel, J.E., Kumar, S., Sarma, S.E.: The future Internet of Things: secure, efficient, and model-based. IEEE Internet of Things J. **5**(4), 2386–2398 (2018)
17. Shahzad, M., Singh, M.P.: Continuous authentication and authorization for the Internet of Things. IEEE Internet Comput. **21**(2), 86–90 (2017)
18. Sarigiannidis, P., Karapistoli, E., Economides, A.A.: Modeling the Internet of Things under attack: a G-network approach. IEEE Internet of Things J. **4**(6), 1964–1977 (2017)
19. Wolf, M., Serpanos, D.: Safety and security in cyber-physical systems and Internet-of-Things systems. Proc. IEEE **106**(1), 9–20 (2018)
20. Wu, H., Wang, W.: A game theory based collaborative security detection method for Internet of Things systems. IEEE Trans. Inf. Forensics Secur. **13**(6), 1432–1445 (2018)
21. Zhang, S., Xu, X., Peng, J., Huang, K., Li, Z.: Physical layer security in massive Internet of Things: delay and security analysis. IET Commun. **13**(1), 93–98 (2018)
22. Zhang, M., Leng, W., Ding, Y., Tang, C.: Tolerating sensitive-leakage with larger plaintext-space and higher leakage-rate in privacy-aware Internet-of-Things. IEEE Access **6**, 33859–33870 (2018)
23. Yang, X., Wang, X., Wu, Y., Qian, L.P., Lu, W., Zhou, H.: Small-cell assisted secure traffic offloading for Narrowband Internet of Thing (NB-IoT) systems. IEEE Internet of Things J. **5**(3), 1516–1526 (2018)
24. Zhang, P., Nagarajan, S.G., Nevat, I.: Secure Location of Things (SLOT): mitigating localization spoofing attacks in the Internet of Things. IEEE Internet of Things J. **4**(6), 2199–2206 (2017)
25. Xu, G., Cao, Y., Ren, Y., Li, X., Feng, Z.: Network security situation awareness based on semantic ontology and user-defined rules for Internet of Things. IEEE Access **5**, 21046–21056 (2017)
26. Ulz, T., Pieber, T., Höller, A., Haas, S., Steger, C.: Secured and easy-to-use NFC-based device configuration for the Internet of Things. IEEE J. Radio Freq. Identif. **1**(1), 75–84 (2017)
27. Wazid, M., Das, A.K., Odelu, V., Kumar, N., Conti, M., Jo, M.: Design of secure user authenticated key management protocol for generic IoT networks. IEEE Internet of Things J. **5**(1), 269–282 (2018)
28. Yang, A., Li, Y., Kong, F., Wang, G., Chen, E.: Security control redundancy allocation technology and security keys based on Internet of Things. IEEE Access **6**, 50187–50196 (2018)
29. Roy, S., Chatterjee, S., Das, A.K., Chattopadhyay, S., Kumari, S., Jo, M.: Chaotic map-based anonymous user authentication scheme with user biometrics and fuzzy extractor for crowdsourcing Internet of Things. IEEE Internet Things J. **5**, 2884–2895 (2017)
30. Salameh, H.B., Almajali, S., Ayyash, M., Elgala, H.: Spectrum assignment in cognitive radio networks for Internet-of-Things delay-sensitive applications under jamming attacks. IEEE Internet of Things J. **5**, 1904–1913 (2018)
31. Manimuthu, A., Ramesh, R.: Privacy and data security for grid-connected home area network using Internet of Things. IET Netw. **7**(6), 445–452 (2018)
32. Mughal, M.A., Luo, X., Ullah, A., Ullah, S., Mahmood, Z.: A lightweight digital signature based security scheme for human-centered Internet of Things. IEEE Access **6**, 31630–31643 (2018)

33. Sajid, A., Abbas, H., Saleem, K.: Cloud-assisted IoT-based SCADA systems security: a review of the state of the art and future challenges. IEEE Access **4**, 1375–1384 (2016)
34. Xu, Q., Ren, P., Song, H., Du, Q.: Security enhancement for IoT communications exposed to eavesdroppers with uncertain locations. IEEE Access **4**, 2840–2853 (2016)
35. Yin, D., Zhang, L., Yang, K.: A DDoS attack detection and mitigation with software-defined Internet of Things framework. IEEE Access **6**, 24694–24705 (2018)
36. Anil, G.N.: FAN: framework for authentication of nodes in mobile adhoc environment of Internet-of-Things. In: Computer Science On-line Conference, pp. 56–65. Springer, Cham (2018)
37. Anil, G.N., Venugopal Reddy, A.: Strategical modelling with virtual competition for analyzing behavior of malicious node in mobile Adhoc network to prevent decamping. Int. J. Comput. Sci. Issues (IJCSI) **8**(6), 173 (2011)
38. Muneshwara, M.S., Swetha, M.S., Thungamani, M., Anil, G.N.: Digital genomics to build a smart franchise in real time applications. In: 2017 International Conference on Circuit, Power and Computing Technologies (ICCPCT), pp. 1–4. IEEE (2017)
39. Swetha, M.S., Muneshwara, M.S., Anil, G.N.: Mobile cloud computing (MCC) enables storage smartness for every smart mobile agent. Int. J. Innov. Res. Comput. Commun. Eng. **3** (11) (2015)
40. Hegde, S., Sonal, T.G., Swetha, M.S., Muneshwara, M.S., Anil, G.N.: Efficient way of managing the data for cloud storage applications depending on the internet speed in cloud environment. IJEIR **3**(1), 8–11 (2014)
41. Rath, P.K., Anil, G.N.: Proposed challenges and areas of concern in operating system research and development (2012). arXiv preprint: arXiv:1205.6423
42. Anil, G.N., Jindal, P., Rafik, J., Barwade, S.S.: Self organized public key certificate in ad hoc networks. Semantic Scholar. Accessed 11 Jan 2019

The Use of Recursive ABC Method
for Warehouse Management

Milan Jemelka$^{(\boxtimes)}$ and Bronislav Chramcov

Faculty of Applied Informatics, Tomas Bata University in Zlín,
Zlín, Czech Republic
{jemelka, chramcov}@utb.cz

Abstract. The paper solves complex warehouse simulation to achieve effective solution. Most attention is focused on the use of recursive ABC method for warehouse management. The aim of the simulation study is to verify whether recursive ABC method yield additional benefits in optimizing the warehouse. The complete simulation and the mathematical calculations are performed in the Witness Lanner simulation tool. The aim of this simulation study is to discover a better solution using recursive ABC method in each part of the model. The basic warehouse is based on the ABC method. Further, the simulation study provides recommendations that can improve warehouse management and thus reduce costs. The Witness simulation environment is used for modeling and experimenting. All mathematical calculations and simulations are evaluated and measured, as well as all settings of input and output values. Description of the proposed simulation experiments and evaluation of achieved results are presented.

Keywords: Witness simulation · ABC method ·
Recursive mathematical calculations · Warehouse management ·
Logistics · Inventory

1 Introduction

In our current society, every company is currently exposed to competitive pressures. Operations efficiency is the key success of every company that processes inventories. Management is under constant pressure to optimize the stock operations every day. Therefore, it is important for any business to pay great attention to the effectiveness of each operation and to continually improve their performance in the light of rapidly changing market conditions. Companies must use effective inventory management process to reduce this time to minimum [1]. When the overall warehouse management efficiency is low, material supply is at risk and the overall primary and secondary costs are high. Companies are constantly finding new ways to optimize processes [2]. Current science uses several ways to increase the efficiency of storage devices using mathematical methods. ABC analysis is one of the basic mathematical methods used in storage management and in some other areas. It is based on the principles of Vilfredo Pareto, who first noted that 20% of Italians own 80% of Italian land [3]. Then he realized that this division 20/80 also applies to the whole economy. Based on Pareto's

© Springer Nature Switzerland AG 2019
R. Silhavy (Ed.): CSOC 2019, AISC 984, pp. 223–229, 2019.
https://doi.org/10.1007/978-3-030-19807-7_22

observations, Paret's law was also defined: "In many projects, 20% of the effort produces 80% of the outputs."

ABC analysis is also based on Pareto's law, and it is often also called Pareto's analysis [4]. This analysis divides the stored units into three categories: A, B and C according to their percentage of the total value of the selected parameter. For example, if it is needed to analyze the total value of a storage location, sorting into categories A, B, C and D [5].

Category A includes products that represent about 15–20% of the assortment and represent about 80% of the value of the warehouse. Category B covers products that represent 30–35% of the assortment and approximately 15% of the total value of the warehouse. Category C includes 50% of products with a 5% share in the value of the warehouse. Sometimes, this classification also includes group D, which has a 0% share. This group includes goods that are no longer used or sold. These goods should be discounted, written off or disposed of, their storage is being a net loss [6].

XYZ analysis is a modification of ABC analysis [7]. The essence of XYZ analysis is the classification of products per the stability of their sales. Sales stability is measured as a variance of sales or consumption over a given time horizon. The XYZ analysis is a way to classify inventory items per variability of their demand. X means "Very little variation". X items are characterized by steady turnover over time. Future demand can be reliably predicted. Y means "Some variation". Although demand for Y items is not steady, variability in demand can be predicted to an extent. This is usually because demand fluctuations are caused by known factors, such as seasonality, product lifecycles, competitor action or economic factors. It's more difficult to forecast demands accurately. Z means "The most variation". Demand for Z items can fluctuate strongly or occur sporadically. There is no trend or predictable causal factors [8].

ABC and XYZ analysis alone may be sufficient, but their combination has a much better predictive value [11]. Therefore, the results of both analyzes should be combined. For this purpose, the ABCXYZ matrix is used, where categories A, B and C are set on one axis, and on the other X, Y and Z. The control of the AX, BX, CX, AY, BY, CY, CY, AY, BY, CY products can be fully or partially automated, the demand forecast for these products is sufficiently accurate. The management of AZ, BZ, CZ products needs to be given additional attention, it cannot be fully automated, the predictions are not reliable in relation to them. AZ and BZ products have a significant share of the company's profit, but are characterized by unstable turnover, meaning that caution should be exercised in handling this product group. CY and CZ products should be discounted or their sales should be guaranteed in some other way. The products of the AZ, BZ, CZ, AY, BY, CY groups could be suitable for JIT supply.

Recursive ABC method is a new approach to the ABC method where the same mathematical method is used for smaller parts of the model [9]. It is a mathematical method, where the effort is to recalculate all areas in the mathematical model recursively. The goal is to re-calculate all these areas again. All the results need to be measured [10]. To measure results, it is best to build a model that is as close to reality as possible. This model is then debugged and multiple random processes are started. All results are measured and entered into tables for further comparison. This work refers to the CoDIT conference, but is focused on another subject of research [10].

2 Problem Formulation

The aim of this paper is to test recursive ABC method for warehouse stock management. The complete simulation and the mathematical calculations are performed in the Witness Lanner simulation tool. The aim of this simulation is to find a better solution using recursive ABC method in each part of the model. The current warehouse operation is based on the ABC method. It is important to create the model of the current state and the model (warehouse according to the ABC method) and the model using the recursive ABC model. Both approaches (models) are compared on the base of some simulation experiments. The model for this comparison has a variable warehouse stock size to make the results as close as possible to reality. The simulation study contains a total of 9 simulation experiments for each model. Each simulation experiment has a different storage area of a different size. The experiments are measured by the total distance traveled by the operator during the storage process. For details on objective function, refer to Sect. 4.

Regarding the logical distribution of the warehouse, the default setting is the ratio 70:20:10 assumed. For this division, a different warehouse size is tested. In the next step, the calculation moves further, and the next ratio is advantageous and overall results of the model are measured. As the unit of measurement are meters, the total distance that an operator must travel to handle 100,000 storage keeping units (SKU). Storage keeping unit is a universal storage unit that is used globally. The best solution in the model is that the total distance traveled by the operator is as low as possible. For all simulations, the rectangular shape of the warehouse is used.

3 Model Construction in Witness Simulation Environment

Modeling and subsequent implementation of the designed experiments, it is possible to use a broad range of simulation programs and systems. The Witness Lanner program is used for this case of mathematical calculation and experiment simulation. This system is granted by the Lanner Group and it includes many elements for automotive and discrete-part manufacturing. This simulation tool has preset specific objects and elements that connect in logical contexts. These objects and elements can be further configured and programmed according to the needs of the experiment. All models can be displayed in 2D or 3D layout. For this simulation, one simple model is created according to the ABC basic theory and for comparison a complex model with recursive calculation and results. The buffer elements and model are used to process individual parts. Each buffer has a limit setting of 10,000. For random simulations, the entire range of random streams is used over the range of one thousand random choices. Using random streams, the simulation approaches the most realistic use. For all simulations, the rectangular shape of the warehouse is used.

The main element of the model is the machine that simulates the manipulation of the SKUs. This element takes the parts from different buffers. Picking from different buffers is based on a special ratio. The special ratio is set according to Pareto's principle in the base model. The basic division is 70:20:10. Pick from these buffers according to a particular division is chosen randomly with full respect of a certain ratio. To prove the calculation expectations, a further change is chosen and that is the size of the warehouse. The size of the warehouse is in different dimensions and the practical experiment confirms that the warehouse dimensions also affect the overall functioning of the warehouse. A random variable is selected to run the model. It has 1000 different streams available. Output data is sent to the SHIP object. An important fact is that for the quality of the model, it must be guaranteed that no buffer will end with a zero supply of parts. Therefore, protective principles are set to guarantee this premise. The appropriate lot size and interval settings ensure the correct course of the simulation.

Recursive simulations differ by having more buffers and variables in the model. Complex programming logic is used for recursive mathematical computations. The reason trying to verify recursively the calculations is finding better, more accurate mathematical calculations to optimize warehouse. The benefits are measured of this solution. In this model, the benefit is measured so that the supply truck path is shorter. All results are recorded in a uniform table.

4   The Objective Function

The objective function is to test the recursive ABC method for warehouse management. The entire experiment is created in the Witness simulation tool. There are two models to explore. One model is created for basic ABC method and the other one is created for a recursive ABC method. The basic element of the model is the operator who manages the fork-lift truck. The simulation experiments are evaluated by the total distance traveled by the operator during the storage process. All the measured results are recorded in the excel tables. For the simulation study, the number of SKU movements must be determined. The total number of the fork-lift truck movements is 100,000 SKUs. For all simulations, the rectangular shape of the warehouse is used. SKU is an international term for storage keeping unit. The total path is determined by the sum of the stock-lift movement tracks in each store section.

(see the formula (1)), where N_A, N_B, N_C are the numbers of SKUs from each store section.

$$D^{total} = \sum_{n_A=1}^{N_A} D_A + \sum_{n_B=1}^{N_B} D_B + \sum_{n_C=1}^{N_C} D_C \qquad (1)$$

The distance, which fork-lift truck needs totally for picking up the SKUs in numerous storage sections is different. Average distance is used for each warehouse area. The average distance of the supply truck path is used and calculated using the Pythagoras theorem. The formula (2) presents the calculation of the average distance of the stock-lift truck path for the A section. This is the average distance, which the operators select in the A section of the warehouse during the storage process of 1 SKU.

$$D_A = \sqrt{\left(\frac{W}{2}\right)^2 + (A * L)^2} \qquad (2)$$

Where:

D is the total average distance in a section.
W is the width of the warehouse.
A is the percentage size of section A (In this simulation study, it is 70%.)
L is the length of the warehouse.

5   Results of Simulation Experiments

The whole ABC model is set to ratio 70:20:10. Total handling is 100,000 SKU. The size of the stock is variable, values are displayed in the table. Nine different experiments with different storage sizes are selected. To prove the calculation expectations, a further change is chosen and that is the size of the warehouse. The size of the warehouse is in different dimensions and the practical experiment confirms that the warehouse dimensions also affect the overall functioning of the warehouse. A random variable is selected to run the model. It has 1000 different streams available. All simulations share the fact that the model is always tested to relocate 100,000 SKUs. The Table 1 presents that handling 100,000 SKUs in a smaller warehouse is more effective. This is not the purpose of verification. The purpose of verification is to discover, if recursive calculations in different warehouses sizes manage higher storage efficiency.

Table 1. Results of simulation experiments

Number	Height [m]	Length [m]	Distance of stock lift for ABC method [m]	Distance of stock lift for recursive ABC method [m]
1	100	120	6 134 995,83	5 334 505,60
2	100	180	7 065 276,25	5 645 899,40
3	100	240	8 094 539,05	6 004 156,90
4	100	300	9 191 938,93	6 394 071,00
5	200	140	10 961 940,57	10 262 475,10
6	260	140	13 799 660,14	13 211 994,40
7	600	800	37 984 350,35	32 390 886,10
8	700	1000	45 323 442,51	38 123 325,70
9	1000	1000	58 563 162,69	52 455 285,80

Recursive ABC method for warehouse management is a new approach to the ABC method where the same mathematical method is used for smaller parts of the model. All smaller areas of the model are recursively recalculated. This slightly modify the method of storing materials in the Warehouse stock. There is a finer definition of the Warehouse efficiency. It can be seen from each of the results, that in all cases the efficiency is increased by the fact that the total distance traveled is lower. This means that a recursive calculation gives us some benefits in storing and manipulating SKUs.

6 Summary of Results

The aim of this paper was to test recursive mathematical calculation for warehouse stock management. The complete simulation and the mathematical calculations were performed in the Witness Lanner simulation tool. All test results were presented in Table 2. This size was calculated by multiplying the length and width of the warehouse. For all simulations, the rectangular shape of the warehouse was used.

Table 2. Comparison of results of simulation experiments

Number	Distance of stock lift for ABC method [m]	Distance of stock lift for recursive ABC method [m]	Difference (m)
1	6 134 995,83	5 334 505,60	−800 490,23
2	7 065 276,25	5 645 899,40	−1 419 376,85
3	8 094 539,05	6 004 156,90	−2 090 382,85
4	9 191 938,93	6 394 071,00	−2 797 867,93
5	10 961 940,57	10 262 475,10	−699 465,47
6	13 799 660,14	13 211 994,40	−587 665,74
7	37 984 350,35	32 390 886,10	−5 593 464,25
8	45 323 442,51	38 123 325,70	−7 200 116,81
9	58 563 162,69	52 455 285,80	−6 107 876,89

About the table description, the first column in the table presents us the experiment number. The second column presents the measured distance in the ABC model. The third column display us the distance measured from the recursive mathematical model. In the fourth column, there is an illustration of basic ABC methods and recursive methods. The negative number represents the total distance savings in favor of the recursive model. The last column presents the total storage area in square meters.

7 Conclusion

The aim of this paper was to test recursive ABC method for warehouse management. The complete simulation and the mathematical calculations were performed in the Witness Lanner simulation tool. The aim of this simulation was to discover a better solution using recursive mathematical calculations in each part of the model. The warehouse was based on the ABC method and simulated in Witness program. The ABC method was compared with recursive mathematical method and the results were compared. This recursive calculation proved to be effective. The model for this comparison had a variable warehouse stock size to make the results as close as possible to reality. All dimensions of the warehouse were displayed including the calculation results. Regarding the logical distribution of the warehouse, the default ratio settings was 70:20:10. As the unit of measurement were used meters, the total distance that an operator had to travel to handle 100,000 storage keeping units (SKU). Storage keeping unit was a global universal storage unit. For all simulations, the rectangular shape of the warehouse was used.

Acknowledgement. This work was supported by the Ministry of Education, Youth and Sports of the Czech Republic within the National Sustainability Program, Project No.: LO1303 (MSMT-7778/2014); as well as the European Regional Development Fund, under CEBIA-Tech Project No.: CZ.1.05/2.1.00/03.0089.

References

1. Banks, J., et al.: Discrete-Event System Simulation, p. 608. Prentice Hall, New Jersey (2005). ISBN 0-13-144679-7
2. Baker, P., Canessa, M.: Warehouse design: a structured approach. Eur. J. Oper. Res. **193**(2), 425–436 (2009)
3. Bottani, E., Montanari, R., Rinaldi, M., Vignali, G.: Intelligent algorithms for warehouse management. Intell. Syst. Ref. Libr. **87**, 645–667 (2015)
4. Curcio, D., Longo, F.: Inventory and internal logistics management as critical factors affecting the chain performances. Int. J. Simul. Process Modell. **5**(4), 278–288 (2009)
5. Gu, J., Goetschalckx, M., McGinnis, L.F.: Research on warehouse operation: a comprehensive review. Eur. J. Oper. Res. **177**(1), 1–21 (2007)
6. Karasek, J.: An overview of warehouse optimization. Int. J. Adv. Telecommun. Electrotech. Signals Syst. **2**(3), 111–117 (2013)
7. Kare, S., Veeramachaneni, R., Rajuldevi, M.K.: Warehousing in theory and practice. University College of Boras dissertation, Sweden (2009)
8. Muppani, V.R., Adil, G.K., Bandyopadhyay, A.: A review of methodologies for class-based storage location assignment in a warehouse. Int. J. Adv. Oper. Manage. **2**(3–4), 274–291 (2010)
9. Petersen, C.G., Aase, G.: A comparison of picking, storage, and routing policies in manual order picking. Int. J. Prod. Econ. **92**(1), 11–19 (2004)
10. Raidl, G., Pferschy, U.: Hybrid optimization methods for warehouse logistics and the reconstruction of destroyed paper documents. Dissertation paper. Vienna University of Technology, Austria (2010)
11. Jirsák, P., Mervart, M., Vinš, M.: Logistics for economists, Wolters Kluwer
12. Jemelka, M., Chramcov, B., Kříž, P.: ABC analyses with recursive method for warehouse. In: CoDIT (2017)

A Robust Modeling of Impainting System to Eliminate Textual Attributes While Restoring the Visual Texture from Image and Video

D. Pushpa[✉]

Maharaja Institute of Technology Mysore, Belavadi, Karnataka, India
pushpa_ise@mitmysore.in

Abstract. The area of video analytics in the context of collaborative networking has gained a lot of attention from the research community owing to its potential applicability in the real life aspects. However, although image and video content which mostly get exchanged in the networking pipelines consist of several significant textual information from the application view-point which often display various confidential textual credentials of a corresponding individual. The realization of this fact that this textual attributes has to be removed for various image forensic requirements, has led to image impainting. The study has addressed this problem and come up with a novel analytical solution which imposes two different methods and further combines this two. In the 1st stage it applies a robust mechanism to detect the region of an image and video frame sequence where textual data representation can be localized and perform extraction of those data it introduces artifact and visual anomalies. On the completion of this stage in the 2nd phase, to eliminate the artifacts from the respective locations, it introduces a novel impainting technique which is computationally efficient and attain higher degree of textual data eliminated recovered image or video sequence which is almost similar like the original image or video sequence, can be visually perceived. The comparative performance analysis show that the proposed technique attain better outcome in terms of textual attributes detection accuracy (%) from specific region of interest (ROI) and also consume very less processing time (Sec) in contrast with the existing system.

Keywords: Image or video impainting ·
Textual data localization and extraction · Textual regions selection ·
Post-processing

1 Introduction

The recent advancement in the technology of video compression, communication and networking provides a wide usage of videos in many walk of life such as surveillance systems, live news reporting and many video content delivery based entertainment channels such as Hot-star, Netflix etc. [1, 2].

© Springer Nature Switzerland AG 2019
R. Silhavy (Ed.): CSOC 2019, AISC 984, pp. 230–239, 2019.
https://doi.org/10.1007/978-3-030-19807-7_23

Due to wider adaptability and popularity of video content, the size of video reposit is growing exponentially that leads to many challenges. One of such challenges is fast and accurate video clip search from video reposit.

For the purpose of easiest search and indexing videos are made labeled and text can be extracted accurately. Though it is a mandatory requirement to embed text into images and videos for seamless process, whereas due to many forensic requirements or application collaboration, the embedded text on the image or video needed to be removed without making it to be noticed and also the process should not affect the visual texture properties of the image and videos to a greater extent. Thereby, the realization of the fact in this context shows why textual data which may contain significant user credentials and attributes are needed to be preserved for various security means from application viewpoint [3–7].

In the process of text removal from the images and videos, it creates many visual artifacts which can be noticed. Therefore, to handle this issue a technique of impainting plays an important role [8–10]. The notion of impainting technique includes local-ization of text region into the image or video scenes, removal of text from this region and then filling the artifacts to make the image and video as original one. Thereby, restoring the visual texture from both image and video objects after applying impainting where the textual localization and extraction is performed is challenging in practice, which also require a large amount of computational effort when image or video frame structure is concerned [11, 12].

This study basically introduces an optimized mechanism which in 1st stage find and detect text attributes from a specific region associated with an image and video frame sequence and further applies a simple and robust analytical design to extract the textual representation from that particular region of image or video in the 2nd stage. The analytical modeling also applies texture synthesis paradigm which ensure good per-formance towards restoring the image attributes after text extraction in the 3rd stage of design and modeling using a novel impainting mechanism. The image texture synthesis is conceptualized and modelled in a way where computational effort is reduced in the post-processing stage. It also attain very negligible computational time complexity when the textual localization takes place in the region of interest. The experimental simulation which is modeled with respect to localizing both image and video text followed by an extraction process. Further, impainting has been applied to minimize the artifacts which get produced while the textual attributes get extracted after localization process.

A comparative performance analysis is performed on the basis of two different performance parameters such as textual attributes detection accuracy (%) in the image or video scenes along with the processing time (Sec), the proposed impainting mechanism yielded. The outcome of the study clearly shows that the proposed method attain superior detection accuracy and negligible analytical design processing time while attaining higher degree of artifact removal which ensures the superior visual perceptual quality of the post processed image and video scenes. The paper is orga-nized in a way where Sect. 2 talk about the existing related works carried out in the similar problem and also extract the open research problems which has defined and given a shape to the research goal set in this paper. The identified research problem is explored and discussed in Sect. 3 from a theoretical view point followed by elaborated discussion on the design methodology of the introduced technique in Sect. 4. Section 5 discusses the computational steps associated with the analytical design implementation strategy where in Sect. 6, the numerical simulation outcome along with the visual outcomes are illustrated. Finally Sect. 7 brings an insight into the concluding remarks.

2 Related Work

This section discusses about the existing literatures being carried out in past towards the similar problem.

Huang et al. [13] have presented the noise extraction and impainting scheme for iris images. It exhibits a result through simulation with high performance in terms of recognition accuracy. Karras and Mertzios [14] have introduced the discretization system and Impainting model using a numerical estimation of partial differential equation (PDE) for MRI reconstruction. The performance of the present scheme decrease the processing time, however, it precedes poor image reconstruction under complex real-world task. Stolojescu-Crisan and Isar [15] have introduced the Classical and Robust Exemplar-based image impainting algorithm for SONAR images. The performance of the method is evaluated in terms of visual superiority by calculating the PSNR value. By the comparison between both the methods Robust Exemplar base technique shows superiority. Chan and Shen [16] have presented the local nontexture Inpainting using a logical and mathematical model. It gives many applications by demonstrating the present method through text removal, image restoration from the scratched photos, edge related image coding, and digital zooming. Perz and Toyama [17] have addressed the Exemplar-based Inpainting in object removal for various kind of image processing application. From the evaluation, the method achieved computational efficiency. Dimiccoli and Salembier [18] have focused on perceptual filtering method for image processing using image inpainting and connected logical operator. The performance is calculated by PSNR and SSPQ, that provides high visual quality and decrease the error visibility compared to a related method. Ruzic and Pizurica [19] have presented the context-aware patch-based image inpainting using Gabor filtering and Markov random field methods. It exhibits high speed compared to other MRF related methods. Buyssens et al. [20] have presented the Depth guided based inpainting method for RGB images in recovering the lost image data. From the demonstration, the method shows the effectiveness in terms of speed, quality and robustness in both color and depth images. Jin et al. [21] have presented the image inpainting identification method for the sparsity-based image using Canonical correlation analysis. The result is compared with the related inpainting method and it shows better performance in the process of Gaussian noise addition and JPEG compression. Li [22] have introduced the localization of images in digital images using image inpainting technique. The method is demonstrated and it shows the effectiveness in identifying diffusion-based image inpainting. Ding and Ram [23] have discussed the image inpainting for image processing using nonlinear filtering technique and nonlocal texture matching. The outcome shows that the method efficient among similar kind of method in inpainting large geometric and texture image missing. Similar kind of image inpainting aware research study for various kind of image processing applications is found in Jin and Ye [24], Li et al. [25], Amrani et al. [26], Guo et al. [27], Cai and Kim [28], Lu et al. [29],

Annunziata et al. [30], Ogawa [31], Zhang et al. [32], Xu [33], Lun [34], Tirer and Giryes [35], Wen et al. [36] and Gao et al. [37].

3 Problem Description

This section basically brings an insight into the fact that, in past a large amount of work has been carried out in the domain of image inpainting where localization of textual attribute in terms of structure inpainting and texture inpainting have been considered for the purpose of localizing and extracting text regions from the image. However, the following are the open research problems which are still unsolved and needed to be addressed.

- Less Emphasize given towards Extraction of Textual Attributes form Video Scenes

- It is observed in the existing significant studies which are explored in the previous section, that very few studies has considered this problem and very limited steps found attempted towards solving the extraction of text attributes from video scenes. On the other hand more emphasize has been given on Image.

- Very less studies found on this problem

- Very less studies found in past which are highly cited and carry significant impact factor. This present form of study found very less archives within the timeline of (2015–2018) pertaining to this similar problem which clearly shows the gap which is needed to be filled up.

- Studies mostly limited towards image inpainting rather impainting

- Existing literatures mostly emphasized towards image inpainting rather impainting to restore the texture properties of image and video.

- No Effective Benchmarking till Date

- Not a single effective benchmarking solution explored till date which addresses both complexity and ROI detection accuracy and textual data extraction problem.

4 Proposed Methodology

The prime aim of the proposed technique is to perform a computationally efficient impainting schema which can perform reduction of textual data attributes from both image and video without compromising its quality. The novelty of this proposed system is that it detects the textual ROI regions from any scene of image frame where the complexity of processing is reduced to a significant extent. Finally it removes the textual characters from image with respect to different complexity levels which is defied with a variable called ε. The proposed study combines a set of processes while structuring the operational modeling of the proposed impainting solution for both image and videos. In the initial phase the analytical design considers an image (I_1) and a video object (V_o) where textual embedded data can be found. Further it performs dimensionality reduction of the data by reading the frame sequence $f_{SEQ} = [f_1, f_2, f_3 \ldots f_n]$. For $I_1, \sum f_{SEQ} = 1$ as an image object contains only 1 frame. After, performing sampling and quantization of the input data followed by dimension reduction, the process further applies edge detection to smoothen textual character identification process. The process gets repeated for a set of iteration for both I_1 and V_o to ensure completeness.

The analytical model is defined in a way where certain hypothetical assumptions also got considered such as the intensity value associated with the textual representation different from the image background where the difference can be calculated with respect to a threshold γ. The following Fig. 1 shows how the conceptual design methodology has been formulated step wise to make the system operational where the PSNR of reconstructed image can be controlled with restrictive dilation process where reference frames (Rf) plays crucial role.

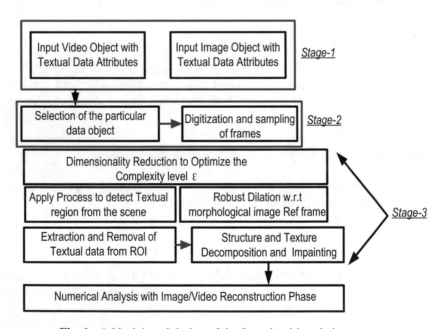

Fig. 1. A block-based design of the formulated impainting system

As mentioned in the above figure, it is clear that the formulated system comprises of multi-stages of computation which are analytically decomposed in the consecutive section. In stage-1 the system applies a computational mechanism to select the intended video or image object from the respective hard-disk drives. These all steps come under the image-preprocessing stage. Further the modeling is designed in a way where the processing of the image starts where initial detection of textual data attributes within a scene region is performed followed by a textual dilation process. During the processing of image and video data the complexity is reduced up to a significant extent by dimensionality reduction where the complexity level as per the corresponding ε plays a significant role. Further the textual data attributes get selected and removed from the highlighted regions which generates artifacts during the processing stage when dilation of image textual attributes are performed with respect to a reference frame intensity threshold value γ. It basically produce a large number of anomalies to make the image or video pixel attributes visually distorted. In this phase the proposed solution applies impainting to fill the missing or broken the textual attribute eliminated distorted image data for the removal of scratches. This post processing stage reconstructs the resultant image with respect to a set of iteration where the prime emphasize given towards making it visually rich in terms of perceptual quality. The numerical outcome demonstrated in Sect. 6 clearly shows that the reconstructed image found as much as similar to the original one with efficient PSNR.

5   Analytical Modeling and Implementation Strategy

The system basically detect the text region from the scene and localize the textual attributes prior extraction. It also performs edge selection and detection mechanism. The character or textual attributes extraction from the image or video attribute if performed using the following mathematical expression.

$$C = \bigcup_{i=1} \overline{|I_i||V_i|}_\rho \tag{1}$$

The Eq. 1 implies that the proposed method extract binary textual character attributes from both image and video frames. The novelty of the proposed system here is it imposes uniform white character pixels for the detection of textual attributes where remaining background remain black.

Here C denotes the textual attributes with character data type, whereas union operator is used to extract the sub image and video frames I and V for $i^{th\ instances}$. It basically applies a thresholding algorithm $\overline{|\cdot|}$ ρ. The prime aim of this thresholding algorithm is to segment the textual region in a manner where characters appear with white color code.

The image impainting and reconstruction phase while applying dilation can be computed by the following mathematical Eq. 2.

$$f_B(\eta|Rf) \underline{\underline{\Delta}} (\eta \oplus B) \cap Rf \tag{2}$$

Equation 2 also refers to a conditional dilation process where image reconstruction takes place with respect to artifacts removal and noise elimination to make it more similar like the original image or video frame sequence where textual attributes will be eliminated. In the above Eq. 2 Rf refers to the reference frame inside which will undergoes through the dilation. η refers to the marker for dilation using prediction attributes and B refers to the unitary disk attributes. The following section presents the numerical simulation outcome and their contributory aspects.

6 Results Discussion

The experimental simulation has been carried out considering the latest version of numerical computing platform where the execution process has been speeded up by dimension reduction of the input data attributes which are images and videos respectively. The complexity level ε is extracted and based on the complexity level the proposed technique performed its cost effective computation. The simulation outcome obtained for different performance metrics such as detection accuracy (%), PSNR (dB) along with Processing Time (Sec) are tabulated in the below Table 1.

Table 1. Numerical outcome assessment of the proposed impainting mechanism

Test Instance	Histogram of the Input Data with complexity Reduction	Complexity Level of Textual Data ε	Dimension Reduced w.r.t to ε	Computed Detection Accuracy (%)	Peak Signal to Noise Ratio Computation (dB)	Processing Time (Sec)
1		8	911×726	90.0999	0.0666	0.00234
2		4	255×255	89.0004	49.00567	0.001124
3		1	269×187	92.00045	57.0098	0.000345
4		2	179× 250	93.00053	56.0456	0.23456
5		5	168×300	94.000678	56.0987	0.456789
6		3	205×246	92.05678	51.0987	0.56789
7		11	179×322	93.0890	53.8796	0.897658

The above Table 1 clearly shows the outcome obtained for the proposed impainting method where computational time complexity has been significantly minimized. It also shows that the trade-off between textual region detection accuracy and reconstructed image quality is also found optimal in the context of texture synthesis and post impainting processing stage. The numerical data which are tabulated in the above Table 1 found superior with respect to the outcome obtained from the existing system Amrani, Serra-Sagristà [26] where image impainting has been considered only for image scenes not for the videos. However, the significant outcome obtained by the proposed technique thereby ensures the effectiveness and its applicability into various practical forensic applications where image and video both are considered.

7 Conclusion

This paper introduces a novel impainting mechanism by means of strengthening the image forensics based application. Here the prime emphasize has been laid towards removal of textual data attributes without compromising the quality aspect of the reconstructed image. The performance analysis of the formulated system shows it achieved 20%, 32% and 40% performance improvement in PSNR, Detection Accuracy and Time complexity as compared to the existing Amrani, Serra-Sagristà [26] method.

References

1. Walden, S.: The Ravished Image. St. Martin's Press, New York (1985)
2. Emile-Male, G.: The Restorer's Handbook of Easel Painting. Van Nostrand Reinhold, New York (1976)
3. King, D.: The Commissar Vanishes. Henry Holt and Company, 1997. 6 Intuitively, it is clear that we can not address the inpainting problem with a single second order PDE, since regularization constraints on vector fields are needed (in [13], the general variational formulation suggested will normally lead to a fourth order gradient descent flow)
4. Joyeux, L., Buisson, O., Besserer, B., Boukir, S.: Detection and removal of line scratches in motion picture films. In: Proceedings of IEEE International Conference on Computer Vision and Pattern Recognition, CVPR 1999, Fort Collins, Colorado, June 1999
5. Kokaram, A.C., Morris, R.D., Fitzgerald, W.J., Rayner, P.J.W.: Detection of missing data in image sequences. IEEE Trans. Image Process. 11(4), 1496–1508 (1995)
6. Kokaram, A.C., Morris, R.D., Fitzgerald, W.J., Rayner, P.J.W.: Interpolation of missing data in image sequences. IEEE Trans. Image Process. 11(4), 1509–1519 (1995)
7. Braverman, C.: Photoshop Retouching Handbook. IDG Books Worlwide, Foster City (1998)
8. Hirani, A., Totsuka, T.: Combining Frequency and spatial domain information for fast interactive image noise removal. Comput. Graph. 96, 269–276 (1996). SIGGRAPH
9. Efros, A., Leung, T.: Texture synthesis by non-parametric sampling. In: Proceedings of IEEE International Conference Computer Vision, Corfu, Greece, pp. 1033–1038, September 1999
10. Heeger, D., Bergen, J.: Pyramid based texture analysis/synthesis. In: Computer Graphics, SIGGRAPH 1995, pp. 229–238 (1995)

11. Simoncelli, E., Portilla, J.: Texture characterization via joint statistics of wavelet coefficient magnitudes. In: 5th IEEE International Conference on Image Processing, Chicago, IL, 4–7 October 1998 (1998)
12. Nitzberg, M., Mumford, D., Shiota, T.: Filtering, Segmentation, and Depth. Springer, Berlin (1993)
13. Huang, J., Wang, Y., Cui, J., Tan, T.: Noise removal and impainting model for iris image. In: 2004 International Conference on Image Processing ICIP 2004, 24 October, vol. 2, pp. 869–872. IEEE (2004)
14. Karras, D.A., Mertzios, G.B.: Discretization schemes and numerical approximations of PDE impainting models and a comparative evaluation on novel real world MRI reconstruction applications. In: 2004 IEEE International Workshop on Imaging Systems and Techniques (IST), 14 May 2004, pp. 153–158. IEEE (2004)
15. Stolojescu-Crisan, C., Isar, A.: Impainting SONAR images by exemplar-based algorithm. In: 2013 International Symposium on Signals, Circuits and Systems (ISSCS), 11 July 2013, pp. 1–4. IEEE (2013)
16. Shen, J., Chan, T.F.: Mathematical models for local nontexture inpaintings. SIAM J. Appl. Math. **62**(3), 1019–1043 (2002)
17. Criminisi, A., Perez, P., Toyama, K.: Object removal by exemplar-based inpainting. In: Proceedings of 2003 IEEE Computer Society Conference on Computer Vision and Pattern Recognition, Jun 2018, vol. 2, p. II. IEEE (2003)
18. Dimiccoli M, Salembier P.: Perceptual filtering with connected operators and image inpainting. In: Proceedings of ISMM 2007 October, vol. 1, pp. 227–238 (2007)
19. Ružić, T., Pižurica, A.: Context-aware patch-based image inpainting using markov random field modeling. IEEE Trans. Image Process. **24**(1), 444–456 (2015)
20. Buyssens, P., Le Meur, O., Daisy, M., Tschumperlé, D., Lézoray, O.: Depth-guided disocclusion inpainting of synthesized RGB-D images. IEEE Trans. Image Process. **26**(2), 525–538 (2017)
21. Jin, X., Su, Y., Zou, L., Wang, Y., Jing, P., Wang, Z.J.: Sparsity-based image inpainting detection via canonical correlation analysis with low-rank constraints. IEEE Access **6**, 49967–49978 (2018)
22. Li, H., Luo, W., Huang, J.: Localization of diffusion-based inpainting in digital images. IEEE Trans. Inf. Forensics Secur. **12**(12), 3050–3064 (2017)
23. Ding, D., Ram, S., Rodríguez, J.J.: Image inpainting using nonlocal texture matching and nonlinear filtering. IEEE Trans. Image Process. **28**(4), 1705–1719 (2019)
24. Jin, K.H., Ye, J.C.: Annihilating filter-based low-rank hankel matrix approach for image inpainting. IEEE Trans. Image Process. **24**(11), 3498–3511 (2015)
25. Li, Z., He, H., Tai, H.M., Yin, Z., Chen, F.: Color-direction patch-sparsity-based image inpainting using multidirection features. IEEE Trans. Image Process. **24**(3), 1138–1152 (2015)
26. Amrani, N., Serra-Sagristà, J., Peter, P., Weickert, J.: Diffusion-based inpainting for coding remote-sensing data. IEEE Geosci. Remote Sens. Lett. **14**(8), 1203–1207 (2017)
27. Guo, Q., Gao, S., Zhang, X., Yin, Y., Zhang, C.: Patch-based image inpainting via two-stage low rank approximation. IEEE Trans. Visual. Comput. Graph. **24**(6), 2023–2036 (2018)
28. Cai, L., Kim, T.: Context-driven hybrid image inpainting. IET Image Process. **9**(10), 866–873 (2015)
29. Lu, H., Zhang, Y., Li, Y., Zhou, Q., Tadoh, R., Uemura, T., Kim, H., Serikawa, S.: Depth map reconstruction for underwater Kinect camera using inpainting and local image mode filtering. IEEE Access **5**, 7115–7122 (2017)

30. Annunziata, R., Garzelli, A., Ballerini, L., Mecocci, A., Trucco, E.: Leveraging multiscale hessian-based enhancement with a novel exudate inpainting technique for retinal vessel segmentation. IEEE J. Biomed. Health Inform. **20**(4), 1129–1138 (2016)

31. Ogawa, T., Haseyama, M.: Adaptive subspace-based inverse projections via division into multiple sub-problems for missing image data restoration. IEEE Trans. Image Process. **25**(12), 5971–5986 (2016)

32. Zhang, H., et al.: Computed tomography sinogram inpainting with compound prior modelling both sinogram and image sparsity. IEEE Trans. Nucl. Sci. **63**(5), 2567–2576 (2016)

33. Xu, Y., Yu, L., Xu, H., Zhang, H., Nguyen, T.: Vector sparse representation of color image using quaternion matrix analysis. IEEE Trans. Image Process. **24**(4), 1315–1329 (2015)

34. Lun, D.P.: Inpainting for fringe projection profilometry based on geometrically guided iterative regularization. IEEE Trans. Image Process. **24**(12), 5531–5542 (2015)

35. Tirer, T., Giryes, R.: Image restoration by iterative denoising and backward projections. IEEE Trans. Image Process. **28**(3), 1220–1234 (2019)

36. Wen, F., Adhikari, L., Pei, L., Marcia, R.F., Liu, P., Qiu, R.C.: Nonconvex regularization-based sparse recovery and demixing with application to color image inpainting. IEEE Access **5**, 11513–11527 (2017)

37. Gao, Z., Li, Q., Zhai, R., Shan, M., Lin, F.: Adaptive and robust sparse coding for laser range data denoising and inpainting. IEEE Trans. Circ. Syst. Video Technol. **26**(12), 2165–2175 (2016)

SkyWay in Zlín

Pavel Pokorný[✉] and Lukáš Laštůvka

Department of Computer and Communication Systems,
Faculty of Applied Informatics, Tomas Bata University in Zlín,
Nad Stráněmi 4511, 760 05 Zlín, Czech Republic
pokorny@utb.cz, lukelastuvka@gmail.com

Abstract. Zlín is a city in the south-east of the Czech Republic. Like other cities, it has to deal with problems arising from population growth. One of these fields is Transport. Local public transport is currently, mainly provided for by trolleybuses, buses and trains in Zlín, and continues to be modernised. One of the public transportation possibilities for the future is the ability to quickly transport people inside the city - using the SkyWay technology. SkyWay is an elevated lightweight transport system, using a pre-loaded one-way rail with cables, ("threads"), and filled with a special concrete, between two or more buildings in an urban area. This paper describes work done at the Faculty of Applied Informatics, TBU in Zlín, representing a visualisation of the appearance of such a SkyWay in Zlín. This work encompasses the combination of real-time city footage of Zlín traffic - captured by camera; and the 3D SkyWay technology models that are imported into the recorded footage. All works were created in the Blender software suite. Its outputs are in the form of rendered images and animations.

Keywords: SkyWay · 3D visualisation · Rendering

1 Introduction

Transport means the movement of humans, and/or animals, goods from one location to another. Transport can be divided into infrastructure: (fixed installations – roads, railways, airways, etc.); vehicles - (bicycles, cars, buses, trains, etc.); or operations: (the way vehicles are operated – private or public).

People mainly need to organise transport in cities in line with its increased density. All over the world, cities are feeling the ever-increasing burden of traffic. A significant element of the problem is the enduring popularity of private cars. Public transport has always seemed to take second-place to the car - and yet, alternative ways of moving around cities are possible [1].

Measures to improve public transport, as well as initiatives to encourage walking and cycling, have been introduced in many large cities to decrease car-use, or at least to persuade people to use their cars in different ways. Public transport systems in cities include bus, trolley-bus, rail, metro and taxi modes. All these technologies have their own characteristics - which predict their use; especially their cost structures and costing methods; transport capacity and energy consumption [2].

© Springer Nature Switzerland AG 2019
R. Silhavy (Ed.): CSOC 2019, AISC 984, pp. 240–248, 2019.
https://doi.org/10.1007/978-3-030-19807-7_24

The Skyway technology can represent one of the modern public transport in cities solutions. This paper describes in short on possible visualisations of this technology in Zlín - a city in the south-eastern part of the Czech Republic [7].

Fig. 1. An example of the SkyWay system; Author: SkyWay Technologies [4]

2 The SkyWay Technology

SkyWay represents an innovative transport technology developed by engineer Yunit-skiy [3], and provides passenger and cargo transportation in a separate space – on the'second level' - above the ground. SkyWay is a complex system that encapsulates an overpass-type information, energy and transport communicator - based on rail-string technologies. The specific feature of this technology is its environmental friendliness; and therefore, the existing natural landscape and surrounding flora and fauna are not disturbed in the vicinity of transport over-pass constructions. An example of the SkyWay system is shown in Fig. 1 [4].

Other benefits include optimised aerodynamics, unprecedented safety, increased speed, comfort, minimised environmental damage caused by transport, and rational land and resources use. In addition, construction and operation costs are significantly lower when compared to existing transport solutions [3].

Yunitskiy's string-transport system can be divided into several types: High-Speed Transport - (between cities, regions, countries or continents – speeds can be up to 500 km/h); Urban Transport - (solutions for towns or big cities – speeds can be up to 150 km/h); Cargo Transport - (primarily intended for mass); or, Sea Transport - (cargo trans-shipments from SkyWay rolling stock) [4].

String-rail represents a conventional, un-split - steel, reinforced concrete, or steel-reinforced concrete beam or truss; equipped with a rail-head, reinforced with pre-stressed, (stretched), strings. The Maximal String Tension per rail (depending on the unibus or unitruck design mass; speed motion limits, and span length), can be 10–1,000 tons - and more - (at an assembly design temperature of ±20 °C). String-rail combines the qualities of a flexible thread - (a large span between supports), and rigid beams, (small spans under the wheel of a rail vehicle, and support above) [5].

The above-mentioned urban SkyWay transport system is designed for short distance passenger transportation inside small or big cities. It can resolve transport problems by developing and constructing a network of high-rise buildings connected with each other by elevated air transportation - with a maximum speed of 150 km/h. three transport unit types currently exist – unibike, unicar and unibus. Unibike represents a light and compact vehicle on steel wheels and the model range includes single, double, three, four and five-seater unibikes. Performance can be up to 20,000 passengers/hour and more [3].

Unicar and unibus are similar – the difference is in passenger capacity. Unicar is designed for 2–6 people - (up to 6–18 in articulated vehicles); the middle-class unibus is for 3–14 people - (up to 84 in articulated vehicles); and large-class unibuses, (Fig. 2), for 7–28 people - (up to 84–168 in articulated vehicles). Performance can be up to 50,000 passengers/hour and more.

Fig. 2. Large-class double-rail unibus; Author: Global Transport Investments [3, 6]

3   Resources and Software Used

The first phases that needed to be done were, to collect all suitable materials of the SkyWay system, and to select adequate software for visualisation creation.

3.1   Acquiring Resources

All the main sources were collected after consultation with Mr. Domanský [8], who has been interested in the SkyWay technology and its implementation in Zlín for a long time. He proposes places that are suitable for departures/arrivals, and a suitable type of the SkyWay technology for Zlín.

Based on the information that was obtained, these places in Zlín were captured with a camera - and saved as animations. Some of them were created from the ground-level, other from nearby buildings' roofs. In order to attain high final visualisation quality, we used high animation resolutions – Full HD (1920 × 1080 pixels).

The next step was to select a specific SkyWay technology. Here, we selected hinged rails because they have less visual impact on the Zlín architectural landscape. The transport unit was chosen in relationship to passenger capacity. Therefore, the Unibus unit was selected as the basic version, whose capacity is up to 28 passengers, (Fig. 2).

The last part of this work was to select suitable images of the supporting technologies - (rails, support pylons, and departure/arrival stations), in order to create 3D models. Sources [3, 4, 6], were used for rails and pylons. However, we were unable to find an image of a suitable station building with its future appearance that would not significantly disrupt the architectural landscape of Zlín. Therefore, an improvisation was made in connection with its main function - (the departure and arrival of people).

3.2   Software Used

In order to select applications, the requisite works that are required were compiled. The whole process includes 3d model creation - (a wide spectrum of modelling tools are suitable for this); to set the material properties for object-faces, (textures are not necessary); to import an animation and to combine these with created models; then, to render images/animation and to support post-processing - (final composition or time-based animation editing, and/or adding a sound track). In addition, we wanted to use free-to-use software.

All these operation can be performed with the Blender software suite [9]. It is a free and open source software, based on the OpenGL technology cross-platform, and offers essential tools for modelling, sculpting, texturing, UV-mapping, lighting, skinning, scripting, camera work, rendering - (with different render engines), animation, rigging, particles and other simulations. Blender also allows one to perform compositing, post-production, and game creation [10].

4   Modelling, Materials and Light

We used polygonal representation for all models, performed with mesh objects in Blender. In transport unit, rails and supporting pylons cases, adequate images were loaded into the Blender environment and placed on the background screen. The stations were create without this image because we improvised, as mentioned above.

The most complex procedure was the creation of the transport unit model. It was divided into several parts - modelled separately, but connected to each other. Where the parts had a square shape, Cube objects were added. Plane objects for Glass Fillings and

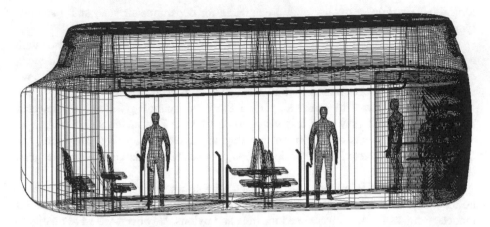

Fig. 3. Mesh model of a transport unit, (wireframe shading)

Cylinder Objects for rounded shapes were used. Using standard modelling tools, (basic editing commands, transformations, modifiers, Extrude, Subdivide, Proportional Editing, Knife, Loop Cut and Slide, etc., [10]), the shapes were gradually improved into their final appearance.

A simple interior was also created - in order to attain more realistic visual outputs. None of the rendered images or animations were generated from a short distance, so this was improvised, and all models were simplified. In this way, seats were created - (using the Extrude Tool and Transformation on a Cube Object), and some low-poly characters made using the Sculpting Modelling Technique. The final transport unit model is shown in Fig. 3. It contains approximately 120 000 vertices.

3D model of rails and supporting pylons were relatively simple to create - a Cube object was added into the 3D scene; Subdivide, Extrude and Transformation tools were then used to model it into the final shape. The supporting reinforcements in the pylons were made from two rectangular objects, (transformed Plane Objects), mutually rotated by 45°.

The problem was to find a model solution for station objects. The decision fell on creating a regular shape. Since these objects represent places where more routes can cross, an octagon was selected. A Circle Object was then inserted into the 3D scene and its vertices were modified to eight.

As usual, final models were created using modelling tools. In order to achieve the modern appearance of these buildings, triangle shapes were used - and every second one of them was horizontally flipped.

A Node Editor in Blender was used to set all objects' material properties. Simple material settings were linked to the objects - (a diffuse colour and glossy parameter with transparency to glass objects). To light the 3D scene, a global light Ambient Occlusion was activated in combination with one local light object – Sun. The direction of its rays was based on camera-orientation in order to achieve the same lighting as in the captured animations.

Figure 4 illustrates all of the created models - (approximately 164 000 vertices), with their material settings. In order to attain better visual appearance, colours are inverted.

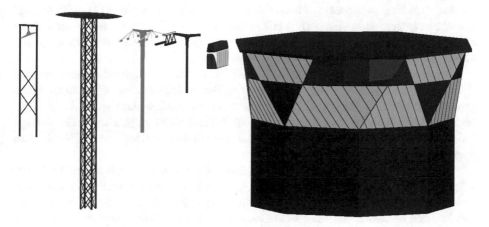

Fig. 4. Mesh models of all created objects with materials - (inverted colours)

5 Motion Tracking, Rendering and Compositing

Blender also offers 2D and 3D Motion Tracking tools used to track objects' motion - and/or, a background scene. Through the compositor, this tracking data can then be applied to any 3D objects. Blender's motion tracker supports camera-tracking as well as object-tracking, including some special features like plane-track for compositing. Tracks can also be used to move and deform masks for rotoscoping in the Mask Editor (a special mode in the Movie Clip Editor) [11].

In order to use these tools – it is necessary to analyse the capture animation, (this procedure is described here for one animation – others were made in the same way). After the animation was loaded, it was necessary to set three parameters – firstly, to specify a device which captured the animation; the focal length; and the "strength" of distortion. In case the focal length is unknown, it can be automatically only be obtained from the camera's settings - or from the EXIF information, (as in our case).

The next step was to specify tracking markers - (pixels are traced in pictures during the animation process). This can be done manually or automatically using the Detect Features function, which plays animation in a memory and detects anomalies between the neighbouring images. We used this automatic detection. After that, we ran the Solve Camera Motion function, which evaluates these tracking markers for further use.

Once this process was complete, we oriented the real scene in the 3D-scene for more convenient compositing - (it is necessary to define the floor, the scene-origin, and the X/Y axes in order to perform scene orientation), and to specify camera-position and view-direction during the animation. These values were obtained in the previous step and where some of them were wrong, they could be manually corrected.

After that, we imported all prepared 3D models into one scene and corrected the transformation. We used curve objects that were shaped as required to make complete rails, and then the Array and Curve Deform modifier was used on one part of the model rail.

For rendering purposes, we used the Cycles Rendering Engine [12], with settings based on captured animation - (full HD resolution with a 50 frame rate per second, the output quality was set to 1,000 samples per frame). The rendered outputs were then saved into PNG graphic format with an alpha channel.

We rendered all of the animations in this way. The final animation encapsulated 3,150 image files, (the total duration was 63 s). For rendering calculation, we used both types of processors, (CPU and GPU). The compute configuration was a CPU AMD Ryzen 7 1700, with 16 threads and an NVIDIA 1060 6 GB graphic card. The average value for rendering one picture was 35 s; i.e., the whole rendering time in Blender was approximately 31 h.

Once the rendering process was completed, the masking phase using the Movie Clip Editor in Blender was begun. Here, we combined the captured animation with rendered images using masks that cover unwanted objects in the images. These masks can be made automatically - but where necessary, they can be modified in a UV/Image Editor window.

All of the obtained resources were then used in the Node Editor Window, where these resources were compiled together. The result was an animation in a sequence of images, as rendered. The final edits were made in the Video Sequence Editor window, where is possible to edit animations - and to add some effects between animation strips.

Fig. 5. The original image of Zlín - captured by a camera

Fig. 6. The final image of Zlín - with implemented SkyWay system

Figure 5 shows one example image of Zlín, captured by a camera. Figure 6 shows the same image with the implemented SkyWay system - (the final result), resulting from the operations described above.

6   Conclusion

This article briefly describes a visualisation method used for the implementation of SkyWay technology into contemporary Zlín. SkyWay is an elevated lightweight transport system that uses a preloaded, one-way rail with cables special concrete filler between two or more buildings in an urban area.

The complete visualisation was created in the Blender software that supports the entirety of the 3D pipeline – including modelling, rigging, animation, simulation, or rendering, compositing and motion-tracking, or even video-editing and game-creation. In this environment - (in one 3D scene), we imported real shots captured the city centre and then created 3D models of the SkyWay technology suitable for public transport in Zlín. These resources were then rendered and composed.

The outputs are provided in the form of rendered images and animations based on Mr. Domanský proposal, as mentioned in the text. This visualisation procedure could be also performed in the same way for other stations of this mass transit system or in other parts of the city. If the SkyWay system can one day be realised in Zlín or not, only time will tell. This visualisation can help with decision-making.

References

1. Richards, B.: Future Transport in Cities, 1st edn. Taylor & Francis, London (2001). ISBN 978-1135159658
2. White, P.: Public Transport – Its Planning, Management and Operation, 6th edn. Taylor & Francis, London (2016). ISBN 978- 1317383185
3. Yunitskiy, A.: UST: New world reality. http://www.yunitskiy.com/author/english.htm. Accessed 12 Jan 2019
4. Home – sw-tech. http://www.sw-tech.by. Accessed 12 Jan 2019
5. Yunitskiy, A.: Transport Complex SkyWay in Questions and Answers. http://www.yunitskiy.com/author/2016/2016_98.pdf. Accessed 12 Jan 2019
6. Official Site of SkyWay group of Companies. http://rsw-systems.com/. Accessed 12 Jan 2019
7. Laštůvka, L.: Camera and Object Tracking in Blender. http://digilib.k.utb.cz/bitstream/handle/10563/43413/la%C5%A1t%C5%AFvka_2018_dp.pdf?sequence=1&isAllowed=y. Accessed 12 Jan 2019
8. Domanský, Z.: ING. http://www.z-domansky.eu/. Accessed 12 Jan 2019
9. Blender.org – Home. http://www.blender.org. Accessed 12 Jan 2019
10. Fisher, G.: Blender 3D Basics: Second Edition, 2nd edn. Packt Publishing, Birmingham (2014). ISBN 978-1783984909
11. Motion Tracking – Blender Manual. https://docs.blender.org/manual/en/latest/editors/movie_clip_editor/tracking/index.html. Accessed 12 Jan 2019
12. Valenza, E.: Blender Cycles: Materials and Textures Cookbook, 3rd edn. Packt Publishing, Birmingham (2015). ISBN 978- 1784399931

Using Multimedia in Blended Learning

Jiri Vojtesek[✉] and Jan Hutak

Faculty of Applied Informatics, Tomas Bata University in Zlin,
Nam. T.G.Masaryka 5555, 760 01 Zlin, Czech Republic
vojtesek@utb.cz
http://www.utb.cz/fai

Abstract. The paper is focused on the use of modern methods, especially computers and the Internet in the educational process. The so-called Blended learning combines printed and electronic materials, online and offline teaching methods, etc. in the class and e-learning is phenomenon nowadays mainly in distance learning. The contribution also mentions Khan Academy as maybe the famous non-profit educational portal in the world. The goal of this contribution is to give an initial overview into this field of education and presents examples of educational videos from the field of Information Technologies that were created, verified and published on the Khan Academy portal.

Keywords: E-learning · Multimedia · Blended learning · Khan Academy · Learning Management System

1 Introduction

Education process varies every time and depends on the time and opportunities especially technical support that have teacher has at his disposal. The great help was the invention of the computer in the last century. As Information Technologies (IT) and computers are the fastest industry concerning innovation, education is also affected very quickly. Also, another milestone was the foundation of the Internet and his rapid development after its commercialization in 1992. This contribution is focused on the use of e-learning and video tutoring over the internet in modern educational process. E-learning has several faces from the simple sharing of the files, online courses through Voice-over-IP (VoIP) services to more sophisticated Learning Management Systems (LMS) [1]. A typical member of this LMS is open-source system Moodle [2] which is very popular in high schools and universities. Moodle is written in PHP under the GNU General Public License and offers mostly all features needed for modern education like a blended, distance of flipped classroom learning. The teacher can create various types of classes depending on the type of learning (full-time, distance learning, etc.). Material sharing is a matter of course; tutor can use Moodle also for testing students, for giving them exercises that can be revised again on-line, etc. The plugins can add other functionalities [3].

© Springer Nature Switzerland AG 2019
R. Silhavy (Ed.): CSOC 2019, AISC 984, pp. 249–258, 2019.
https://doi.org/10.1007/978-3-030-19807-7_25

Another modern teaching methods are the use of educational videos that are uploaded to one of the online video sharing services like YouTube [4], Vimeo [5] etc. Teachers can use these sites directly and distribute only a link among students. A more sophisticated solution can be found on sites like Dumy.cz [6] or Khan Academy [7] that uses YouTube as a video-sharing service but offers some extra functionalities like a clear grouping of the videos, adding exercises, testing of students etc. Maybe the greatest public free online education sites. KA was founded by Salman Khan in 2008. Khan gets his degrees in mathematics, electrical engineering computer sciences at MIT and MBA on Harvard Business School. He was very good on tutoring especially on-line and decided to create a non-profit learning platform named Khan Academy which motto is To accelerate learning for students of all ages [7]. The site offers more than 20 000 videos which are at most 15 min long sorted in eight main groups. The popularity of this site still grows, and the main donators of this non-profit project are Google, AT&T, Bill Gates foundation, etc. which gives reliability to this site. The use of Khan Academy is very wide from the courses at basic schools, through high schools, and of course universities [8–10] etc. The site is open to everybody who can help with the translations to local languages or even creating own videos. The only condition is to follow suggestions and style that was defined by Salman Khan. The local Khan Academy Council checks this.

2   Teaching Methods

Teaching method could be in the concept of the topic of this article divided into classical teaching methods, activating teaching methods and their combinations in the complex teaching methods. All approaches will be discussed in the next sub-chapters.

2.1   Classical Teaching Methods

Classical teaching methods are basics of didactic categories that are integrated into the education system. We can divide them into three main groups [11]:

- *Word methods* – since Ancient time, speaking methods belongs the most versatile and most widespread methods. Spoken and written words are both very important, and the role of the teacher and student is essentials. Typical forms are narrations, explanations, lectures, interviews, etc.
- *Demonstration methods* improve ideas and concepts about the subject at the students. Emphasis is placed on linking real experience with the practice. Demonstration and observation, work with the picture and training belong to these methods.
- *Practical methods* are focused on action-oriented learning. Attention is put on student activities connected with real life. As a result, teachers use skills building, imitation, manipulation and experimentation and production methods are used here.

Classical teaching methods are still necessary for teaching specific subjects such as practical subjects, arts etc. On the other hand, lessons of a wide range of subject may be complemented by modern teaching methods mentioned in the next chapter.

2.2 Activating Teaching Methods

As every method is evolving, also classical teaching methods are changing in time thanks to new concepts, options and especially thanks to computers and information technologies generally. Activating methods increase the role of the students and put emphases at the teamwork that teaches a student responsibility. As a result, the student is more active in the classroom and not only listen to the teacher who could be more attractive for the student.

For example, the discussion, problem-solving, situation methods or didactic games belong to this group of activating teaching methods [11].

2.3 Complex Teaching Methods

Combinations of classical and complex approaches could result in so-called complex teaching methods as a special tool that spread nowadays more widely. Brainstorming that improves critical discussion for solving a specific problem is one of the complex methods. The Information and Communication Technologies (ICT) are more and more included in the learning process in this lets call it Internet Age. This contribution is focused on this last group of methods as you will see in next chapters.

3 E-learning

Probably the most common use of ICT in education is in e-learning which can be defined as learning using electronic media, especially Internet and computer networks generally to access the learning curricula outside the classroom worldwide from every computer connected to the Internet [1]. As a result, e-learning is an ideal tool, especially for distance learning.

E-learning is universal, and it can be used for various situations, not only for the education of the students on the schools and universities but also for the different training sessions and educations of seniors, etc.

We can say, that E-learning is generally not a computer program or system, but a tool for deepening the quality the learning. It is a complex system including teachers and users (students) that communicates using computers and the Internet.

If we combine e-learning with classical teaching methods, achieved goals and results are gain more effectively. This combination is called *Blended learning* [12], and we can combine electronic and printed learning materials, online and offline teaching methods, individual and group teaching, structured and non-structured study materials, etc.

Online teaching courses that are not fixed the student attendance in the classroom are very popular. Technical equipment for online courses is not very money demanding – it needs the only computer with a connection to the Internet and special communication program installed. The big advantage is that a lot of these programs are free, for example, Skype, Microsoft Teams, etc.

Like everything in this World, also use of ICT has its advantages and disadvantages that are concluded from the students and teachers point of view [13] are shown in Table 1.

Table 1. Advantages and disadvantages of the e-learning

From the students view	
Advantages	Disadvantages
Availability of the information	Hardware and software costs
Backup of study materials	Internet connection
Possibility to learn everywhere	Complexity of ICT
Flexibility in learning	Information overflow
Better information and computer literacy	Plagiarism
Communication with other students	Addiction and other health problems
From the teachers view	
Advantages	Disadvantages
Creation, archiving, distribution of multimedia materials	Insufficient teachers know-how
Support and improvement of the ICT literacy	Poor quality or lack of ICT resources
Cooperative teaching	More demanding preparation
Support for innovative ICT didactics	Lack of natural communication
Classroom administration	

3.1   Learning Management System (LMS)

As e-learning has many forms, one of the most common is Learning Management Systems (LMS) as a software tool offering management of the online courses. LMS systems are used for the administration of the courses, distribution of the study materials, questioning students throughout on-line test in various forms. They are widely used at universities especially in those that offer distance courses but they are often deployed as support for full-time students and also on lower education levels – at high school especially.

A typical member of LMS is Moodle which is definitely favorite LMS not only among Czech universities but worldwide [2]. The biggest advantage is that it is written in PHP under the open-source licensing. Moodle offers all standard and special tools through add-ons that are also mostly free [3].

Also all big IT companies like Google, Microsoft and Apple have their tools for blended learning like Google Classroom, Apple Classroom or Microsoft Classroom but some of them are paid or oven platform-depended.

3.2 Educational Portal

Another very progressive part of the e-learning is educational portals that provide study materials in text, audio or video forms. One of the most known and famous is Khan Academy (KA) [7] which is a non-profit organization founded in 2008 by Salman Khan with a goal to educate student using online tools, mainly videos published on todays phenomenon Googles portal YouTube. The motto of KA is *"A free, world-class education for anyone, anywhere."*

The Salman Khans educational KA YouTube channel includes in January 2019 over than 6800 educational videos with more than 4,5 million followers. The educational videos are mainly from the areas: mathematics, physics, chemistry, biology, economy, informatics, medicine, and astronomy. Most of the educational videos are in English, but some of them are localized to other 15 languages, including the Czech language. Some videos are in the local language or English with local subtitles. KA was founded in 2008, and it is registered as a non-profit organization supported for example by Google or Bill Gates. The KA offers not only educational videos on YouTube channel, but it also has very professional web pages https://www.khanacademy.org/ with online courses that offer a connection to the educational videos on YouTube and also exercises that verifies gained information. Everything is structured into groups and subgroups by subject.

KA web pages are actually (in January 2019) divided into eight main groups

1. Math – subdivided into groups by theme – geometry, algebra, statis-tics, etc.
2. Math by grade – subdivided by the age of the student
3. Science & Engineering – deals with physics, astronomy, biology, etc.
4. Computing – including programming, animation, science, etc.
5. Arts & Humanities – with art, world and US history, grammar, etc.
6. Economics & Finance – microeconomics, macroeconomics etc.
7. Test prep
8. College, careers, & more

Each subject has web page structured in sections including teaching videos, that explains the problem, revision exercises with quizzes and test where a student can get points that shows his progress in learning. The topic is explained from the basics to the more complex examples which means that the information does not overload the student. Some courses, for example, programming, also offers an online editor that shows the results in real time which is also very demonstrative. The concept of the KA offers three types of registration – student, teacher, and parent. At teacher login, a mentor can create a personalized course for specific students and revise their progress. Similarly, a parent can create the account for his child and watch the progress in selected topics.

KA also aims to the very young children in age 2–6 with the mobile application "Khan Academy Kids."

Of course, there are a lot of other educational portals like Dumy.cz [6] which primarily focused on the online support for teachers, students, schools and also parents. A server has among teaching videos also e-books, presentations, etc. Users can upload materials which are then verified and rated by other users. Ad-vantage can be found in the wide range of topics, materials are under the license Creative Commons, and teachers can create own e-classroom with specific con-tent. The disadvantage is that the materials and web pages are mainly in the Czech language.

Specific portal for teaching programming can be found at www.codecademy.com. The server is focused on teaching languages: Javascript, HTML, Python, Ruby, etc. [14]. It was founded in 2011, and most of the lectures are free of charge. The negative feature can be found in the limited just to the programming and not to the other subjects.

Lastly, the Google famous YouTube portal offers tons of education channels that can be used for teaching. YouTube is the worlds biggest video-sharing portal founded in 2005 [4]. It belongs to the Google which bought this service on 2006 and integrates it to his services. It means that one login can be used for Gmail, YouTube and also Google docs, etc. Since then, the popularity grows rapidly to the second-most popular web pages on the Internet. YouTube can serve education videos itself or acts as a support for above mentioned educational portals.

There are also a lot of other sites with the specific subjects of learning mostly connected with IT like Academic Earth, Coursera, FutureLearn, etc. Dealing with all of them is out of the range of this paper.

4   Practical Part

Practical part will be focused on the procedure of creating and publishing teaching video on the KA. Next sub-chapters will describe the preparation process firstly, then the software needed for creating this video and finally the process of uploading the video to Khan Academys web portal. There were created teaching videos from the four fields:

1. Numeral system conversion,
2. Hardware, software and peripheral,
3. Introduction to HTML language and CSS and
4. Introduction to Algorithms [15].

4.1   Preparation Stage

Firstly, as we will prepare material for KA, it is good to choose a topic that suits one of the eight groups mentioned above. It is good to check, if the material still does not exist because it makes no sense to create it again. The second way is to contact representatives of KA and ask them for topics which they need to prepare.

It is also good to prepare material and scenario of the video in advance. Good preparation saves a lot of time during the creation of the material. The timing of the video is also very important; it is good to prepare the length no longer than 10 min which is standard on the KA. Longer videos could act boringly. If we still need more time, it is possible to divide the topic into several chapters. The right voice and legible written text are also very important.

If we think that our preparation is finished, we can move on to the creation of the video.

4.2   Tools for Creating Teaching Videos

Tools that are important for the creation of the teaching video can be divided into two groups – hardware and software tools.

Hardware Tools. Into this group, we belong of course the computer that could in the form of a notebook or desktop computer with an appropriate microphone for recording our speech. It is common that videos at KA have a specific form so-called writing hand where student see black background and teacher writes, illustrates and comment on the topic like it is classical blackboard at school. This style was introduced by the founder of the KA Mr. Salman Khan in his first teaching videos. This effect could be achieved with the use of a graphical tablet that is connected to the computer and offers transmission of our written ideas and drawings directly into the digital form.

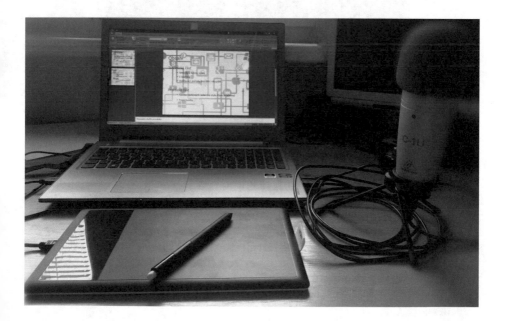

Fig. 1. Hardware tools used for preparing educational video

Sample hardware assembly used for recording of educational videos in this work is shown in Fig. 1. There were used an ordinary notebook with an external microphone for recording of audio tracks and external graphical tablet Wacom Bamboo 3 Pen.

Software Tools. Once we prepared and checked the hardware, it is time to introduce software that will be used. As we want to minimize costs, let us focus on free software.

Transmission of the manual writings and drawings from the graphical table in-to the digital form can serve program *SmoothDraw*. This program is freeware and offers everything we need for our task – we can write by pencil, pen or brush using a wide range of colours. Moreover, it collaborates with the graphical tablet, and Salman Khan recommends it. Sample screen from one of the educational videos made by SmoothDraw is shown in Fig. 2. The text inside the picture comes from the teaching videos that is why they are in Czech language.

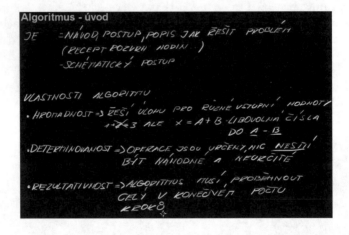

Fig. 2. Sample scren from SmoothDraw drawing software

We also need a tool for recording our desktop. This could be done for example by the program *Active Presenter* which also has freeware version with limited features that do not limit us when creating the video. The program offers to record of the desktop with additional commenting by the voice through the microphone. The resulting video could be later edited with other sources like pictures etc. Final export could be done into AVI or MP4 video format.

One of the best audio editors is *Audacity* which offers recording of the audio, conversion from the analog to the digital form and finally edit of the audio track. Audacity can help us also with the noise reduction, editing, and mixing of recordings, etc.

Finally, simple, free and suitable video editing tool for Windows platform called *Windows Movie Maker* can be used for finalization of our recording. A program can effectively cut, edit or change audio and video tracks; we can add new audio track or video and audio effect. The resulted project can be exported into video or directly to YouTube.

4.3 Publishing of the Teaching Video

The final stage is publishing of the video on the Internet. The best way is to upload the video to the portal YouTube which is the biggest video portal in the world.

A user can upload the video with the maximal length 15 min that can be prolonged to the 10 hours, there is supported a wide range of the video formats, and the video could be uploaded directly from the mobile phone. Video can have captured and it could be placed into one of 15 categories, one of them is Education.

Authors of this contribution were created 13 teaching videos in the Czech language from the field of Informatics aimed mainly on the basic school and high school students [15] (Table 2).

Table 2. List of author's published educational videos on KA and their length in min.

Numerical System Conversion			
Binary to decadic	0:05:44	Decadic to binary	0:03:50
Binary to hexadecimal	0:03:11	Decadic to binary	0:03:40
Hardware, Software, peripherals			
Parts of computer	0:05:44	Basic types of software	0:04:33
Periphcrals	0:03:53		
Introduction to HTML and CSS			
Introduction to HTML	0:06:20	Images in HTML	0:04:38
Links in HTML	0:04:50	HTML + CSS	0:06:17
Introduction to Algorithms			
Introduction to Algorithms	0:03:51	Written forms of algorithms	0:05:05
Total time of educational videos			1:00:56

All videos were verified by Czech representatives of the KA and uploaded on the Czech Khan Academy YouTube canal available on the web address https://www.youtube.com/playlist?list=PL_0PIu4A1-CJVsuU51kYTaNWYypTNpqt3

5   Conclusion

The goal of this contribution was to show new trends in education. The main attention was focused on the computer support of the teaching process. This support is very popular, and we can meet computers in every step of the education process. Computers and the Internet can be used in various ways, but the main help is in the sharing of knowledge, questioning of students, testing their knowledge and teaching remotely. These features are very important mainly in the distant form of study. A practical part of this contribution shows the procedure of the creating teaching video. We focused on necessary hardware and software and described the process of creating and publishing the video on Khan Academy, which special worldwide online teaching portal. The contribution could be then used as a starting source of information for teachers that wants to use computers and the Internet in the educational process. This time certainly offers plenty of possibilities of usage of computers and Internet as a support of the education and it is up to us how we use this offer.

References

1. Clark, R.C., Mayer, R.E.: E-Learning and the Science of Instruction: Proven Guidelines for Consumers and Designers of Multimedia Learning, 4th edn. Wiley, Hoboken (2016). ISBN 978-1119158660
2. Official website of the LMS Moodle. https://moodle.org/
3. LMS Moodle plugins directory. https://moodle.org/plugins/
4. Google's YouTube official website. https://www.youtube.com/
5. Official website of video sharing service Vimeo. https://vimeo.com/
6. Digital educational materials website Dumy.cz (in Czech). http://dumy.cz/
7. Khan Academy's official webside. https://www.khanacademy.org/
8. Cargile, L.A., Harkness, S.S.: Flip or flop: are math teachers using Khan Academy as envisionedby Sal Khan? Tech. Trends **59**, 21 (2015). https://doi.org/10.1007/s11528-015-0900-8
9. O'Donnell, J.F.: The Khan Academy: a great opportunity for cancer education. J. Cancer Educ. **27**(3), 395–396 (2012). https://doi.org/10.1007/s13187-012-0405-5
10. Henrquez, V., Scheihing, E., Silva, M.: Incorporating blended learning processes in K12 mathematics education through BA-Khan platform. In: Pammer-Schindler, V., Prez-Sanagustn, M., Drachsler, H., Elferink, R., Scheffel, M. (eds.) Lifelong Technology-Enhanced Learning. EC-TEL 2018. Lecture Notes in Computer Science, vol. 11082. Springer, Cham (2018). https://doi.org/10.1007/978-3-319-98572-5_26
11. Manak, J., Svec, V.: Teaching methods (in Czech). Paido, Brno (2003). 219 p. ISBN 80-7315-039-5
12. Ifenthaler, D.: Blended Learning. In: Seel, N.M. (ed.) Encyclopedia of the Sciences of Learning. Springer, Boston (2012)
13. Zounek, J., Juhanak, L., Staudkova, H., Polacek, J.: E-learning: teaching with digital technologies (in Czech). Wolters Kluwer, Praha (2016). ISBN 978-807-5522-177
14. Codecademy on Wikipedia.org. https://en.wikipedia.org/wiki/Codecademy
15. Hutak, J.: The Use of Multimedia for Teaching Information Technologies (in Czech). Tomas Bata University in Zlin, Faculty of Applied Informatics, Master Theses (2018)

Modeling the Behavior of Objects in the Pursuit Problem

A. A. Dubanov[(✉)]

Buryat State University, 24 A, Smolin-Street, Ulan-Ude, Russian Federation
alandubanov@mail.ru

Abstract. This article provides a description of the developed behavior models of objects in the persecution task, the objects is the pursuer and pursued. The idea of the research is to development an algorithm for autonomous robotic systems. In the proposed behavior models, local dynamic coordinate systems are introduced, which are formed by the direction of movement of objects. For a certain interval of time, the object must decide in which direction it should move depending on the result of the analysis of the coordinates of the second object. Due to the fact that an object cannot move instantly when moving in space, in our problems, "inertia" is modeled using the angular velocity of rotation. According to the proposed models of the behavior of objects in the pursuit problem, programs are written in the computer math system "MathCAD", which can be found on the website of the author. The results of the programs obtained animated images of the movement of objects, references to which are given in the text of the article #CSOC1120.

Keywords: Behavior model · Pursuit · Object movement · Velocity · Rotation

1 Introduction

Consider the problem of pursuit on rough terrain, when the pursued object (Rabbit) and the pursuing object (Fox) move in three-dimensional space with constant velocities. Earlier, in the works of Isaacs R., Pontryagin LS considered the theoretical aspects of the problem of pursuit-evasion (boat - torpedo). At present, with the development of the material and technical base responsible for providing information to objects in the pursuit task (laser rangefinders, optical station for circular viewing, etc.), it becomes necessary to build behavioral models with the development of original algorithms for each type of task. In article [1] the model was considered when the trajectory of the "Rabbit" movement was predetermined. That is, the position coordinates of the Rabbit at a certain point were known. In such a model, it was suggested that the velocity vector of the Fox would be directed to the Rabbit or its intended position at a certain point in time. This article proposes different behaviors of both the Rabbit and the Fox. In the proposed Rabbit model of behavior, an analysis of the spatial position of the Fox will be made, its speed and, depending on this, the "Rabbit" will decide on the direction of its movement. Exactly the same reasoning can be carried out with respect to the behavior of the Fox. In our article, we rely on the results of the creators of the theory of differential games Isaacs [2], Pontryagin [3], Krasovsky [4], and others. We believe that

© Springer Nature Switzerland AG 2019
R. Silhavy (Ed.): CSOC 2019, AISC 984, pp. 259–274, 2019.
https://doi.org/10.1007/978-3-030-19807-7_26

the theoretical questions presented in the works of the founders of the theory need practical realization. Taking into account the progress of IT-technologies, it seems possible to model the processes of pursuit, evasion, evasion and support in computer mathematics systems. We consider our work as a development and continuation of work [5] in terms of transferring the task to an intersected surface, with giving our model "inertness". The development of the pursuit and evasion ideas follows from [6–8]. The results of this article can be applied by the developers of robotic complexes in the implementation of some tasks of pursuit.

2   To the Choice of the Coordinate System of the Object of Pursuit of the "Rabbit"

In this article, the main thing is that the choice of the "solution" (direction of motion) follows after analyzing the current position in the dynamic coordinate system. In the Rabbit model of behavior, it is proposed to choose a dynamic local coordinate system, where:

- The center of the local coordinate system coincides with the current coordinates of the horizontal projection of the rabbit's position (Fig. 1).

$$\overrightarrow{Center}_{dynamic}(T) = \begin{bmatrix} X_{rabbit}(T) \\ Y_{rabbit}(T) \\ 0 \end{bmatrix} \qquad (1)$$

- The abscissa axis of the local coordinate system coincides with the horizontal projection of the speed of the Rabbit at the current time.

$$\vec{E}_1(T) = \frac{1}{\sqrt{\left(\frac{d}{dT}X_{rabbit}(T)\right)^2 + \left(\frac{d}{dT}Y_{rabbit}(T)\right)^2}} \cdot \begin{bmatrix} \frac{d}{dT}X_{rabbit}(T) \\ \frac{d}{dT}Y_{rabbit}(T) \\ 0 \end{bmatrix}$$

$$\vec{E}_2(T) = \frac{1}{\sqrt{\left(\frac{d}{dT}X_{rabbit}(T)\right)^2 + \left(\frac{d}{dT}Y_{rabbit}(T)\right)^2}} \cdot \begin{bmatrix} -\frac{d}{dT}Y_{rabbit}(T) \\ \frac{d}{dT}X_{rabbit}(T) \\ 0 \end{bmatrix} \qquad (2)$$

- The ordinate axis belongs to the horizontal plane of the projections. The local coordinate system is orthogonal, respectively, the axis of applicate is the product of the vector product of vectors $\vec{E}_1(T)$ and $\vec{E}_2(T)$.

In our task explicit pronounced dependencies $X(T)$, $Y(T)$ don't exist. In the program posted on the website [9], for the model of the "Rabbit" trajectory, the ordered sets of points are given: X_i, Y_i, $Z_i = Surface(X_i, Y_i)$, $i \in [0..N]$, where $Surface(X_i, Y_i)$ – is a function describing the terrain based on a point basis, after performing polynomial regression and spline interpolation procedures. This procedure is described in detail in the comments to the program code on the author's website [9], as well as in article [1].

Further, to build a functional dependence of the "Rabbit" trajectory on time T, a formal parameter is introduced t, $t_i = i$, $i \in [0..N]$. From the formula for the arc length segment $dS^2 = dX_{rabbit}^2 + dY_{rabbit}^2 + dZ_{rabbit}^2$ we have: $\frac{dt}{dS} = \dfrac{1}{\sqrt{\left(\frac{d}{dt}X_{rabbit}(t)\right)^2 + \left(\frac{d}{dt}Y_{rabbit}(t)\right)^2 + \left(\frac{d}{dt}Z_{rabbit}(t)\right)^2}}$.

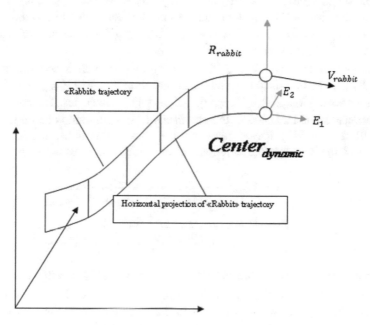

Fig. 1. Local coordinate system of "Rabbit"

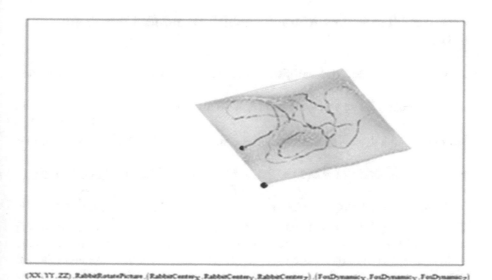

$(XX, YY, ZZ), RabbetRotatePicture, (RabbetCenter_X, RabbetCenter_Y, RabbetCenter_Z), (FoxDynamic_X, FoxDynamic_Y, FoxDynamic_Z)$

Fig. 2. Trajectories of "Rabbit", "Foxes" in the dynamic coordinate system

In Fig. 2 shows the first frame of the animated image [10]. This video clearly shows that the dynamic coordinate system is tied to the Rabbit system. Since it is motionless in the video. The program code executed in the "MathCAD" system can be downloaded from the author's website [9] by the link «Archive of the program "Rabbit" and "Fox" on the plane. "Rabbit" runs away, "Fox" is catching up». The built-in solver of the MathCAD system allows us to solve a differential equation: $\frac{dt}{dS} = \frac{1}{\sqrt{\left(\frac{d}{dt}X_{rabbit}(t)\right)^2 + \left(\frac{d}{dt}Y_{rabbit}(t)\right)^2 + \left(\frac{d}{dt}Z_{rabbit}(t)\right)^2}}$, in the form of a definition of dependence $t(S)$. Considering that we viewing the case of motion with a constant in module velocity V_{rabbit}, we introduce calculation in real time T, $t = t(V_{rabbit} \cdot T)$. In the calculations we need $\frac{d}{dt}X_{rabbit}(t)$, $\frac{d}{dt}Y_{rabbit}(t)$, $\frac{d}{dt}Z_{rabbit}(t)$ dependencies. The construction of such dependencies is indicated in the comments to the program code on the author's website [9, 11]. Taking into account all the above conditions, we find the time derivatives of the T coordinate functions of the Rabbit. As a result, we have:

$$\frac{d}{dT}X_{rabbit}(T) = V_{rabbit} \cdot \frac{\frac{d}{dt}X_{rabbit}(t)}{\sqrt{\left(\frac{d}{dt}X_{rabbit}(t)\right)^2 + \left(\frac{d}{dt}Y_{rabbit}(t)\right)^2 + \left(\frac{d}{dt}Z_{rabbit}(t)\right)^2}}$$

$$\frac{d}{dT}Y_{rabbit}(T) = V_{rabbit} \cdot \frac{\frac{d}{dt}Y_{rabbit}(t)}{\sqrt{\left(\frac{d}{dt}X_{rabbit}(t)\right)^2 + \left(\frac{d}{dt}Y_{rabbit}(t)\right)^2 + \left(\frac{d}{dt}Z_{rabbit}(t)\right)^2}} \quad (3)$$

The resulting formulas (3) are substituted into formulas (2) to define a new dynamic local basis. $\vec{E}_1(T)$, $\vec{E}_2(T)$, $\vec{E}_3(T)$. The coordinates $\overrightarrow{Center}_{dynamic}(T)$ of the center of the local coordinate system are also known to us. Therefore, preliminary work on the construction of a dynamic coordinate system can be considered complete.

3 "Foxes" Trajectory in the "Rabbit" Dynamic Coordinate System

Now we need to find out what the "Foxes" trajectory will look like on a surface defined as $z = Surface(x, y)$. The surface specified in the form $\vec{R}_{land}(x, y) = \begin{bmatrix} x \\ y \\ Surface(x, y) \end{bmatrix}$, when moving to a dynamic local coordinate system $\vec{E}_1(T)$, $\vec{E}_2(T)$, $\vec{E}_3(T)$ with a center at a point $\overrightarrow{Center}_{dynamic}(T)$ $\vec{R}_{rot}(T) = \begin{bmatrix} X_{rot}(T) \\ Y_{rot}(T) \\ Z_{rot}(T) \end{bmatrix}$ will look like the figure (Fig. 2).

All points of the basis on which the base surface was built were built in the basis $\vec{E}_1(T)$, $\vec{E}_2(T)$, $\vec{E}_3(T)$ with a center at a point $\overrightarrow{Center}_{dynamic}(T)$ (Fig. 2). Referring to [12] you can see how the original surface transforms over time. Also, pay attention to the coordinates of the current position of the "Rabbit". As expected, the abscissa and ordinate are equal zero, only the applicate changes over time. The coordinates of the current position of the "Rabbit" and "Foxes", as well as the coordinates of the points of

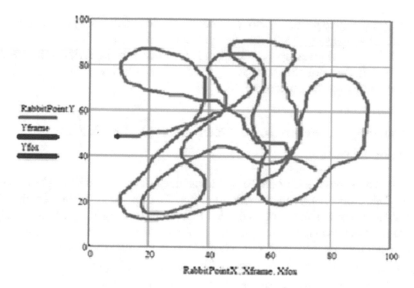

Fig. 3. Persecution in the world Coordinate System

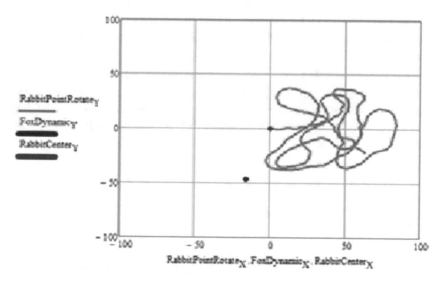

Fig. 4. Persecution in a dynamic Coordinate System

the "Rabbit" trajectory, were calculated in the same way as in formula (4). Referring to [10] You can see a pursuit picture of the Fox and Rabbit's, the points of the Rabbit's trajectory, and also the points of the original surface. For further work, we will need to see the picture of the pursuit on the plane, in the horizontal projection. Referring to [13] presents a picture, both in the dynamic coordinate system and in the world. In Figs. 3 and 4, we can see fragments of the first frame of the video that shown by reference [13].

We see a comparative picture of motion from the world coordinate system and from the Rabbit coordinate system on the horizontal plane of projections.

Analysis (Fig. 4) gives us a mechanism for modeling the behavior of the "Rabbit" and "Foxes". In the sense that now "Rabbit" can choose which way to move to the right or left depending on the position of "Fox".

4   Modeling the Behavior of "Rabbit" and "Fox". Situation on the Plane

We now need to consider the "flat" case for visibility, demonstrating an approach to the choice of the Rabbit movement model. Let the "Fox" pursue "Rabbit" on the plane. "Rabbit" moves in a plane with a constant in module velocity V_{rabbit}. Figure 5 shows schematically the change in the dynamic local coordinate system associated with the Rabbit at the time points T and $T + \Delta T$. At time T, the dynamic coordinate system is given by the vectors $\vec{E}_1$ и $\vec{E}_2$, the center of coordinates is given by point $\vec{C}$. At the time, $T + \Delta T$ the coordinate system of the "rabbit" moves to the position $\vec{E}_1'$, $\vec{E}_2'$ centered at $\vec{C}'$. Figure 5 presents the situation in the global coordinate system $\vec{H}_1$, $\vec{H}_2$. Next, we need to go to the Rabbit coordinate system $\vec{E}_1$, $\vec{E}_2$, $\vec{C}$ (Fig. 6).

We assumed that the Fox is approaching the Rabbit from the **II** quarter of the plane (Fig. 6). As already assumed, "Rabbit" is located at $\vec{C}$. Our time is divided into fairly small intervals. ΔT. During this time, "Rabbit" goes from the position $\vec{C}$ to position $\vec{C}'$: $|\vec{C} - \vec{C}'| = V_{rabbit} \cdot \Delta T$. The line connecting the points $\vec{R}_{fox}$ and $\vec{C}$ makes up the direction of the vector $\vec{E}_1$ and angle α. But in our model, "Rabbit" cannot change the direction of movement by an angle α over a period of time ΔT. We assume that the angle $\alpha > \alpha_0$, where α_0 - some threshold value. If "Rabbit" will have an angular

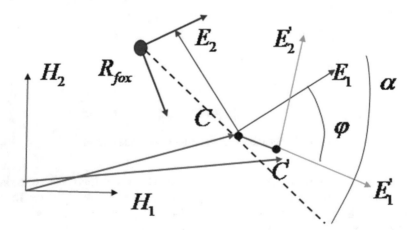

Fig. 5. Changing the Coordinate System in the next point in time

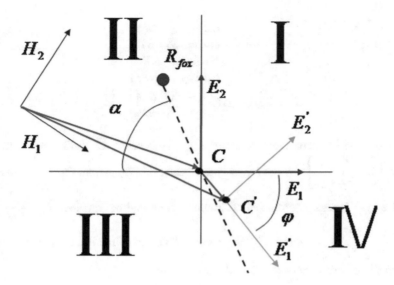

Fig. 6. "Rabbit" coordinate system

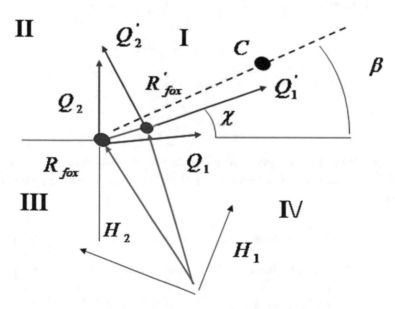

Fig. 7. Coordinate system "Foxes"

velocity of rotation ω_{rabbit}, then the velocity vector will turn through an angle $\varphi = \omega_{rabbit} \cdot \Delta T$. We have that in the coordinate system $\vec{E}_1, \vec{E}_2, \vec{C}$, coordinates of the vectors $\vec{E}_1', \vec{E}_2', \vec{C}'$ will be expressed as follows:

$$\vec{E}_1 = \begin{bmatrix} \cos(\omega_{rabbit} \cdot \Delta T) \\ -\sin(\omega_{rabbit} \cdot \Delta T) \end{bmatrix}$$

$$\vec{E}_2 = \begin{bmatrix} \sin(\omega_{rabbit} \cdot \Delta T) \\ \cos(\omega_{rabbit} \cdot \Delta T) \end{bmatrix} \tag{4}$$

$$\vec{C'} = V_{rabbit} \cdot \Delta T \cdot \begin{bmatrix} \cos(\omega_{rabbit} \cdot \Delta T) \\ -\sin(\omega_{rabbit} \cdot \Delta T) \end{bmatrix}$$

In the world coordinate system $\vec{H}_1$, $\vec{H}_2$ vector $\vec{C'}$ will look like this: $\vec{C}_{new} = \begin{bmatrix} \vec{C'} \cdot \vec{H}_1' + x_r \\ \vec{C'} \cdot \vec{H}_2' + y_r \end{bmatrix}$. Where $\vec{H}_1'$, $\vec{H}_2'$ basis vector $\vec{H}_1$, $\vec{H}_2$ in coordinate system $\vec{E}_1$, $\vec{E}_2$, $\vec{C}$, and x_r, y_r - current coordinates of "Rabbit". It should be clarified: $\vec{E}_1 = \frac{1}{V_{rabbit}} \begin{bmatrix} V_x \\ V_y \end{bmatrix}$, $\vec{E}_2 = \frac{1}{V_{rabbit}} \begin{bmatrix} -V_y \\ V_x \end{bmatrix}$, are formed by the current direction of the Rabbit's speed. And the components of the vectors $\vec{H}_1'$, $\vec{H}_2'$ after calculations:

$$\vec{H}_1' = \begin{bmatrix} \vec{H}_1 \cdot \vec{E}_1 \\ \vec{H}_1 \cdot \vec{E}_2 \end{bmatrix} = \frac{1}{V_{rabbit}} \begin{bmatrix} [1 \ 0] \begin{bmatrix} V_x \\ V_y \end{bmatrix} \\ [1 \ 0] \begin{bmatrix} -V_y \\ V_x \end{bmatrix} \end{bmatrix} = \frac{1}{V_{rabbit}} \begin{bmatrix} V_x \\ -V_y \end{bmatrix}$$

$$\vec{H}_2' = \begin{bmatrix} \vec{H}_2 \cdot \vec{E}_1 \\ \vec{H}_2 \cdot \vec{E}_2 \end{bmatrix} = \frac{1}{V_{rabbit}} \begin{bmatrix} [0 \ 1] \begin{bmatrix} V_x \\ V_y \end{bmatrix} \\ [0 \ 1] \begin{bmatrix} -V_y \\ V_x \end{bmatrix} \end{bmatrix} = \frac{1}{V_{rabbit}} \begin{bmatrix} V_y \\ V_x \end{bmatrix}$$

$$\vec{V}_{rabbit} = \begin{bmatrix} V_x \\ V_y \end{bmatrix}$$

In more detail this moment is described on the resource [14]. In the final form, vectors $\vec{E}_1'$, $\vec{E}_2'$ and a new coordinate center $\vec{C'}$ in the world coordinate system $\vec{H}_1$, $\vec{H}_2$ will look like:

$$\vec{C}_{new} = \begin{bmatrix} \Delta T \cdot [\cos(\omega_{rabbit} \cdot \Delta T) \quad -\sin(\omega_{rabbit} \cdot \Delta T)] \cdot \begin{bmatrix} V_x \\ -V_y \end{bmatrix} + x_r \\ \Delta T \cdot [\cos(\omega_{rabbit} \cdot \Delta T) \quad -\sin(\omega_{rabbit} \cdot \Delta T)] \cdot \begin{bmatrix} V_y \\ V_x \end{bmatrix} + y_r \end{bmatrix}$$

$$\vec{E}_{1_{new}} = \begin{bmatrix} \frac{1}{V_{rabbit}} \cdot [\cos(\omega_{rabbit} \cdot \Delta T) \quad -\sin(\omega_{rabbit} \cdot \Delta T)] \cdot \begin{bmatrix} V_x \\ -V_y \end{bmatrix} \\ \frac{1}{V_{rabbit}} \cdot [\cos(\omega_{rabbit} \cdot \Delta T) \quad -\sin(\omega_{rabbit} \cdot \Delta T)] \cdot \begin{bmatrix} V_y \\ V_x \end{bmatrix} \end{bmatrix} \tag{5}$$

$$\vec{E}_{2_{new}} = \begin{bmatrix} \frac{1}{V_{rabbit}} \cdot [\sin(\omega_{rabbit} \cdot \Delta T) \quad \cos(\omega_{rabbit} \cdot \Delta T)] \cdot \begin{bmatrix} V_x \\ -V_y \end{bmatrix} \\ \frac{1}{V_{rabbit}} \cdot [\sin(\omega_{rabbit} \cdot \Delta T) \quad \cos(\omega_{rabbit} \cdot \Delta T)] \cdot \begin{bmatrix} V_y \\ V_x \end{bmatrix} \end{bmatrix}$$

Formulas (5) are applicable in the event that the "Fox" $\vec{R}_{fox} = \begin{bmatrix} x_{fox} \\ y_{fox} \end{bmatrix}$ in dynamic coordinate system $\vec{E}_1$, $\vec{E}_2$, $\vec{C}$ approaching from the **II** quarter (See Fig. 6), provided $\propto \, > \propto_0$ some threshold

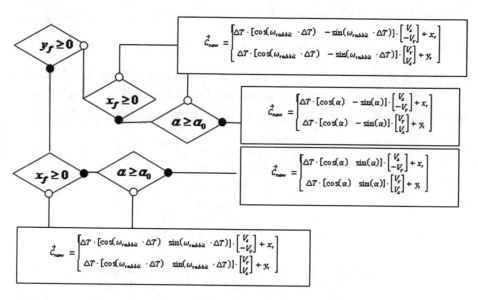

Fig. 8. Coordinates of the new center of coordinates $\vec{C}_{new}$ in the world coordinate system

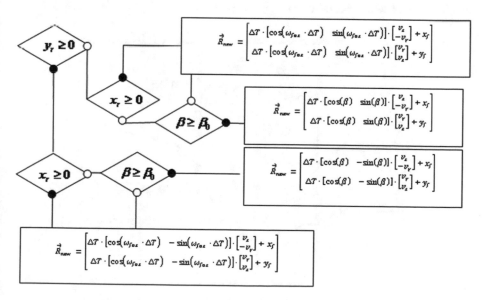

Fig. 9. Coordinates of the new position "Foxes" $\vec{R}_{new}$ in the world coordinate system

Figure 8 represents possible values $\vec{C}_{new}$ depending on which quarter the Foxes approach comes from and the angle $\propto$.

Similar reasoning can be carried out in relation to the behavior of "Fox". Let's create a coordinate system $\vec{Q}_1$, $\vec{Q}_2$, $\vec{R}_{fox}$ (Fig. 7). As before, the vectors of the "Fox" basis are formed by its direction of movement: $\vec{Q}_1 = \frac{1}{v_{fox}} \cdot \begin{bmatrix} v_x \\ v_y \end{bmatrix}$, $\vec{Q}_2 = \frac{1}{v_{fox}} \cdot \begin{bmatrix} -v_y \\ v_x \end{bmatrix}$, $\vec{R}_{fox} = \begin{bmatrix} x_f \\ y_f \end{bmatrix}$, $\vec{v}_{fox} = \begin{bmatrix} v_x \\ v_y \end{bmatrix}$. In the case shown in Picture (Fig. 7), in coordinate system $\vec{Q}_1$, $\vec{Q}_2$, $\vec{R}_{fox}$ Rabbit is in the first quarter. The line connecting "Fox" and "Rabbit" is with the vector $\vec{Q}_1$ angle β. At the expiration of the interval ΔT, "Fox" will move a distance $\left| \vec{R}_{fox} - \vec{R}'_{fox} \right| = v_{fox} \cdot \Delta T$. In the direction of the vector $\vec{Q}'_1$:

$$
\vec{Q}'_1 = \begin{bmatrix} cos(\omega_{fox} \cdot \Delta T) \\ sin(\omega_{fox} \cdot \Delta T) \end{bmatrix}
$$
$$
\vec{Q}'_2 = \begin{bmatrix} -sin(\omega_{fox} \cdot \Delta T) \\ cos(\omega_{fox} \cdot \Delta T) \end{bmatrix} \tag{6}
$$
$$
\vec{R}'_{fox} = v_{fox} \cdot \Delta T \cdot \begin{bmatrix} cos(\omega_{fox} \cdot \Delta T) \\ sin(\omega_{fox} \cdot \Delta T) \end{bmatrix}
$$

Reminder that formulas (6) are expressed in the coordinate system $\vec{Q}_1$, $\vec{Q}_2$, $\vec{R}_{fox}$. Then, as in the case of the "Rabbit", we can present (See Fig. 9) the coordinates of the new position of the Fox $\vec{R}_{new}$ in the world coordinate system (β_0 - threshold value for angle β, Fig. 7). Figure 10 shows the results of the program on algorithms, the development of which is shown in Figs. 8 and 9.

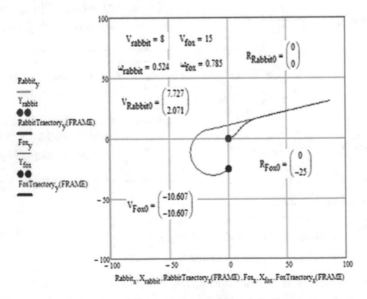

Fig. 10. The results of the program. Modeling behavior on the plane

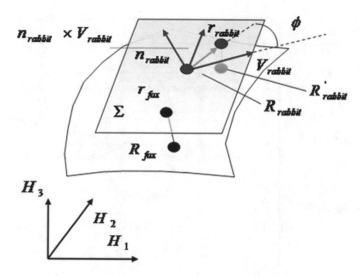

Fig. 11. Model movement on the surface ("Rabbit")

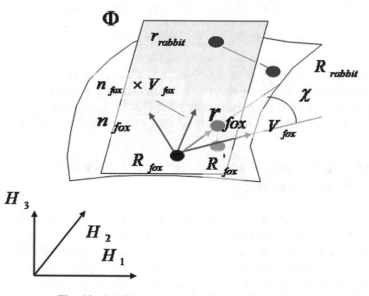

Fig. 12. Model movement on the surface ("Fox")

In addition, we created a playlist [15], in which the dependence of the situation of the pursuit on the initial data is partially displayed, be it the relative position, the angles of the speeds and their absolute values.

Figure 13 show the first frame of one of the video playlists [15], where the "Fox" speed exceeds the "Rabbit" speed. And the angular velocity of the Fox is greater than the angular velocity of the Rabbit. Note that the angular velocity in the proposed

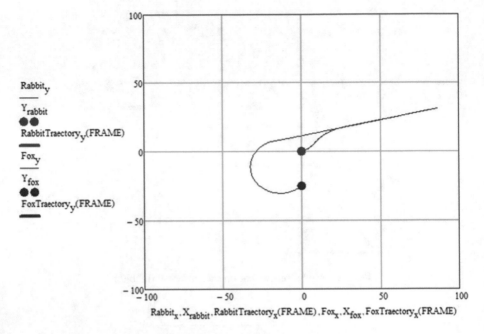

Fig. 13. Model of motion on a plane

model, the magnitudes are constant. With this pattern of behavior, "Rabbit" and "Foxes", we see that over time, this turns into pursuit in a straight line. The program code can be downloaded from the author's site [13] by link «Archive of the program "Rabbit" and "Fox" on the plane. "Rabbit" runs away, "Fox" is catching up.

5   Modeling the Behavior in the Problem of Pursuit on a Rugged Surface

We created models of behavior for Fox and Rabbit on the plane. Similar reasoning can be carried out in three-dimensional space, when modeling the problem of pursuit on the surface. The simplest thing to do is to project the process of solving the problem on the plane onto the surface, which was done in the next video [16].

Figure 14 shows the first frame of the video [16]. As you can see, there was a simple projection of the trajectory shown in Fig. 13, to the surface. The program code itself with a projection on the horizontal plane can be downloaded from the author's website [9] by the link «Archive of the program "Rabbit" and "Fox" on the surface. Vertical projection from plane to surface».

But a simple projection from the plane to the surface will not satisfy the conditions for moving objects along the surface at a constant speed.

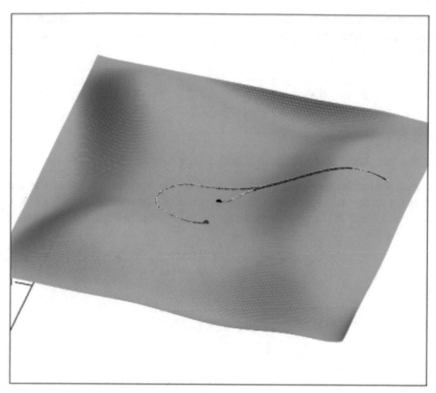

SurfacePicture$_1$ · (Rabbit$_x$ · Rabbit$_y$ · Rabbit$_z$) · (Fox$_x$ · Fox$_y$ · Fox$_z$) · (X$_{rabbit}$ · Y$_{rabbit}$ · Z$_{rabbit}$) · (X$_{fox}$ · Y$_{fox}$ · Z$_{fox}$)

Fig. 14. Vertical projection of the pursuit task

Consider the movement of the "Rabbit" on the surface. The figure (Fig. 11) shows the tangent plane Σ to the point of the surface, where the Rabbit is at a certain point in time. At this point, the local dynamic coordinate system is entered: $\left[\dfrac{\vec{V}_{rabbit}}{|\vec{V}_{rabbit}|} \quad \dfrac{\vec{n}_{rabbit} \times \vec{V}_{rabbit}}{|\vec{n}_{rabbit} \times \vec{V}_{rabbit}|} \quad \vec{n}_{rabbit} \right]$, where $\vec{V}_{rabbit}$ - Rabbit speed, and $\vec{n}_{rabbit}$ − normal to plane Σ. As before, Rabbit will analyze the position of the Fox to take the next step. For this, an orthogonal projection $\vec{r}_{fox}$ is constructed from the position point $\vec{R}_{fox}$ of the Fox on the plane Σ (Fig. 11). The regarding point $\vec{r}_{fox}$ uses the same logic as in the flowchart shown in Fig. 8. The orthogonal projection of the Fox position $\vec{r}_{fox}$ on the plane Σ calculated as follows:

$$\vec{r}_{fox} = \vec{R}_{fox} + \frac{\left(\vec{R}_{rabbit} - \vec{R}_{fox} \right) \cdot \vec{n}_{rabbit}}{\vec{n}_{rabbit} \cdot \vec{n}_{rabbit}} \cdot \vec{n}_{rabbit} \tag{7}$$

In order to analyze the components of the vector "Fox" (7), the projection of the vector $\vec{r}_{fox}$ it is necessary to convert to the local dynamic coordinate system of the Rabbit:

$$\begin{bmatrix} x_f \\ y_f \end{bmatrix} = \begin{bmatrix} \left(\vec{r}_{fox} - \vec{R}_{rabbit} \right) \cdot \frac{\vec{V}_{rabbit}}{\left[\vec{V}_{rabbit} \right]} \\ \left(\vec{r}_{fox} - \vec{R}_{rabbit} \right) \cdot \frac{\vec{n}_{rabbit} \times \vec{V}_{rabbit}}{\left| \vec{n}_{rabbit} \times \vec{V}_{rabbit} \right|} \end{bmatrix} \qquad (8)$$

After this transformation, we can analyze the coordinates $\begin{bmatrix} x_f \\ y_f \end{bmatrix}$ according to the algorithm shown in the figure (Fig. 8), to select the next movement of the Rabbit. Similar reasoning we can produce in relation to the movement "Foxes". Orthogonal projection of the Rabbit's position point $\vec{r}_{rabbit}$ (9) on plane Φ (Fig. 12) is calculated as:

$$\vec{r}_{rabbit} = \vec{R}_{rabbit} + \frac{\left(\vec{R}_{fox} - \vec{R}_{rabbit} \right) \cdot \vec{n}_{fox}}{\vec{n}_{fox} \cdot \vec{n}_{fox}} \cdot \vec{n}_{fox} \qquad (9)$$

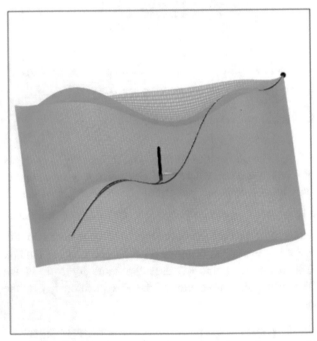

$SurfacePicture_2, NormalPicture_2, SpeedPicture_2, OrdinataPicture_2, (X_{rabbit}, Y_{rabbit}, Z_{rabbit}), (X_{fox}, Y_{fox}, Z_{fox}), (FrFox_X, FrFox_Y, FrFox_Z)$

Fig. 15. Model of motion on the surface

We can transform the projection of the Rabbit position (9) into the Fox system of coordinates (10):

$$\begin{bmatrix} x_r \\ y_r \end{bmatrix} = \begin{bmatrix} \left(\vec{r}_{rabbit} - \vec{R}_{fox} \right) \cdot \frac{\vec{V}_{fox}}{\left[\vec{V}_{fox} \right]} \\ \left(\vec{r}_{rabbit} - \vec{R}_{fox} \right) \cdot \frac{\vec{n}_{fox} \times \vec{V}_{fox}}{\left| \vec{n}_{fox} \times \vec{V}_{fox} \right|} \end{bmatrix} \tag{10}$$

Analyze the position of "Rabbit" and make the appropriate decision regarding the direction of movement "Fox" will be able according to the algorithm described in Fig. 9. On video [17] you can see the result of the program in the system «MathCAD», written according to the algorithm outlined in this paragraph. Figure 15 shows the first frame video [17]. The pursuit according to the algorithm of this paragraph also comes down to pursuing along the "Rabbit" trajectory, and "Rabbit" simply turns its back on the "Fox". The program code itself is posted on the author's website [9] in the section "Patterns of behavior in the pursuit task" - "Patterns of behavior of the Rabbit and Foxes. It can also be downloaded from the link. «Archive of the program Models of behavior "Rabbit" and "Foxes" on the surface».

6 Conclusions

This article proposes mathematical models of the behavior of the pursuing object ("Foxes") and the pursued ("Rabbit"). Models of object behavior on the plane and on the surface are proposed. The basis of the behavior models is the input of dynamic coordinate systems, both for the Fox and for the Rabbit. Coordinate systems are formed by velocity vectors. The main point is to enter the angular velocity of rotation as the "inertness" of the movement. A further prospect for the development of the proposed models is the introduction of the dependence between the angular velocity and the velocity of motion. In other words, the higher the speed, the greater the inertia. We believe that the results proposed in this article can be claimed by the developers of autonomous robotic systems that perform the pursuit tasks.

When working on this article, mathematical modeling in the computer mathematics system "MathCAD" was chosen as the research methods. The mathematical models of the movement of the "Rabbit" and "Foxes" [4] on the plane and on the surface were revised. And for these adapted models for the systems of computer mathematics, algorithms of motion, analysis of the coordinates of the opponent and decision-making were compiled. The results, in the form of animated images, were posted on the author's channel in http://www.youtube.com.

Acknowledgement. The article was funded by innovative grant of the Buryat State University in 2019 "Hardware and software complex of persecution".

References

1. Dubanov, A.A.: The task of pursuit. Solution in the system of computational mathematics MathCAD. Inf. Technol. **4**(24), 251–255 (2018)
2. Isaacs, R.: Differential Games. World, Moscow, Russia (1967)
3. Pontryagin, L.S.: Linear differential runaway game. Tr. MIAN USSR **112**, 30–63 (1971)
4. Krasovsky, N.N., Subbotin, A.I.: Positional Differential Games. Science, Moscow, Russia (1974)
5. Burdakov, S.V., Sizov, P.A.: Algorithms for controlling the movement of a mobile robot in the pursuit task. Scientific and technical statements of the St. Petersburg State Polytechnic University. Comput. Sci. Telecommun. Control **6**(210), 49–58 (2014)
6. Zhelnin, Yu.N.: Linearized pursuit and evasion problem on the plane. TsAGI Rec. **3**(8), 88–98 (1977)
7. Simakova, E.N.: On a differential pursuit game. Automation and Remote Control 2 (1967)
8. Merz, A.W.: The game of two identical cars. J. Optim. Appl. **5**(9), 324–343 (1972)
9. Section "Model of Behavior in the Task Prosecution". http://dubanov.exponenta.ru
10. Video "Change of Base, Prosecution". https://www.youtube.com/watch?v=L5Z0MCDIlEs
11. Section "Fox and Rabbit". http://dubanov.exponenta.ru
12. Video "Dynamic Basis". https://youtu.be/KsMZ9Zy8XRs
13. Video "Change of Base, Pursuit, Plane". https://youtu.be/1nNlN-U8WyY
14. Section "Base Conversion". http://dubanov.exponenta.ru/russian/book_rus.htm
15. Video playlist: "The Task of Prosecution. Behavior Model (adaptive)". https://www.youtube.com/playlist?list=PLfGCUhhiz5wuPh-rKHWYvi-RZf9M215bq
16. Video "A Simple Projection of Pursuit From Plane to Surface". https://youtu.be/BAew7xTw-iU
17. Video "Models of behavior" Rabbit "and" Foxes "on the Surface". https://www.youtube.com/watch?v=QrKpJoiyB0o

Studying the Schrödinger Equation, Implementation of Qubits and Visualization of Matrices Using Matlab

Alexey Samoylov, Sergey Gushanskiy,
and Natalia Korobeynikova[(⊠)]

Department of Computer Engineering, Southern Federal University,
Taganrog, Russia
{asamoylov, smgushanskiy}@sfedu.ru, natali_dce@mail.ru

Abstract. In this paper, we consider two key concepts that form the basis of quantum physics, in particular, the Schrödinger equation and the Born probability equation. A mental experiment that allows one to show the incompleteness of quantum mechanics in the transition from subatomic systems to macroscopic systems is also analyzed. It is proposed to use the MATLAB package, which makes it possible to solve the time-dependent Schrödinger equation for the infinite rectangular potential well of a harmonic oscillator, as well as for potentials that can be solved only numerically. The relevance of these studies is to find methods for the eigenvalues definition of the Schrödinger equation. The topicality of this research is explained either by the large number of both experimental and theoretical works devoted to these issues. The aim of the work is to determine the methods for finding the eigenvalues of the Schrödinger equation using the MATLAB package, which will allow to visualize the density matrix for a different number of qubits of the quantum scheme.

Keywords: Schrödinger equation · Quantum computing · Simulation ·
Density matrix · MATLAB

1 Introduction

The Schrödinger equation and the Born equation of probability are two key concepts that form the basis of quantum physics. In 1930, Dirac introduced the concept of sconces and ket for the quantum states description, which emphasizes and explains the role of inner products and linear functions of the space. These two equations are fundamental to understanding modern quantum mechanics [3]. The Schrödinger equation demonstrates how the state of a particle ψ changes with time. In the description of the bra-ket it looks like

$$i\hbar \frac{d}{dt}|\psi\rangle = H|\psi\rangle, \tag{1}$$

where H is the Hamiltonian operator, and $|\psi\rangle$ is a ket or a column vector represented by the quantum state of the particle.

© Springer Nature Switzerland AG 2019
R. Silhavy (Ed.): CSOC 2019, AISC 984, pp. 275–282, 2019.
https://doi.org/10.1007/978-3-030-19807-7_27

The Hamiltonian in quantum theory is the operator of the total energy of the system. Its spectrum is the set of possible values when measuring the total energy of a system. The Hamiltonian spectrum can be discrete or continuous. There may also be a situation (for example, for the Coulomb potential) when the spectrum consists of a discrete and continuous part. Since energy is a real quantity, the Hamiltonian is a self-adjoint operator.

2 Materials and Methods

The famous mental experiment is also associated with the name of Schrödinger. With this fictional experience, the scientist wanted to show the incompleteness of quantum mechanics in the transition from subatomic systems to macroscopic systems.

According to quantum mechanics, if no observation is made on the atomic nucleus, then its state is described by mixing two states – a decaying nucleus and a non-decaying nucleus, therefore, the cat sitting in the box and personifying the atomic nucleus is both alive and dead (Fig. 1).

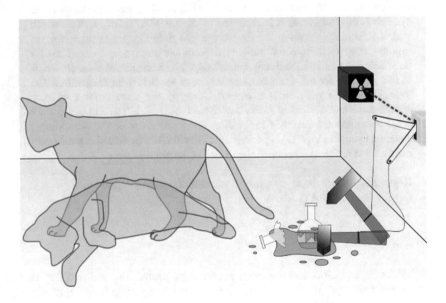

Fig. 1. Interpretation of the experiment with Schrödinger's cat

If the box is opened, then the experimenter can see only one specific state – "the nucleus disintegrated, the cat is dead" or "the nucleus has not disintegrated, the cat is alive". In other words:

1. There are a box and a cat. In the box there is a mechanism containing a radioactive atomic nucleus and a container with a poisonous gas. The parameters of the experiment are chosen so that the probability of disintegration of the nucleus in 1 h is 50%. If the nucleus disintegrates, the gas tank opens and the cat dies. If the decay of the nucleus does not occur – the cat remains alive.

2. Close the cat in the box, wait an hour and ask ourselves: is the cat alive or dead?
3. Quantum mechanics as it tells us that the atomic nucleus is in all possible states at the same time. Before we opened the box, the cat – nucleus system is in the "decayed, the cat is dead" state with a probability of 50% and the "core has not disintegrated, the cat is alive" with a 50% probability. It turns out that the cat sitting in the box is both alive and dead at the same time.
4. According to the modern Copenhagen interpretation, the cat is alive/dead without any intermediate states. And the choice of the state of decay of the nucleus occurs not at the moment of opening the box, but also when the nucleus enters the detector. Because the reduction of the wave function of the "cat-detector-nucleus" system is not connected with the human observer of the box, but is connected with the detector-observer of the nucleus.

The MATLAB package makes it possible to solve the time-dependent Schrödinger equation for an infinite rectangular well potential of a harmonic oscillator, as well as for potentials that can be solved only numerically. In order to minimize the amount of RAM, we will use sparse matrices, which store only nonzero elements [4, 5].

When performing numerical calculations, it is important to minimize the effect of rounding errors by selecting modules so that the variables used in the simulation are of the order of unity [1].

An alternative approach used in research is to convert the equation to a dimensionless variable, for example, by re-scaling the energy. To make an equation dimensionless, you need to express it in terms of some characteristic energy [2].

Solution $|\psi_E(t)\rangle$ is a subset of all possible solutions of Eq. (1). Fortunately, we can build a general solution using linear superposition

$$|\psi(t)\rangle = \sum_E a_E |\psi_E(0)\rangle \exp(-iEt/\hbar) \tag{2}$$

where a_E is the sum of the constants of all possible values of E. The main difference of this solution is that the probability density does not depend on time, in contrast to the density obtained from general solutions, which depends on time [5]. A quantum transformation is a unitary state transformation of a quantum system, which uniquely describes the evolution of a system and in which orthogonal states remain orthogonal [18]. In other words, a linear transformation U is unitary if and only if it is a linear operator preserving the scalar product:

$$\langle x \mid y \rangle = \langle Ux|U|y\rangle = \langle x|U^\dagger U|y\rangle, \tag{3}$$

where $|x\rangle$ and $|y\rangle$ arbitrary states, $U^\dagger$ is the linear adjoint operator U and

$$U^\dagger U = I, \tag{4}$$

where I is the identity operator.

In a finite-dimensional linear complex space, any linear operator can be represented by a matrix. In these terms

$$U^{\dagger} = U^{T*}, \tag{5}$$

and the identity operator will be described by the identity matrix. An important consequence of the fact that quantum transformations are unitary is their reversibility. The repeated action of a unitary operator on an arbitrary quantum system transfers it to the initial state. An important consequence of the fact that quantum transformations are linear is that their action is completely determined by the action on the vectors of an orthonormal basis representing the selected states of the system.

The tensor product of operators gives the operator acting in the tensor product of the spaces of its factors, that is,

$$U_0 \otimes U_1 |x\rangle \otimes |y\rangle = U_0 |x\rangle \otimes U_1 |y\rangle. \tag{6}$$

Thus, we obtain operators expressing simultaneous, heterogeneous effects on different parts of a quantum register. Next the elementary quantum operators most commonly used in quantum computing algorithms will be given (Table 1).

Table 1. Elementary quantum operators

Operator	Matrix	Action
Identical transform I	$\begin{pmatrix} 1 & 0 \\ 0 & 1 \end{pmatrix}$	$\begin{cases} \|0\rangle \to \|0\rangle \\ \|1\rangle \to \|1\rangle \end{cases}$
Invert X	$\begin{pmatrix} 0 & 1 \\ 1 & 0 \end{pmatrix}$	$\begin{cases} \|0\rangle \to \|1\rangle \\ \|1\rangle \to \|0\rangle \end{cases}$
Phase shift Y with inversion	$\begin{pmatrix} 0 & 1 \\ -1 & 0 \end{pmatrix}$	$\begin{cases} \|0\rangle \to -\|1\rangle \\ \|1\rangle \to \|0\rangle \end{cases}$
Phase shift Z	$\begin{pmatrix} 1 & 0 \\ 0 & -1 \end{pmatrix}$	$\begin{cases} \|0\rangle \to \|0\rangle \\ \|1\rangle \to -\|1\rangle \end{cases}$
Operator Walsh-Hadamard H	$\frac{1}{\sqrt{2}} \begin{pmatrix} 1 & 1 \\ 1 & -1 \end{pmatrix}$	$\begin{cases} \|0\rangle \to \dfrac{1}{\sqrt{2}} \\ \|1\rangle \to \dfrac{1}{\sqrt{2}} \end{cases}$
Controlled Inversion CN	$H \otimes I \cdot Z_2 \cdot H \otimes I = \begin{pmatrix} 1 & 0 & 0 & 0 \\ 0 & 1 & 0 & 0 \\ 0 & 0 & 0 & 1 \\ 0 & 0 & 1 & 0 \end{pmatrix}$	$\begin{cases} \|00\rangle \to \|00\rangle \\ \|01\rangle \to \|01\rangle \\ \|10\rangle \to \|11\rangle \\ \|11\rangle \to \|10\rangle \end{cases}$
Controlled Inversion CN	$I \otimes H \cdot Z_2 \cdot I \otimes H = \begin{pmatrix} 1 & 0 & 0 & 0 \\ 0 & 0 & 0 & 1 \\ 0 & 0 & 1 & 0 \\ 0 & 1 & 0 & 0 \end{pmatrix}$	$\begin{cases} \|00\rangle \to \|00\rangle \\ \|01\rangle \to \|11\rangle \\ \|10\rangle \to \|10\rangle \\ \|11\rangle \to \|01\rangle \end{cases}$
Controlled inversion	$H \otimes I \otimes I \cdot Z_3 \cdot H \otimes I \otimes I =$ $\begin{pmatrix} 1 & 0 & 0 & 0 & 0 & 0 & 0 & 0 \\ 0 & 1 & 0 & 0 & 0 & 0 & 0 & 0 \\ 0 & 0 & 1 & 0 & 0 & 0 & 0 & 0 \\ 0 & 0 & 0 & 1 & 0 & 0 & 0 & 0 \\ 0 & 0 & 0 & 0 & 1 & 0 & 0 & 0 \\ 0 & 0 & 0 & 0 & 0 & 1 & 0 & 0 \\ 0 & 0 & 0 & 0 & 0 & 0 & 0 & 1 \\ 0 & 0 & 0 & 0 & 0 & 0 & 1 & 0 \end{pmatrix}$	$\begin{cases} \|000\rangle \to \|000\rangle \\ \|001\rangle \to \|011\rangle \\ \|010\rangle \to \|010\rangle \\ \|011\rangle \to \|001\rangle \\ \|100\rangle \to \|100\rangle \\ \|101\rangle \to \|111\rangle \\ \|110\rangle \to \|111\rangle \\ \|111\rangle \to \|110\rangle \end{cases}$

The operators discussed above and many other operators can be successfully simulated in MATLAB environment. Let us give an example of setting qubits and matrix visualization.

Explicit definition of the first qubit:

$Q1 = [1; 0]$

Explicit definition of the second qubit:

$Q2 = [0; one]$

Visualization of two qubit density matrices (Fig. 2, upper left diagrams):

RD1 = 10 00; RD2 = 0001
»Subplot (1,2,1); bar3 (RD1);
»Subplot (1,2,2); bar3 (RD2);

Applying the Hadamard transducer to both qubits gives the following result:

HQ1 = 0.7071, 0.7071; HQ2 = 0.7071, −0.7071

Visualization of the result (Fig. 2, upper right diagrams)

RD1_H = 0.5000, 0.5000, 0.5000, 0.5000
RD2_H = 0.5000, −0.5000, −0.5000, 0.5000

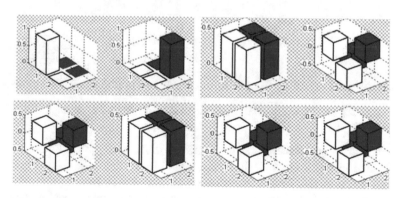

Fig. 2. Visualization of two qubit density matrices (upper left diagrams); visualization of the result of applying the Hadamard transducer to both qubits (upper right diagrams); visualization of the use of controlled exchange (lower left diagrams); visualization of the application of controlled NOT (lower right diagrams)

The use of controlled exchange with visualization of the result (Fig. 2, lower left diagrams)

RD1_H = 0.5000, −0.5000, −0.5000, 0.5000
RD2_H = 0.5000, 0.5000, 0.5000, 0.5000
»Subplot (1,2,1); bar3 (RD1_H);
»Subplot (1,2,2); bar3 (RD2_H);

The use of a controlled NOT result with visualization (Fig. 2, lower right diagrams)

RD1_H = 0.5000, −0.5000, −0.5000, 0.5000
RD2_H = 0.5000, −0.5000, −0.5000, 0.5000
»Subplot (1,2,1); bar3 (RD1_H);
»Subplot (1,2,2); bar3 (RD2_H).

A special place among the instrumental applications is occupied by the Simulink visual modeling system. Models development in Simulink (so called s-models) is based on the Drag-and-Drop technology. The blocks included in the model being created can be connected to each other both in terms of information and control. The type of communication depends on the type of block and the logic of the model. Data can be scalar values, vectors or matrices of arbitrary dimension. S-model can have a hierarchical structure, the number of hierarchy levels is almost unlimited. The view of the CNOT converter is shown in Fig. 3.

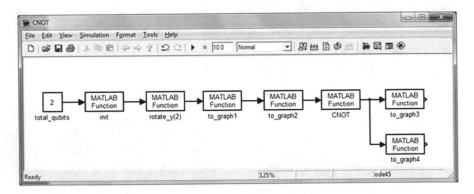

Fig. 3. CNOT converter implemented in the Simulink package

The number of qubits to be created is entered in the total_qubits block. The initial state is formed in the init block. The rotate_y (2) block rotates the second qubit along the y axis so that the CNOT transformation can be observed. The to_Graph1, to_Graph2, to_Graph3 and to_Graph4 blocks display the state of qubits before and after the conversion. The CNOT block performs a NOT controlled. An example of the model is shown in Fig. 4. The first two diagrams show the initial states of the qubits, and the third and fourth states show the states after the CNOT transformation.

In this case, the second qubit is the controlling one, so no changes have occurred to it. The first qubit is controlled; it has changed its state.

Traditionally, to solve the problem of finding the eigenvalues of the Schrödinger equation, the false position method is used. The idea of the false position method is as follows. Suppose one of the bound states is searched for as the desired value, therefore, the negative eigenvalue is chosen as the trial initial energy value. We integrate the Schrödinger equation by any known numerical method on the interval. In the course of integration in the direction of large values, the solution $\psi(x)$, which grows exponentially

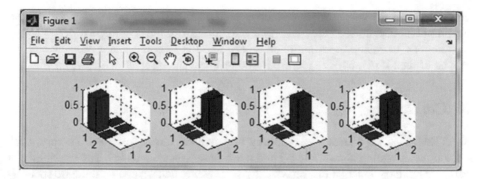

Fig. 4. Four diagrams in the Simulink package

within the classically forbidden region, is first calculated. After passing through the turning point, which limits the area of motion permitted by classical mechanics to the left, the solution of the equation becomes oscillating. If we continue the integration further beyond the right turning point, then the solution becomes numerically unstable [5, 6].

It is valid due to the fact that even with the exact choice of an eigenvalue, it can always contain some admixture of an exponentially growing solution that does not have physical content. The noted circumstance is a general rule: integration inward direction of a field prohibited by classical mechanics will be inaccurate. Therefore, for each energy value, it is more reasonable to calculate one more solution integrating Eq. (2) from downwards. The criterion for the coincidence of this energy value is the

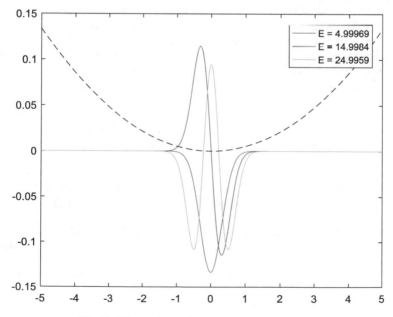

Fig. 5. Eigenvalues of the Schrödinger equation

coincidence of the values of the functions at some intermediate point. Usually, the left turning point is chosen as the given point. In addition to the coincidence of the values of functions, to ensure the smoothness of the solution stitching, the coincidence of the values of their derivatives is required. The result of the research is shown in Fig. 5.

3   Conclusion

MATLAB is a high-level programming system that allows to reduce sharply work content when testing algorithms [7] and performing calculations [8]. The possibility of carrying out large calculations on MATLAB is mainly determined by the time spent by the user: one has to select between ease and clarity of programming [9] and presentation of results, on the one hand, and time spent on the calculation, on the other. The system is very convenient for mastering and testing numerical methods, which we have shown above. That is why it is recommended as one of the main research areas for quantum processes by many scientific laboratories and universities.

Acknowledgments. The reported study was funded by RFBR according to the research project № 19-07-01082.

References

1. Viamontes, G.F.: Efficient quantum circuit simulation. A Ph.D. dissertation (Computer Science and Engineering). http://citeseerx.ist.psu.edu/viewdoc/download?doi=10.1.1.126.3074&rep=rep1&type=pdf
2. Potapov, V., Gushanskiy, S., Samoylov, A., Polenov, M.: The quantum computer model structure and estimation of the quantum algorithms complexity. In: Advances in Intelligent Systems and Computing, vol. 859. pp. 307–315. Springer (2019)
3. Adler, S.L., Brody, D.C., Brun, T.A., Hughston, L.P.: Martingale models for quantum state reduction. J. Phys. Math. Gen. **34**(42), 8795–8820 (2001)
4. Buri'c, N.: Hamiltonian quantum dynamics with separability constraints. In: arXiv:0704.1359v1 (2007)
5. Deutsch, D.: Quantum theory, the church-turing principle and the universal quantum computer. Proc. R. Soc. Lond. Ser. A, Math. Phys. Sci. **400**(1818), 97–117 (1985)
6. Hughston, L.P.: Geometry of stochastic state vector reduction. Proc. R. Soc. Lond. A **452** (1947), 953–979 (1996)
7. Potapov, V., Gushanskiy, S., Guzik, V., Polenov, M.: The computational structure of the quantum computer simulator and its performance evaluation in: software engineering perspectives and application in intelligent systems. In: Advances in Intelligent Systems and Computing, vol. 763, pp. 198–207. Springer (2019)
8. Potapov, V., Gushansky, S., Guzik, V., Polenov, M.: Architecture and Software Implementation of a quantum computer model. In: Advances in Intelligent Systems and Computing, vol. 465, pp. 59–68. Springer (2016)
9. Raedt, K.D., Michielsen, K., De Raedt, H., Trieu, B., Arnold G., Marcus Richter, Th Lippert, Watanabe, H., Ito, N.: Massively parallel quantum computer simulator. Computer Physics Communications, 176, pp. 121–136 (2007)

Performance Analysis of Weighted Priority Queuing Systems

Dariusz Strzęciwilk[✉]

Department of Applied Informatics, University of Life Sciences,
Nowoursynowska Street 159, 02-787 Warsaw, Poland
dariusz_strzeciwilk@sggw.pl

Abstract. The following article analyzes and studies the impact of traffic variability on the efficiency of WPQS (*Weighted Priority Queuing System*). Net models have been designed and tested taking into account the architecture supporting data transmission with different priorities. Priority queue models consisting of three different priority queues were examined. In order to examine the significant qualitative features of the studied models, TPN (*Temporal Petri Nets*) models were used. All performance features of the tested systems shown in the following article were obtained by simulating discrete events of TPN models. The use of simulation tools by using TPN aimed at verification of traffic shaping mechanisms with selected queuing algorithms. The obtained results are a perfect illustration of efficiency of the tested WPQS. The research has shown that TPN models can be effectively used in modeling and analyzing the performance of queuing systems.

Keywords: Weighted Priority Queuing System · Petri nets ·
Performance analysis · Modeling · QoS data

1 Introduction

PQS (*Priority Queuing System*) with various types of access restrictions to service stations are currently being intensively examined, due to their modeling applications in many phenomena occurring in technical sciences and economics. They are used in areas such as cellular automata [1], systems biology [2], business [3], telecommunications [4], computing [5], automated manufacture [6], queuing systems with impatient customers [7] etc. Thanks to rapid technical progress, the role of such systems in the area of operating systems or computer networks has also increased [8, 9]. Computer network is currently the most popular information exchange medium, whereas queuing is one of the key mechanisms in traffic management systems of computer networks. The quality of QoS (*Quality of Service*) is one of the most challenging tasks during the design and maintenance of both modern computer networks and next generation networks [10, 11]. The queuing theory is intensively used in modeling and research of QoS transmission quality in computer networks [12, 13]. The following theory uses a mathematical apparatus related to the theory of stochastic processes, Markov processes [14] in particular. The queuing system can be characterized by queue regulations, i.e.

© Springer Nature Switzerland AG 2019
R. Silhavy (Ed.): CSOC 2019, AISC 984, pp. 283–292, 2019.
https://doi.org/10.1007/978-3-030-19807-7_28

the manner in which the order of handling queues placed in the system is managed [15]. The most common queue disciplines are: FIFO (*First In First Out*), LIFO (*Last In First Out*), SIRO (*Select In Random Order*), PQ (*Priority Queuing*). Priority queueing [16] is the simplest mechanism that provides preferential service to some classes of traffic; in the priority queueing, lower priority traffic can be serviced only when all queues of higher priority classes are empty. Typical examples of queuing algorithm applications are routers which manage subsequent arriving packets. Each router in the network must have an implemented queuing discipline, which determines the way of buffering packets while queuing their further transmission. Excess packets can be dropped into the FIFO buffer in the order in which they arrived at the system. Packets are buffered until the transmission medium is released, that can be sent further. The disadvantage of the following solution is that sources with uneven traffic adversely affect the delays of other traffic flows. In particular, such action is unfavorable in case of real-time applications, such as VoIP, Video or online games [17]. The advantage of FIFO queues is that there is no need to configure them, and planning operations are managed in a small amount of time. An excellent way, in which the FIFO queue fulfills its tasks, are interfaces on which there is not much traffic. However, FIFO scheduling does not support good quality of transmission, once the packets arrive from different traffic flows, one of them can easily disrupt the flow of data remaining in other streams. The processing of packets in the order of their arrival means that an aggressive data stream can acquire a larger FIFO queue capacity. This may result in deterioration of the transmission quality causing, for example, a sudden increase in the delay of sent packets. A number of packet scheduling algorithms have been developed that show better memory isolation between flows [18]. In scheduling algorithms with PQS priority, some entries may be handled before others, regardless of when they appeared in the system. Ensuring the desired quality of service for packets along the entire route from the sender to the recipient is the subject of many research works [19–21]. In the literature, there have been a number of contributions with respect to priority scheduling. An overview of some basic priority queueing models in continuous time can be found in [22, 23] Although there are several well know basic QoS algorithms, migration to NGN (*Next Generation Networks*) imply investigation of new algorithms or their modifications to provide better behavior to time critical applications [24]. The analysis of available research shows that analytical methods usually include relatively simple queuing systems that require meeting many stochastic assumptions of traffic stream. The research was carried out in search of a solution for algebraic or differential equations that link the probabilities of occurrence of events in the system. In simulation methods, implantation of queuing algorithms was most often used, which were then subjected to statistical analysis. It turns out, that constructing a sufficiently accurate model is not easy, and the waiting time for results can be discouragingly long. Complex queuing systems are very difficult to analyze, however their functioning can be effectively investigated by simulation methods. In the following study, an attempt was made to compare the WPQS systems, which was conducted for a system with three queues with different priorities (high, medium and low). The assessment may be useful in designing and analysis of data transmission in computer networks, distributed systems or multiprocessor systems. The built-in queue models allowed to estimate significant features and parameters of the tested model. This work is a continuation of previous studies on priority queuing systems using TPN models [25], which allowed to evaluate the efficiency and effectiveness of priority queuing mechanisms. The remainder of this paper is organized in the following manner. Section 2 recalls basic concepts of Petri Nets. Section 3 describes the net model of priority queueing system while Sect. 4 uses the developed model to analyze the performance of weighted priority queueing systems. Section 5 concludes the paper.

2 Petri Nets and Network Models

Petri nets are perceived as a mathematical tool for modeling concurrent systems [26]. There are many different varieties of Petri nets [27–29]. Stochastic Petri nets [30] and timed Petri nets [25] have many similarities but deal with temporal properties of models in a different way, so sometimes the similarities may be misleading. A common feature of these networks is a structure based on a bigraph directed with two types of vertices, alternately connected by directed edges (i.e. arcs). These two types of vertices represent, in a general sense, the conditions and events occurring in the modeled system, where each event can occur, once all conditions connected with it are fulfilled. What's important is that the interpretation of the system introducing events dependent on the conditions related to them (condition-event systems) is very convenient and helpful in modeling complex discrete systems [31]. For a more complete presentation of the subject, one should first introduce the basic definitions.

Definition 2.1. In classic terms, the Petri net is defined as an ordered four:

$$C = (P, T, A, m_0) \tag{1}$$

where: $P = \{p_1, p_2, \ldots, p_n\}$ non-empty, finite set of places, $T = \{t_1, t_2, \ldots, t_m\}$ non-empty, finite set of transitions and $P \cap T = \varnothing$, $A-$ non-empty, finite set of directed arcs, such as: $A \subset (P \times T) \cup (T \times P)$, m_0 – initial marking function $m_0 : P \rightarrow \mathcal{N}\{0\}$.

Definition 2.2. Petri net with enabling and inhibitor arcs:

$$C_A = (P, T, A, m_0) \tag{2}$$

where: A is a set of arcs such that: $A = A_o \cup A_e \cup A_i, A_o : A_o \subset (P \times T) \cup (T \times P), A_e : A_e \subset (P \times T), A_i : A_i \subset (P \times T)$ other symbols as in (1). In addition, we can define auxiliary concepts.

Definition 2.3. The sets of enabling, inhibitor and output transitions of places are respectively called sets $P_t^{in(o)}$, $P_t^{in(e)}$, $P_t^{in(i)}$ and P_t^{out},
    where: $P_t^{in(o)} = \{p \in P : (p, t) \in A_o\}, P_t^{in(e)} = \{p \in P : (p, t) \in A_e\}, P_t^{in(i)} = \{p \in P : (p, t) \in A_i\}, P_t^{out} = \{p \in P : (p, t) \in A_o\}$.

Definition 2.4. The sets of input and output transitions of p places are called T_p^{in} and T_p^{out}, respectively,
    where: $T_p^{in} = \{t \in T : (t, p) \in A_o\}, T_p^{out} = \{t \in T : (t, p) \in A_o\}$.

Definition 2.5. The place's activity function is called the $ac(p) \rightarrow \{true, false\}$ function, which assigns *true* value to every place of the net, if the place has a token, and *false* in the opposite situation.

Definition 2.6. The occurrence of events is represented by the so-called tokens assigned to the p-elements of the network, usually the p-element to which at least one token is associated, means that the condition represented by this element is met.

Definition 2.7. Marking m is a function that assigns a number to every number of tokens present at that point at a given point in time. The placement of tokens in the p-elements can be described by marking function $m : P \rightarrow \{0, 1, 2 ...\}$ or as a vector describing the number of tokens assigned to the next p-elements of nets $m = [m_{(p1)}, m_{(p2)}, ...]$.

Definition 2.8. The C net together with the (initial) marking function m_0 is called the marked net $M = (C, m_0) = (P, T, A, m_0)$.

To assess the performance of the modeled systems, i.e. to determine how quickly certain events may occur after each other, the duration of the modeled event should be additionally taken into account. The temporal model should be supplemented with times in the relevant states (holding times) and transition probabilities or transition rates. The completed graph becomes the Markov chain, representing the real behavior of the model. The evaluation of system performance using its Markov chains is well described in the literature [25]. Broadening the net with time definitions allows to use them in real-time system modeling [32]. The temporal net T is thus defined as the system $T = (M, c, f)$ where: M is a marked network, $M = (P, T, A, m_0)$, c- is a function of conflict resolution $c : T \rightarrow [0, 1]$, which for each decision class gives the probabilities of individual events belonging to this class, and for other conflict events gives their relative rates used for random conflict resolution, and f - defines event occurrence times $f : T \rightarrow \mathbb{R}_+$, where $\mathbb{R}_+$ - represents a set of nonnegative real numbers. The duration of events can be deterministic, determined by the value $f(t)$ or stochastic, described by the corresponding probability density function with the parameter $f(t)$.

3   The Model

To assess the efficiency of the modeled systems, the Petri net model presented schematically in Fig. 1 was used. Based on the prepared models, the mechanisms of shaping traffic in systems based on priority queues were studied.

The models assumed that the PQ model will consist of three priority queues (places *p1, p2, p3*). The three places, *p1, p2* and *p3*, are queues for class-1, class-2 and class-3 packets, respectively. It is assumed that all queues have infinite capacities. This will allow you to visualize traffic for data with high, medium and low priority. In the tested models, high priority data was marked as class-1, data with medium priority was marked as class-2, and low priority data was class-3. In the transit model, *t1, t2, t3* represent the transmission channel through which data is transmitted. To simplify, it was assumed that the transmission time is deterministic and amounts to 1 time unit for all 3 classes. The transmission channel is connected to *p4* squares to ensure that only

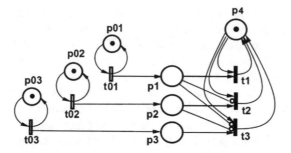

Fig. 1. Petri net model of priority queuing system

one data type is transported in the transmission channel at any given time. Inhibitors arc $p1–t2$ connected to the $t2$ transitions do not allow for data transmission in case when class-1 data being in place $p1$ is waiting for transmission. The arc inhibitors $p1–t3$ and $p2–t3$ connected to the $t3$ transition do not allow for data transmission in case data class-1 and class-2 are waiting for transmission. A characteristic feature is that as long as any packets are in a queue with a higher priority, packets waiting in queues with a lower priority cannot be sent. In order to eliminate the problem of queuing, a limitation of the amount of data transferred in a single cycle was introduced. The problem was eliminated by introducing WPQS. The round-robin mechanism introduced in the model first retrieves 5 data waiting in the class-1 queue, then 4 from the class-2 queue and at the end 2 from the lowest priority queue (class-3). Basing on such prepared models, the mechanisms of shaping traffic in systems based on 5-4-2 priority queues were examined. *M*-timed nets (*Markovian nets*) were used as the source of data generation (place *p01, p02, p03*).

4   Research Results and Discussion

In order to compare the performance of systems made was a comparative analysis of parameters such as traffic intensity, average waiting time, throughput and utilization. Based on the simulations prepared were charts showing the behaviour of the studied systems. Detailed analysis concerned the behaviour of the system on the verge of its stability.

Average waiting times for classes-1, 2 and 3 as functions of traffic intensity ρ are shown in Fig. 2. The conducted research has shown that for the same values of generated data ($\rho_1 = \rho_2 = \rho_3$) and channel utilization, average waiting times are different (Figs. 2 and 3). It was found that ρ ($\rho = \rho_1 + \rho_2 + \rho_3$) in the average waiting times function is linear in the range of $\rho < 0.80$, but after exceeding this value in the model there are dynamic conversions in the class with the lowest priority (class-3). The value of average waiting times class-3 increases exponentially, and changes in other classes are insignificant. Significant differences were found for the average waiting times characteristics when the data in two classes take constants, and the data generated by the third source is exchanged (Figs. 2 and 3). In case when the amount of data increases in class-1 (increase of ρ_1), and data class-2 and class-3 take the values const, i.e. for

$\rho_2 = 0.25$ and $\rho_3 = 0.25$, then the waiting time class-3 increases exponentially for $\rho_1 > 0.4$, while class-1 waiting time increases exponentially for $\rho_1 > 0.5$. A similar effect was achieved in case when the amount of data generated in class-2 increases (increase of ρ_2), and class-1 and class-3 data take a constant value, i.e. $\rho_1 = 0.25$ and $\rho_3 = 0.25$. As expected, the increase in the amount of data in the class with the lowest priority (class-3) at constant values of class-1 and class-3, i.e. $\rho_1 = 0.25$ and $\rho_2 = 0.25$ does not significantly affect the waiting time of calss-1 and class-2 (Fig. 3). It should be noted, however, that the performance of the tested model 5-4-2 is limited to 5/11, 4/11 and 2/11 (0.45, 0.36 and 0.18) for class-1, class-2 and class- 3. Hence, for some values ρ_1, ρ_2, ρ_3 the model falls within the non-stationary range. Studies on the dynamics of the queue length showed that when the traffic intensity ρ approaches 0.90, the most affected class is class–3, the lowest priority class (Fig. 4). In addition, the average

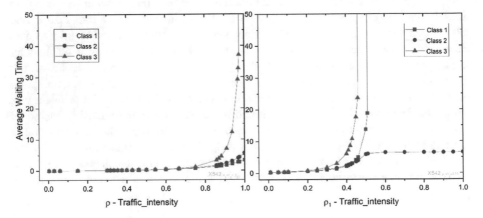

Fig. 2. Average waiting times diagram for model 542 in function ρ (left). Average waiting times diagram in function ρ_1 with constants of $\rho_2 = \rho_3 = 0.25$ (right)

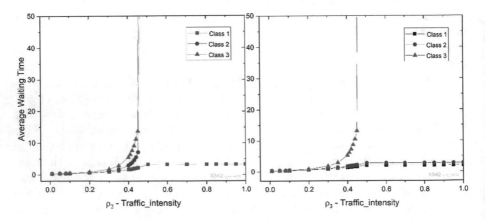

Fig. 3. Average waiting times diagram in the ρ_2 function with constants of $\rho_1 = \rho_3 = 0.25$ (left). Average waiting times diagram in function ρ_3 with constants of $\rho_1 = \rho_2 = 0.25$ (right)

queue length studies in the function ρ_1 with $\rho_2 = 0.25$ and $\rho_3 = 0.25$ have revealed that the class-3 queue length increases sharply for $\rho_1 > 0.40$, while the class-1 class queue increases exponentially for the value of $\rho_1 > 0.50$. Similar behavior has been observed in model 542 for increasing values of ρ_2 and ρ_3.

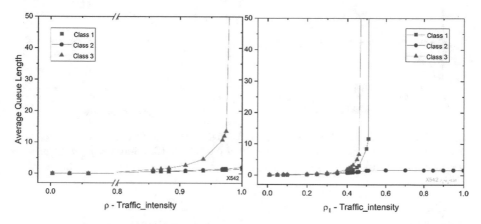

Fig. 4. The average queue length chart of model 542 in the ρ (left) function. The average queue length chart in the ρ_1 function with constants $\rho_2 = \rho_3 = 0.25$ (right).

In the next stage, two parameters of the channel utilization and throughput model were examined. The channel utilization study have shown slight changes in the tested model. Figure 5 shows channel utilization values as functions ρ_1 with $\rho_2 = 0.25$ and $\rho_3 = 0.25$ for weighted priority queuing with weights 5-4-2. It has been found that the maximum channel utilization value for class-1 is 0.54, and for class-2 and class-3 0.25 and 0.21, respectively. Studies show that in case ρ_2 with $\rho_1 = 0.25$ and $\rho_3 = 0.25$ the maximum value of channel utilization for class-2 is 0.50 while the value of channel utilization class-2 and class-3 does not change (Fig. 5). A similar effect is observed for ρ_3 with $\rho_2 = 0.25$ and $\rho_3 = 0.25$. The research also proves that the maximum throughput for class-1 data is 0.45 and remains at this level for the value of $\rho_1 > = 0.47$. For class-2 data, the maximum throughput occurs for $\rho_1 = 0.39$, but increasing the value of ρ_1 above 0.4 results in "blocking effects" created by priority queuing class-1. Throughput for class-2 stabilizes only at 0.36, when $\rho_1 = 0.45$. Similarly to class-2 data, "blocking effects" are also observed for class-3 data. However, for class-3 data, this effect occurs in two stages (Fig. 6). The maximum throughput for class-3 data is 0.33 at $\rho_1 = 0.33$. After reaching the maximum value of class-3 bandwidth, it drops sharply as a result of "blocking effects" created by priority queuing class-1 and class-2. For $\rho_1 > 0.45$, the throughput value stabilizes at 0.18. Above the value of $\rho_1 = 0.45$, the characteristics for all classes are uniform and rectilinear (Fig. 6). As expected, the fraction packed served characteristics of the 542 model are dependent on the value of the traffic intensity function. In cases when the value ρ approaches the non-stationary range (arrival rates are greater than departure rates) then we can observe a decrease in the value of fraction packed served (Fig. 6).

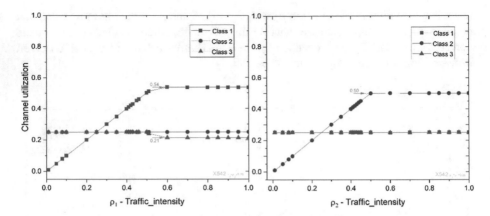

Fig. 5. The channel utilization diagram of the 542 model in function ρ_1 with constants $\rho_2 = \rho_3 = 0.25$ (left). Channel utilization model of the 542 model in function ρ_2 with constants $\rho_1 = \rho_3 = 0.25$ (right).

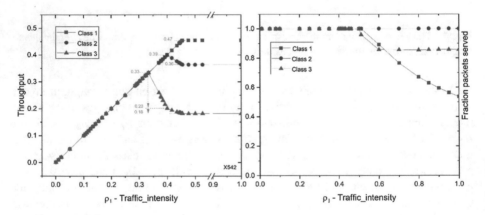

Fig. 6. Throughput diagram of the 542 model in function ρ_1 with constant values of $\rho_2 = \rho_3 = 0.25$ (left). The fraction packed served chart of model 542 in function ρ_1 with constants $\rho_2 = \rho_3 = 0.25$ (right).

5 Summary

With the increase in the number of network users and the rapid development of various network services, the amount of data transferred within the network also increases. Queuing is one of the key mechanisms in traffic management systems, while effective use of modern computer networks and next generation networks require detailed information on the characteristics and type of traffic. It's possible to calculate the data transfer time from node to node, as well as throughput, but the time and waiting period in the given node is unknown. Detailed information can be obtained on the basis of simulation models. Planning and development of computer networks can be supported using mathematical modeling and computer simulation. The work shows that TPN networks can be effectively used for modeling of WPQS. Based on the models built, the characteristics of the 542 priority model has been obtained and demonstrated, which eliminates traffic blocking effect with lower priorities. The obtained results may allow to determine the optimal band intended for both critical traffic and to determine the best network loads. Obtained models or parts of it can be easily subjected to further, detailed analysis. Because in real-life telecommunication networks the characteristics of traffic often change, it is worth considering to examine the impact of the changing nature of traffic on priority queuing models in further studies.

References

1. Zaitsev, D.A.: Simulating cellular automata by Infinite Petri Nets. J. Cell. Automata **13**, 121–144 (2018)
2. Koch, I., Reisig, W., Schreiber, F. (eds.): Modeling in Systems Biology: The Petri Net Approach, vol. 16. Springer, Cham (2010)
3. Rogge-Solti, A., Weske, M.: Prediction of business process durations using non-markovian stochastic Petri nets. Inf. Syst. **54**, 1–14 (2015)
4. Girault, C., Valk, R.: Petri Nets for Systems Engineering: a Guide to Modeling, Verification, and Applications. Springer, Heidelberg (2013)
5. Best, E., Devillers, R., Koutny, M.: Petri Net Algebra. Springer, Heidelberg (2013)
6. Li, Z. (ed.): Formal Methods in Manufacturing Systems: Recent Advances: Recent Advances. IGI Global, Hershey (2013)
7. Wang, K., Li, N., Jiang, Z.: Queueing system with impatient customers: a review. In: 2010 IEEE International Conference on Service Operations and Logistics and Informatics (SOLI). IEEE (2010)
8. Strzeciwilk, D., Zuberek, W.: Modeling and performance analysis of weighted priority queueing for packet-switched networks. J. Comput. Commun. **6**, 195–208 (2018). https://doi.org/10.4236/jcc.2018.611019
9. Zuberek, W.M., Strzeciwilk, D.: Modeling traffic shaping and traffic policing in packet-switched networks. J. Comput. Sci. Appl. **6**(2), 75–81 (2018)
10. Strzęciwilk, D.: Examination of transmission quality in the IP Multi-Protocol label switching corporate networks. Int. J. Electron. Telecommun. **58**(3), 267–272 (2012)
11. Strzęciwilk, D., Zuberek. W.M.: Modeling and performance analysis of QoS data. In: Photonics Applications in Astronomy, Communications, Industry, and High-Energy Physics Experiments 2016. International Society for Optics and Photonics (2016)
12. Czachórski, T., Domański, A., Domańska, J., Rataj, A.: A study of IP router queues with the use of Markov models. In: International Conference on Computer Networks. Springer (2016)
13. Strzęciwilk, D., Pękala, R., Kwater, T.: Performance analysis of priority queueing systems using timed petri nets. In: 19th International Conference Computational Problems of Electrical Engineering. IEEE (2018). https://doi.org/10.1109/cpee.2018.8507124
14. Puterman, M.L.: Markov Decision Processes: Discrete Stochastic Dynamic Programming. Wiley, New York (2014)
15. Bose, S.K.: An Introduction to Queueing Systems. Springer, Boston (2013)
16. Georges, J.-P., Divoux, T., Rondeau, E.: Strict priority versus weighted fair queuing in switched ethernet networks for time critical applications. In: Proceedings of the 19th IEEE International Parallel and Distributed Processing Symposium. IEEE (2005)

17. Kim, H.J., Choi, S.G.: A study on a QoS/QoE correlation model for QoE evaluation on IPTV service. In: 2010 The 12th International Conference on Advanced Communication Technology (ICACT), vol. 2. IEEE (2010)
18. Zhang, Q., et al.: Early drop scheme for providing absolute QoS differentiation in optical burst-switched networks. In: Workshop on High Performance Switching and Routing. HPSR. IEEE (2003)
19. Tarasiuk, H., et al.: Performance evaluation of signaling in the IP QoS system. J. Telecommun. Inf. Technol. **3**, 12–20 (2011)
20. Zuberek, W., Strzęciwilk, D.: Modeling Quality of Service Techniques for Packet–Switched Networks. Dependability Engineering (2018). ISBN 978-953-51-5592-8
21. Shortle, J.F., et al.: Fundamentals of Queueing Theory, vol. 399. Wiley, Hoboken (2018)
22. Pinedo, M.L.: Scheduling: Theory, Algorithms, and Systems. Springer, Heidelberg (2016)
23. Carmona-Murillo, J., et al.: QoS in next generation mobile networks: an analytical study. In: Resource Management in Mobile Computing Environments, pp. 25–41. Springer, Cham (2014)
24. Strzęciwilk, D., Zuberk, W.M.: Modeling and performance analysis of priority queuing systems. In: Silhavy, R. (eds) Software Engineering and Algorithms in Intelligent Systems. CSOC2018 2018. Advances in Intelligent Systems and Computing, vol 763. Springer, Cham (2019)
25. Zuberek, W.M.: Timed Petri nets definitions, properties, and applications. Microelectron. Reliab. **31**(4), 627–644 (1991)
26. Balbo, G.: Introduction to stochastic Petri nets. In: Lectures on Formal Methods and Performance Analysis, pp. 84–155. Springer, Heidelberg (2001)
27. Peterson, J.L.: Petri net theory and the modeling of systems (1981)
28. Jensen, K., Kristensen, L.M.: Coloured Petri Nets: Modelling And Validation of Concurrent Systems. Springer, Heidelberg (2009)
29. Reisig, W.: Petri Nets: An Introduction, vol. 4. Springer, Heidelberg (2012)
30. Marsan, M.A.: Stochastic Petri nets: an elementary introduction. In: European Workshop on Applications and Theory in Petri Nets. Springer, Heidelberg (1988)
31. Pedrycz, W., Gomide, F.: A generalized fuzzy Petri net model. IEEE Trans. Fuzzy Syst. **2**(4), 295–301 (1994)
32. Cheng, A.M.K.: Real-Time Systems: Scheduling, Analysis, and Verification. Wiley, Hoboken (2003)

Topic-Enriched Word Embeddings
for Sarcasm Identification

Aytuğ Onan[(✉)]

Department of Computer Engineering, Faculty of Engineering and Architecture,
İzmir Katip Çelebi University, 35620 İzmir, Turkey
aytug.onan@ikc.edu.tr

Abstract. Sarcasm is a type of nonliteral language, where people may express their negative sentiments with the use of words with positive literal meaning, and, conversely, negative meaning words may be utilized to indicate positive sentiment. User-generated text messages on social platforms may contain sarcasm. Sarcastic utterance may change the sentiment orientation of text documents from positive to negative, or vice versa. Hence, the predictive performance of sentiment classification schemes may be degraded if sarcasm cannot be properly handled. In this paper, we present a deep learning based approach to sarcasm identification. In this regard, the predictive performance of topic-enriched word embedding scheme has been compared to conventional word-embedding schemes (such as, word2vec, fastText and GloVe). In addition to word-embedding based feature sets, conventional lexical, pragmatic, implicit incongruity and explicit incongruity based feature sets are considered. In the experimental analysis, six subsets of Twitter messages have been taken into account, ranging from 5000 to 30.000. The experimental analysis indicate that topic-enriched word embedding schemes utilized in conjunction with conventional feature sets can yield promising results for sarcasm identification.

Keywords: Sarcasm detection · Word-embedding based features ·
Deep learning

1 Introduction

Sarcasm can be defined as a cutting, ironic remark that is intended to express contempt or ridicule [1]. Sarcasm is a type of nonliteral language, where people may express their negative sentiments with the use of words with positive literal meaning, and, conversely, negative meaning words may be utilized to indicate positive sentiment. With the advances in information and communication technologies, the immense quantity of user-generated text documents have been available on the Web. As a result, sentiment analysis has emerged as an important research direction. User-generated content on social platforms can serve as an essential source of information. The identification of sentiments towards entities, products or services can be important for business organizations, governments and individual decision makers [2]. The identification of subjective information from online content can be utilized to generate structured knowledge to construct decision support systems.

© Springer Nature Switzerland AG 2019
R. Silhavy (Ed.): CSOC 2019, AISC 984, pp. 293–304, 2019.
https://doi.org/10.1007/978-3-030-19807-7_29

The social content available on the Web generally contains nonliteral language, such as irony and sarcasm. There are a number of challenges encountered in sentiment analysis, such as negation, domain dependence, polysemy and sarcasm [3]. Sarcastic utterance can change the sentiment orientation of text documents from positive to negative, or vice versa. In text documents with sarcasm, the expressed text utterances and the intension of the person utilizing sarcasm can be completely opposite [4]. Hence, the predictive performance of sentiment classification schemes may be degraded if sarcasm cannot be handled properly.

Automatic identification of sarcasm is a challenging problem in sentiment analysis. First of all, sarcasm is a difficult concept to define. Since is a difficult concept to define, it is even difficult to precisely identify for people whether a particular statement is sarcastic or not [5]. In addition to that, there is no lots of accurately-labeled naturally occurring utterances labeled for machine learning based sarcasm identification.

Sarcasm may be encountered in short text documents, which is characterized by the limited length (such as, Twitter messages), in long text documents (such as, review posts and forum posts) and transcripts of TV shows [1]. Twitter is a popular microblogging platform, where people can express their opinions, feelings and ideas in short messages, called tweets, within 140-character limit. With the use of Twitter, people may interact to each other in a faster way. Twitter may be utilized, as daily chatter, conversation, sharing information and reading breaking news [6]. Twitter serves as an essential source of information for practitioners and researchers. Twitter is an important source for sarcasm identification and sarcasm identification on Twitter is a promising research direction. The earlier supervised learning schemes for sarcasm identification on Twitter have been utilized linguistic feature sets, such as lexical unigrams, bigrams, sentiments, punctuation marks, emoticons, character n-grams, quotes and pronunciations [7]. Sarcasm identification approaches may be broadly divided into three schemes, as rule-based approaches, statistical methods and deep learning based approaches [1]. Rule-based schemes to sarcasm identification seek to identify sarcastic text with the use of rules based on the indicators for sarcasm. Statistical methods to sarcasm identification utilize supervised learning algorithms, such as, support vector machines, logistic regression, Naïve Bayes and decision trees. Deep learning is a promising research direction in machine learning, which can be applied in a wide range of applications, including computer vision, speech recognition and natural language processing, with high predictive performance.

In this paper, we present a deep learning based approach to sarcasm identification. In this regard, the predictive performance of topic-enriched word embedding scheme has been compared to conventional word-embedding schemes (such as, word2vec, fastText and GloVe). In addition to word-embedding based feature sets, conventional lexical, pragmatic, implicit incongruity and explicit incongruity based feature sets are considered.

The rest of this paper is organized as follows: In Sect. 2, related work on sarcasm identification has been presented. Section 3 presents word-embedding schemes utilized in the empirical analysis, Sect. 4 briefly presents convolutional neural network. In Sect. 5, experimental procedure and empirical results are presented. Finally, Sect. 6 presents the concluding remarks.

2 Related Work

This section briefly discusses the existing works on machine learning and deep learning based schemes for sarcasm identification.

Gonzalez-Ibanez et al. [8] examined the predictive performance of unigrams, dictionary based feature sets and pragmatic factors for sarcasm identification. In the presented scheme, support vector machines and logistic regression classifiers have been utilized. In another study, Reyes et al. [9] utilized properties of figurative languages (such as, ambiguity, polarity, unexpectedness and emotional scenarios) for sarcasm identification on Twitter. In another study, Reyes et al. [10] examined the predictive performance of conceptual features (such as, signatures, unexpectedness, style and emotional scenarios) for sarcasm identification on Twitter. The presented scheme obtained a predictive performance of 0.72 in terms of F-measure.

Ptacek et al. [11] presented a machine learning based approach to identify sarcasm on Twitter messages written in English and Czech language. In the presented scheme, n-gram based features (such as, character n-grams, n-grams, skip-bigram), pattern based features (such as, word-shape pattern), part of speech based features (such as, POS characteristics, POS n-grams and POS word-shape) and other features (such as, emoticons, punctuations, pointedness and word-case) have been utilized as the feature sets. In another study, Barbieri et al. [12] examined the predictive performance of feature sets, such as frequency-based features, written-spoken style uses, intensity of adjectives and adverbs, length, punctuation, emoticons, sentiments, synonyms and ambiguities for sarcasm identification. In another study, Rajadesingan et al. [13] presented a behavioral modeling scheme for sarcasm identification.

Farias et al. [14] utilized the predictive performance of affective features for irony detection on Twitter. In this regard, structural features (such as, punctuation marks, length of words, emoticons, discourse markers, part of speech and semantic similarity), affective features (such as, sentiment lexicons, sentiment-related features, and emotional categories) have been taken into consideration. Bouazizi and Otsuki [15] presented a pattern-based scheme for sarcasm identification on Twitter. In the presented scheme, sentiment-related features, punctuation-related features, syntactic and semantic features and pattern-related features have been considered. In another study, Kumar et al. [16] presented a machine learning based scheme for sarcasm identification in numbers. In another study, Mishra et al. [17] utilized lexical, implicit incongruity based features, explicit incongruity based features, textual features, simple gaze based features and complex gaze based features for sarcasm identification.

In addition to machine learning based schemes, the paradigm of deep learning has been recently utilized for sarcasm identification. For instance, Ghosh et al. [18] examined different word-embedding based schemes (such as, weighted textual matrix factorization, word2vec and GloVe) for sarcasm identification. Similarly, Joshi et al. [19] utilized four word-embedding based schemes (namely, latent semantic analysis, GloVe, dependency weights and word2vec) for sarcasm detection. In another study, Poria et al. [20] presented a deep learning based approach for sarcasm identification based on pre-trained sentiment, emotion and personality models based on convolutional neural networks.

3 Word-Embedding Schemes

Bag-of-words model is a typical representation scheme for natural language processing tasks, where words are represented in a vector space such that the more similar words are located to nearby positions. One of the problems regarding bag-of-words scheme is that the model regards words within the same context have the same semantic meaning. In order to build deep learning based schemes for natural language processing tasks, word-embeddings based representation schemes is an important language modelling technique, in which semantic and syntactic relations among the words have been captured based on large unsupervised sets of documents [21]. In this study, the predictive performance of LDA2vec has been compared to three well-known word-embedding schemes, namely, word2vec, global vectors (GloVe) and fastText have been considered. The rest of this section briefly presents word-embedding schemes utilized in the empirical analysis.

The word2vec is an unsupervised and efficient method to construct word embeddings from text documents. The word2vec model consists of two different architectures, namely, continuous bag of words model (CBOW) and continuous skip-gram model [22]. CBOW model takes the context of each word as the input and seeks to identify the word related to the context. In contrast, skip-gram model predicts the particular context words based on the center word. Skip-gram model works well with small amount of data and can effectively represent infrequent words, whereas CBOW model is faster and can effectively represent more frequent words. Let we denote a sequence of training words $w_1, w_2, \ldots, w_T$ with length T, the objective of skip-gram model is determined based on Eq. 1 [23]:

$$argmax_\theta \frac{1}{T} \sum_{t=1}^{T} \sum_{-C \leq j \leq C, j \neq 0} logP_\theta\left(w_{t+j}|w_t\right) \tag{1}$$

where C denote the size of training context, $P\left(w_{t+j}|w_t\right)$ represents a neural network with a set of parameters denoted by θ.

The fastText model is an extension of word2vec model, which represent each word by breaking words into several character n-grams (sub-words) [24]. With the use of fastText based word-embedding, rare words can be represented in a more efficient way. The fastText model is a computationally efficient model and since it takes character n-grams into account, good representation schemes can be obtained for rare words.

The global vectors (GloVe) is a global log-bilinear regression model for word embeddings based on global matrix factorization and local context window methods [25]. The objective of Glove model is determined based on Eq. 2:

$$J = \sum_{i,j=1}^{V} f\left(X_{ij}\right) \left(w_i^T \omega_j + b_i + b_j - logX_{ij}\right)^2 \tag{2}$$

where V denotes the vocabulary size, $w \in R^d$ represent word vectors, $\omega \in R^d$ represent context word vectors, X denote co-occurrence matrix and X_{ij} denotes the number of times word j occurs in the context of word i. $f\left(X_{ij}\right)$ denotes a weighting function and b_i, b_j are bias parameters [25].

The LDA2vec model is a word-embedding scheme based on word2vec and the latent Dirichlet allocation, which extracts dense word vectors from the latent document-level mixture vectors jointly from Dirichlet-distribution [26]. LDA2vec model enables to identify topics from texts and obtain topic-adjusted word vectors. In this way, the interpretability of the word vectors has been enhanced by linking each word to the corresponding topic. The LDA2vec model utilizes skip-gram negative sampling as the objective function to identify topic weights for the documents.

4 Convolutional Neural Networks

Convolutional neural networks (CNNs) are a type of neural networks which process data with a grid-like topology. Convolutional neural networks have been successfully utilized in a number of applications, including image recognition, computer vision and natural language processing [27, 28]. Convolutional neural networks are characterized by the convolution operation in their layers. A typical convolutional neural network architecture consists of input layer, output layer and hidden layers. The hidden layers of the architecture may be convolutional layer, pooling layer, normalization layer or fully connected layers. In convolutional layers, convolution operation has been applied on the input data to obtain feature maps. In order to add nonlinearity to the architecture, each feature map has been also subject to the activation functions [29]. In convolutional neural networks, the rectified linear unit has been widely utilized as the activation function. In pooling layers, the number of parameters and operations for obtaining the output has been reduced in order to eliminate overfitting. In convolutional neural networks, max pooling scheme has been widely utilized as the pooling function. After convolution and pooling layers, the final output of the architecture has been identified by fully connected layers [30].

5 Experimental Procedure and Results

In this section, dataset collection and preprocessing methodology, experimental procedure, evaluation measures and empirical results of the study have been presented.

5.1 Dataset Collection and Preprocessing

In the dataset collection, we adopted the framework presented in [31]. To build the sarcasm dataset with sarcastic and non-sarcastic tweets, self-annotated tweets of Twitter users have been utilized. Twitter messages with hashtags of "sarcasm" or "sarcastic" are taken as sarcastic tweets, whereas hashtags about positive and negative sentiments are regarded as non-sarcastic tweets. In this way, we obtained approximately 40.000 tweets written in English. Twitter4J, an open-source Java library for utilizing Twitter Streaming API, has been utilized to collect dataset. In order to eliminate duplicated tweets, retweets, ambiguous, irrelevant and redundant tweets, automatic filtering has been utilized. Each tweet has been manually annotated with the use of a single-class label, as either sarcastic or non-sarcastic. In this way, we obtained a collection of roughly 15.000 sarcastic tweets and roughly 24.000 non-sarcastic tweets. In order to obtain a balanced corpus, our final corpus contains Twitter messages with 15.000 sarcastic tweets and 15.000 non-sarcastic tweets. To preprocess our corpus, we have adopted the framework presented in [32]. First, tokenization has been utilized on the corpus to divide tweets into tokens, such as words and punctuation marks. To handle with the tokenization process, Twokenize tool has been utilized. At the end of the tokenization process, unnecessary items generated by Twokenize has been eliminated. In addition, mentions, replies to other users' tweets, URLs and special characters have been eliminated. In Table 1, the distribution of the dataset and the basic descriptive information for the dataset has been given.

Table 1. Descriptive information regarding the corpus utilized in empirical analysis

Set	Positive	Negative	Total number
Training set	12.000	12.000	24.000
Testing set	3.000	3.000	6.000

5.2 Experimental Procedure

In the empirical analysis, the predictive performance of topic-enriched word embedding scheme has been compared to conventional word-embedding schemes (such as, word2vec, fastText and GloVe). In addition to word-embedding based feature sets, conventional lexical, pragmatic, implicit incongruity and explicit incongruity based feature sets are considered. In the experimental analysis, six subsets of Twitter messages have been taken into account, ranging from 5000 to 30.000. For each subset of the corpus, 80% of data has been utilized as the training set, while the rest of data has been utilized as the testing set. In order to implement convolutional neural network, TensorFlow has been utilized. For the LDA2vec based word embedding, a number of parameters (including, the number of topics and the negative sampling exponent) have been taken into consideration. Since the best predictive performance is obtained by the number of topics ($N=25$) and the negative sampling exponent ($\beta \in 0.75$), we have reported the results for these parameter sets in the empirical results. For the word-embedding schemes, vector size has been set to 200.

5.3 Evaluation Measures

In order to empirical analysis, we have utilized F-measure, as the evaluation measure.

Precision (PRE) is the proportion of the true positives against the true positives and false positives as given by Eq. 3:

$$PRE = \frac{TP}{TP + FP} \tag{3}$$

Recall (REC) is the proportion of the true positives against the true positives and false negatives as given by Eq. 4:

$$REC = \frac{TP}{TP + FN} \tag{4}$$

F-measure takes values between 0 and 1. It is the harmonic mean of precision and recall as determined by Eq. 5:

$$F - measure = \frac{2 * PRE * REC}{PRE + REC} \tag{5}$$

5.4   Experimental Results

In Table 2, the predictive performance of four word embedding based feature representation schemes, namely, word2vec, global vectors (GloVe), fastText and LDA2vec has been given in terms of F-measure values. In addition to word-embedding based feature sets, conventional lexical, pragmatic, implicit incongruity and explicit incongruity based feature sets are considered [33, 34]. For the F-measure values given in Table 2, convolutional neural network has been utilized. For word2vec and fastText models, continuous skip-gram and continuous bag of words (CBOW) algorithms are taken into consideration. In addition, different subsets of Twitter messages, ranging from 5000 to 30.000 are taken into consideration. In Table 2, subset1 corresponds to subset of corpus with 5000 tweets, subset2 corresponds to subset of corpus with 10000 tweets, subset3 corresponds to subset of corpus with 15000 tweets, subset4 corresponds to subset of corpus with 20000 tweets, subset5 corresponds to subset of corpus with 25000 tweets and subset6 corresponds to the entire corpus with 30000 tweets.

As it can be observed from F-measure values presented in Table 2, LDA2Vec based word embedding scheme yield higher predictive performance compared to other word-embedding based schemes, such as word2vec, fastText and global vectors (GloVe). The second highest predictive performance in terms of F-measure is generally obtained by GloVe based word embedding and the lowest predictive performance is generally obtained by word2vec based word embedding schemes. Regarding the predictive performance of continuous skip-gram and continuous bag of words (CBOW) algorithms utilized in word2vec and fastText models, continuous bag of words (CBOW) models generally outperform continuous skip-gram models. In addition to word-embedding based feature representation, the empirical analysis also examines whether integration of feature sets, such as lexical, pragmatic, implicit incongruity and explicit incongruity based features can enhance the predictive performance of deep learning based sarcasm identification schemes. As it can be observed from the results listed in Table 2, the utilization of lexical, pragmatic, implicit incongruity and explicit incongruity based features in conjunction with word-embedding based representation schemes can enhance the predictive performance of classification schemes. The highest predictive performance among the compared schemes is obtained by the integration of LDA2Vec based word embedding with explicit incongruity based feature sets and the second highest predictive performance is generally obtained by LDA2Vec based word embedding with implicit incongruity based feature sets.

Table 2. F-measure values obtained by different word-embedding schemes

Representation Scheme	Subset#1	Subset#2	Subset#3	Subset#4	Subset#5	Subset#6
word2vec (Skip-gram)	0.734	0.749	0.740	0.772	0.786	0.868
word2vec (CBOW)	0.737	0.762	0.741	0.790	0.796	0.839
fastText (Skip-gram)	0.738	0.769	0.759	0.791	0.806	0.842
fastText (CBOW)	0.740	0.771	0.768	0.793	0.806	0.838
GloVe	0.745	0.774	0.770	0.794	0.807	0.877
LDA2Vec	0.745	0.775	0.775	0.794	0.812	0.847
word2vec (Skip-gram) + Lexical	0.749	0.776	0.775	0.796	0.814	0.881
word2vec (Skip-gram) + Pragmatic	0.749	0.782	0.778	0.801	0.818	0.894
word2vec (Skip-gram) + Implicit incongruity	0.754	0.782	0.780	0.801	0.818	0.848
word2vec (Skip-gram) + Explicit incongruity	0.755	0.783	0.782	0.803	0.819	0.865
word2vec (CBOW) + Lexical	0.761	0.784	0.783	0.805	0.820	0.878
word2vec (CBOW) + Pragmatic	0.762	0.786	0.785	0.807	0.821	0.868
word2vec (CBOW) + Implicit incongruity	0.766	0.788	0.787	0.807	0.823	0.846
word2vec (CBOW) + Explicit incongruity	0.768	0.791	0.789	0.807	0.825	0.879
fastText (Skip-gram) + Lexical	0.770	0.792	0.791	0.811	0.830	0.865
fastText (Skip-gram) + Pragmatic	0.780	0.796	0.792	0.813	0.832	0.886
fastText (Skip-gram) + Implicit incongruity	0.781	0.797	0.796	0.813	0.835	0.863
fastText (Skip-gram) + Explicit incongruity	0.781	0.799	0.798	0.814	0.836	0.872
fastText (CBOW) + Lexical	0.782	0.800	0.798	0.814	0.837	0.871
fastText (CBOW) + Pragmatic	0.782	0.801	0.802	0.815	0.838	0.871
fastText (CBOW) + Implicit incongruity	0.782	0.802	0.806	0.817	0.841	0.907
fastText (CBOW) + Explicit incongruity	0.784	0.807	0.809	0.824	0.843	0.855
GloVe + Lexical	0.787	0.809	0.810	0.826	0.844	0.863
GloVe + Pragmatic	0.788	0.811	0.817	0.831	0.849	0.847
GloVe + Implicit incongruity	0.792	0.811	0.818	0.832	0.853	0.853
GloVe + Explicit incongruity	0.793	0.812	0.820	0.833	0.856	0.858
LDA2Vec + Lexical	0.802	0.813	0.822	0.835	0.858	0.857
LDA2Vec + Pragmatic	0.811	0.815	0.827	0.841	0.861	0.872
LDA2Vec + Implicit incongruity	0.815	0.818	0.828	0.849	0.862	0.858
LDA2Vec + Explicit incongruity	**0.816**	**0.823**	**0.845**	**0.853**	**0.873**	**0.876**

In order to summarize the main findings of the empirical analysis, Fig. 1 presents the main effects plot for average F-measure values of compared representation schemes and Fig. 2 presents the main effects plot for different subsets of dataset. Among the compared subsets of the corpus, ranging from 5000 to 30000, the highest predictive performance is obtained by utilizing the entire corpus, denoted as Subset#6.

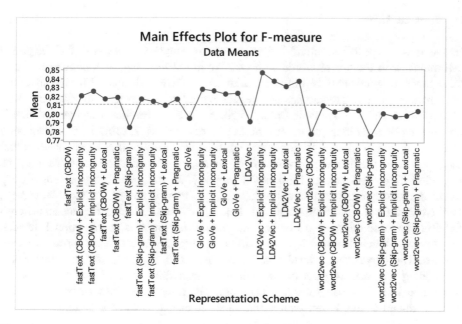

Fig. 1. The main effects plot for average F-measure values of compared representation schemes

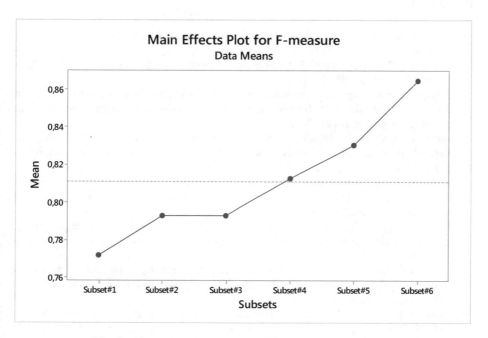

Fig. 2. The main effects plot for different subsets of dataset.

6 Conclusion

Sarcasm is a type of nonliteral language, where people may express their negative sentiments with the use of words with positive literal meaning, and, conversely, negative meaning words may be utilized to indicate positive sentiment. With the advances in information and communication technologies, the immense quantity of user-generated information available on the Web. As a result, sentiment analysis, which is the process of extracting public sentiment towards entities or subjects, is a promising research direction. Much online content contains sarcasm or other forms of nonliteral language. The predictive performance of sentiment classification schemes may be degraded if sarcasm cannot be handled properly. In this paper, we present a deep learning based approach to sarcasm detection. In this scheme, LDA2vec, word2vec, fastText and GloVe word embedding schemes have been utilized for sarcasm identification. In addition to word-embedding based feature sets, conventional lexical, pragmatic, implicit incongruity and explicit incongruity based feature sets are considered. The experimental results indicated that LDA2vec outperforms other word-embedding schemes for sarcasm identification. In addition, the utilization of lexical, pragmatic, implicit incongruity and explicit incongruity based features in conjunction with word-embedding based representation schemes can yield promising results.

References

1. Joshi, A., Bhattacharyya, P., Carman, M.J.: Automatic sarcasm detection: a survey. ACM Comput. Surv. **50**, 73 (2017)
2. Fersini, E., Messina, E., Pozzi, F.A.: Sentiment analysis: Bayesian ensemble learning. Decis. Support Syst. **68**, 26–38 (2014)
3. Joshi, A., Bhattacharyya, P., Carman, M.J.: Understanding the phenomenon of sarcasm. In: Joshi, A., Bhattacharyya, P., Carman, M.J. (eds.) Investigations in Computational Sarcasm, pp. 33–57. Springer, Berlin (2018)
4. Onan, A.: Sarcasm identification on twitter: a machine learning approach. In: Silhavy, R., Senkerik, R., Kominkova, Z., Prokopova, Z., Silhavy, P. (eds.) Artificial Intelligence Trends in Intelligent Systems, pp. 374–383. Springer, Berlin (2017)
5. Muresan, S., Gonzalez-Ibanez, R., Ghosh, D., Wacholder, N.: Identification of nonliteral language in social media: a case study on sarcasm. J. Assoc. Inf. Sci. Technol. (2016). https://doi.org/10.1002/asi.23624
6. Java, A., Song, X., Finin, T., Tseng, B.: Why we twitter: understanding microblogging usage and communities. In: Proceedings of the 9th WebKDD Conference, pp. 56–65. ACM, New York (2007)
7. Zhang, M., Zhang, Y., Fu, G.: Tweet sarcasm detection using deep neural network. In: Proceedings of the 26th International Conference on Computational Linguistics, pp. 2449–2460. COLING, New York (2016)
8. Gonzalez-Ibanez, R., Muresan, S., Wacholder, N.: Identifying sarcasm in twitter: a closer look. In: Proceedings of the 49th Annual Meeting of the Association for Computational Linguistics, pp. 581–586. ACL, New York (2011)
9. Reyes, A., Rosso, P., Buscaldi, D.: From humar recognition to irony detection: the figurative language of social media. Data Knowl. Eng. **74**, 1–12 (2012)

10. Reyes, A., Rosso, P., Veale, T.: A multidimensional approach for detecting irony in twitter. Lang. Resour. Eval. **47**(1), 239–268 (2013)
11. Ptacek, T., Habernal, I., Hong, J.: Sarcasm detection on czech and english twitter. In: Proceedings of COLING 2014, pp. 213–223. COLING, New York (2014)
12. Barbieri, F., Saggion, H., Ronzano, F.: Modelling sarcasm in twitter a novel approach. In: Proceedings of the 5th Workshop on Computational Approaches to Subjectivity, Sentiment and Social Media Analysis, pp. 50–58. ACL, New York (2014)
13. Rajadesingan, A., Zafarani, R., Liu, H.: Sarcasm detection on twitter: a behavioural modelling approach. In: Proceedings of the Eight ACM International Conference on Web Search and Data Mining, pp. 97–106. ACM, New York (2015)
14. Hernandez-Faria, D., Patti, V., Rosso, P.: Irony detection in twitter: the role of affective content. ACM Trans. Internet Technol. **16**(3), 1–19 (2016)
15. Bouazizi, M., Ohtsuki, T.O.: A pattern-based approach for sarcasm detection on Twitter. IEEE Access **4**, 5477–5488 (2016)
16. Kumar, L., Somani, A., Bhattacharyya, P.: Having 2 hours to write a paper is fun: detecting sarcasm in numerical portions of text. arXiv preprint arXiv:1709.01950 (2017)
17. Mishra, A., Kanojia, D., Nagar, S., Dey, K., Bhattacharyya, P.: Harnessing cognitive features for sarcasm detection. arXiv preprint arXiv:1701.05574 (2017)
18. Ghosh, D., Guo, W., Muresan, S.: Sarcastic or not: word embeddings to predict the literal or sarcastic meaning of words. In: Proceedings of the 2015 Conference on Empirical Methods in Natural Language Processing, pp. 1003–1012. ACL, New York (2015)
19. Joshi, A., Tripathi, V., Patel, K., Bhattacharyya, P., Carman, M.: Are word embedding-based features useful for sarcasm detection. arXiv preprint arXiv:1610.00883 (2016)
20. Poria, S., Cambria, E., Hazarika, D., Vij, P.: A deeper look into sarcastic tweets using deep convolutional neural networks. arXiv preprint arXiv:1610.08815 (2016)
21. Rezaeinia, S.M., Ghodsi, A., Rahmani, R.: Improving the accuracy of pre-trained word embeddings for sentiment analysis. arXiv preprint arXiv:1711.08609 (2017)
22. Mikolov, T., Chen, K., Corrado, G., Dean, J.: Efficient estimation of word representations in vector space. arXiv preprint arXiv:1301.3781 (2013)
23. Bairong, Z., Wenbo, W., Zhiyu, L., Chonghui, Z., Shinozaki, T.: Comparative analysis of word embedding methods for DSTC6 end-to-end conversation modelling track. In: Proceedings of the 6th Dialog System Technology Challenges Workshop (2017)
24. Bojanowski, P., Grave, E., Joulin, A., Mikolov, T.: Enriching word vectors with subword information. arXiv preprint arXiv:1607.04606 (2016)
25. Pennington, J., Socher, R., Manning, C.: Glove: global vectors for word representation. In: Proceedings of the 2014 Conference on Empirical Methods in Natural Language Processing, pp. 1532–1543. ACL, New York (2014)
26. Moody, C.E., Johnson, R., Zhang, T.: Mixing Dirichlet Topic Models and Word Embeddings to Make Lda2vec (2014). https://www.datacamp.com/community/tutorials/lda2vec-topic-model
27. Johnson, R., Zhang, T.: Effective use of word order for text categorization with convolutional neural networks. arXiv preprint arXiv:1412.1058 (2014)
28. Young, T., Hazarika, D., Poria, S., Cambria, E.: Recent trends in deep learning based natural language processing. IEEE Comput. Intell. Mag. **13**(3), 55–75 (2018)
29. Kilimci, Z., Akyokus, S.: Deep learning and word embedding-based heterogeneous classifier ensembles for text classification. Complexity **2018**, 1–10 (2018)
30. Cireşan, D., Meier, U., Schmidhuber, J.: Multi-column deep neural networks for image classification. arXiv preprint arXiv:1202.2745 (2012)

31. Gonzalez-Ibanez, R., Muresan, S., Wacholder, N.: Identifying sarcasm in Twitter: a closer look. In: Proceedings of the 49th Annual Meeting of the Association for Computation Linguistics, pp. 581–586. ACL, New York (2011)
32. Paredes-Valverde, M.A., Colomo-Palacios, R., Salas-Zarate, M., Valencia-Garcia, R.: Sentiment analysis in Spanish for improvement of product and services: a deep learning approach. Sci. Program. **2017**, 1–12 (2017)
33. Riloff, E., Qadir, A., Surve, P., De Silva, L., Gilbert, N., Huang, R.: Sarcasm as contrast between a positive sentiment and negative situation. In: Proceedings of the 2013 Conference on Empirical Methods in Natural Language Processing, pp. 704–714. ACL, New York (2013)
34. Ramteke, A., Malu, A., Bhattacharyya, P., Nath, J.S.: Detecting turnarounds in sentiment analysis: thwarting. In: Proceedings of the 51st Annual Meeting of the Association for Computational Linguistics, pp. 860–865. ACL, New York (2013)

Validating the Conceptual Framework with Exploratory Testing

Shuib Basri[1,2], Thangiah Murugan[1,2(✉)],
and Dhanapal Durai Dominic[1,2]

[1] Computer and Information Science Department,
Universiti Teknologi PETRONAS,
Bandar Sri Iskandar, 32610 Seri Iskandar, Perak, Malaysia
{shuib_basri,dhanapal_d}@utp.edu.my,
tm_gun@hotmail.com
[2] Software Quality and Quality Engineering (SQ2E) Research Group,
Universiti Teknologi PETRONAS,
Bandar Sri Iskandar, 32610 Seri Iskandar, Perak, Malaysia

Abstract. The Small and Medium Enterprises (SMEs') in the IT industry plays a major role towards the growth of country's economy. SME makes a substantial contribution towards development of the country in the form of GDP and it is widely recognized around the world. SMEs are dependent on software to accomplish their objectives and maintain survivability and sustainability in their businesses, but they often face many challenges due to various factors. Most of the SME's aim to produce valuable software in a short duration of time, and the traditional method of software development is not viable to provide satisfactory answer with regards to scope, time and resources they need and hence the quality of the product is not guaranteed. Though several approaches towards testing are available, adopting Exploratory testing has so far not been investigated in the context of SME, where software quality is often associated to business-critical values. In this research study, a conceptual framework using Exploratory Testing (ET) has been developed and it is validated by using Partial Least Square Structural Equation Modelling (PLS_SEM).

Keywords: Exploratory testing · PLS-SEM · SMEs' · Software quality

1 Introduction

The unprecedented development in the technology and the competitive driven market strategies have forced the SME's to adopt the changes in technology where they lacked in many factors and are experiencing challenges. Some of the challenges are the projects cannot be completed on time and exceeds the budget, failure to learn from the past mistake, lack of software development skill and poor software quality and collapse in the organizational intelligence [1, 2]. Most of the SME's aim to produce valuable software in a short duration of time, however, due to the above challenges the traditional method of software development is not viable to provide satisfactory answer with regards to scope, time and resources they need [3]. Frequent changes in the requirement

© Springer Nature Switzerland AG 2019
R. Silhavy (Ed.): CSOC 2019, AISC 984, pp. 305–317, 2019.
https://doi.org/10.1007/978-3-030-19807-7_30

and changes in the technology have created the need for more effective approaches in the software development activities [4] beyond the traditional approaches. Hence, the SME's tend to adopt agile development process.

There are several agile methods available and each method has its own significance. Agile method is intended to be a collection of values, principles, and practices applied in a project in a more flexible manner than traditional development methods that can be applied to a software development process. Though agile method has many advantages but it is also having some challenges, especially, when SME's tend to adopt this method. In agile methodologies, due to the frequent interaction with customers and frequent changes in the requirement, it is difficult to maintain a proper documentation for the ongoing projects. Besides that, the success of a project is depending on the user involvement with better understanding and communication process [5]. In addition to that, standard testing procedure is not integrated into the agile development process and most of the organization has to perform manual testing.

Software Testing involves huge cost and majority of the project development lifecycle focus towards testing activities. There are several instances available that major companies have still had to recall products that lead to expensive consequences and companies have been forced to compensate customers due to poor quality. For example, Microsoft received so many complaints about the quality, performance, and security of Windows Vista that it was forced to extend support for its predecessor, Windows XP (http://www.microsoft.com/windows/letter.html).

Since software development practices have moved from sequential development models (waterfall) to iterative models (agile or lean), tester's must provide quicker and accurate status updates more frequently. However, the ability to provide updates using traditional scenario-based testing is limited and so an alternate solution to support this fast-moving new reality is needed. Exploratory Testing (ET) is an important activity in an agile environment which it can help software testers to keep up with the rapid development pace of agile software projects [6]. However, there is not enough research study is available using Exploratory Testing is employed as a testing activity in the SME's practicing agile development.

As far as SME is concerned which is operating under highly competitive and economic pressure due to the dynamic business requirements, quality assurance often helps to focus on testing activities on critical areas of a software product [7]. Though several approaches to testing are available, adopting Exploratory testing has so far not been investigated in the context of SME, where software quality are often associated to business-critical values. Organization may do testing process well even without using the testing tool and methods. Hence, it poses a problem if the organization has the characteristics that does not support and embrace testing due to lack of well-defined testing process and tend to adopt ad-hoc testing methods [2].

Various factors affecting the software quality, which is the essential requirement for customer satisfaction in the SMEs were identified, examined and grouped together. Based on the research gap identified from the literature by examining the weakness in the testing methods and techniques adopted by the SME's, a conceptual framework has been developed [8]. The aim of the study is to propose a conceptual framework using Exploratory Testing that has not been adopted in the SMEs so far, which helps to identify critical bugs more than that of doing with traditional script-based or automated testing.

This research study focuses on to validate the conceptual framework on ET using PLS_SEM and how this testing approach influences to improve the testing methods in order to produce quality software in SME's.

2 Theoretical Background and Hypothesis

Software testing is the process of not only finding the errors, but it also reveals that at what level the quality has been achieved. Hence, software testing provides information about defects and problems in an application and simultaneously evaluates the achieved quality [9]. Much effort is devoted to the improvement in areas of analysis and requirements, design and code reviews. However, in the spirit of continuous improvement in software quality, there is not much effort to improve its testing techniques to reduce customer found defects [10].

2.1 Exploratory Testing

In the traditional method, software testing is defined as test cases are planned prior to the execution and compare their outputs against the expected results. This type of document driven, planned approach towards testing is called as test-case based testing (TCT). TCT emphasis on detailed documentation of test-cases and directed from the requirements that verify the correct implementation of functional specification [11]. In contrast to TCT, exploratory testing needs a tester to listen, read, think and report, effectively and rigorously, thus making this approach as productive in terms of revealing critical information, especially vulnerabilities, as against the scripted approach.

There is common perception and misunderstanding that exploratory testing will simply execute random actions, however, it is important to emphasize that this is a very rigorous approach. Sometimes automation is not sufficient to ensure that the system does everything we expect and there are possibilities that the system may goes wrong while in use and in these circumstances, Exploratory testing is a perfect fit [12]. According to Dunlop, the three reasons why ET is useful when compared to other testing methods are: (1) it helps to expose the defects that manual and automated testing could miss, (2) it lets to find any functional defects when automated test is not viable, and (3) helps to expose a broader variety of issues and reduce the risk that a critical issue goes unnoticed [13]. There is not sufficient theory available with the use of Exploratory Testing in SMEs and there is no standard guidance available to implement ET.

The aim of the study is to develop and validate the conceptual framework using Exploratory Testing that has not been adopted in the SMEs so far, which helps to identify critical bugs more than that of doing with traditional script based or automated testing.

2.2   Research Model and Hypothesis Development

In this research paper, developing the conceptual framework subsequent to the development of the research hypothesis are discussed briefly. A detailed literature review has been discussed prior to this research work [14] and the conceptual framework was developed after analyzing the various factors that influenced the software quality were identified and outlined as shown in Fig. 1 was found in [8].

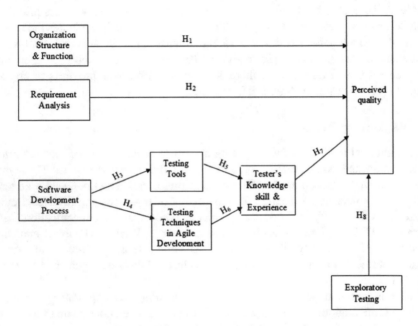

Fig. 1. Proposed conceptual framework

There is a significant relationship exist between these factors and are established by using the research model and framework. This research framework which later translated to become conceptual model was the main driver of this research because this framework will become the main input throughout the other stages of the methodology.

Table 1 represents the detailed description of the relevant construct adopted in the conceptual model. The literature review was not sufficient enough for theoretical interpretations and theory development. Thus, qualitative and quantitative research was deployed with the hopes to get a better understanding of the subject matter. The study extends the basis of using Exploratory Testing in SME's to measure the effectiveness in the software testing process which is an essential requirement for the software quality and therefore the following hypothesizes are proposed:

H1. *Organization structure and function has a significant positive influence on the perceived quality.*

H2. *Requirement analysis has a significant impact and positive influences on perceived quality.*

H3. *Agile development process has significant positive influence on testing tools.*
H4. *Agile development process has a significant positive influence on different testing techniques.*
H5. *Testing tools have a significant positive influence on tester's knowledge, skill and experience.*
H6: *Tester's knowledge skill and experience is positively influenced by the testing techniques adopted in the agile development process.*
H7: *There is a positive influence on tester's knowledge, skills and experience towards achieving the perceived quality of the software project.*
H8: *Exploratory testing has a significant and positive direct influence on the perceived software quality.*

Table 1. Latent variables

Variables	References
Organization structure and function (OS)	Mishra and Mishra [37]; Almomani et al. [34]; Musa and Chinniah [42]; Wang [41]; Rahman et al. [36]
Requirement analysis (RA)	Vasconcelos et al. [47]; Higgins [48]
Software development process (SDP)	Basri and O'Connor [38, 39]; Almomani et al. [35]; Sharif et al. [46]
Testing in agile (TADM)	van der Broek et al. [49]; Hellman et al. [50]; Suresh Kumar and Sambasiva Rao [9]
Testing methods, tools and techniques (TMT)	Thangiah and Basri [14]; Felderer and Ramler [7]; Austin and Williams [32]; Akimoto et al. [40]
Tester's skill, knowledge and experience (TKES)	Musa and Chinniah [42]; Gruner and van Zyl [2]; Almomani et al. [35]
Exploratory testing (ET)	Itkonen et al. [43]; Itkonen and Mäntylä [44]; Hellmann and Maurer [45]; Afzal et al. [11]; Prakash and Gopalakrishnan [33]

3   Methodology

For this research study, a mixed method approach has been adopted which focuses on collecting, analyzing, and mixing both quantitative and qualitative data in a single study. Combining the use of quantitative and qualitative approaches provides a better understanding of research problems than either approach alone. This approach is most suitable to explain, interpret and to explore a new phenomenon which serves a theoretical perspective. Also, this method is most appropriate on designing and validating an instrument.

Data Collection and Sampling Method. Data collection method was governed by the development of conceptual framework and elaborated in the previous study [8]. For this study, the data was collected by conducting structured questionnaire survey and was distributed to organization related to the research study. Stratified random sampling method was employed in this research study for a large scale quantitative survey by dividing the population into a set of smaller non-overlapping sub-groups (strata), then doing a simple random sampling for each sub-group. The survey was conducted in the SME organizations associated with only software development and a total of 280 questionnaire forms were distributed in Kuala Lumpur, Penang, Selangor and Klang Valley. As a response, only 200 samples were considered and selected for the data analysis process. The remaining questionnaire sets were incomplete and considered as inappropriate. The analysis used 200 completed questionnaire sets which are sufficient enough based on the rule of thumb specified in the PLS-SEM [18].

4   Partial Least Square Structural Equation Modelling

In recent years, there have been substantial increases in the applications of structural equation modeling (SEM) in various fields of studies [16, 17]. The primary reason for using this structural equation modeling is because of its improved ability to assess the reliability and validity of the constructs which have multiple item measures and to test the relationship between the constructs created in the structural model [19, 24]. According to Hair Jr. [20], there are two methods of SEM available for the researchers that dominate the literature: (1) covariance-based SEM (CB-SEM), and (2) variance-based partial least squares (PLS-SEM) [29–31]. The fundamental difference between the methods is that CB-SEM is based on common factor model which assumes that the analysis is only based on the common variance in the data and the PLS-SEM is based on composite model which includes common, specific and error variance [21]. This will help to uses all the variance from the independent variable(s) that can predict the variance in the dependent variable(s).

4.1   Measurement Model Assessment

The measurement model assessment in PLS-SEM comprised of two steps, they are the assessment of convergent validity and assessment of discriminant validity. The convergent validity is measured by construct reliability and validity which measures Composite Reliability (CR), Average Variance Extracted (AVE). The purpose of the measurement model assessment is to evaluate the relationship between the indicators and the constructs as well as to identify the correlational relationship between constructs to ensure that each of the constructs in the study is distinct from the other. Validation of the measurement model is a requirement for assessing the structural model. The threshold value for CR must be higher than 0.7 and in this research study all the constructs had satisfied the minimum value which is given in Table 2. Similarly, the threshold values for AVE should be greater than 0.5 and for this assessment also the proposed model satisfied its requirements.

4.2   Structural Model Assessment Criteria

The structural model assessment is the second step of evaluation criteria and this will be done after the measurement model indicates satisfactory quality. This step is mainly focused on learning about the predictive capabilities of the research model such as coefficient of determination (R^2), the path coefficients and the effect size (f^2).

Table 2. Convergent validity: construct reliability and validity

Constructs	CR	AVE	Alpha
Organizational structure	0.858	0.550	0.793
Requirement analysis	0.836	0.563	0.776
Software development process	0.891	0.804	0.756
Testing in agile development	0.850	0.532	0.794
Testing technique and methods	0.871	0.575	0.824
Tester's knowledge, skill, experience	0.824	0.547	0.766
Exploratory testing	0.861	0.559	0.797

Path Coefficient. Path coefficient represents the hypothesized relationship among the constructs and has a standard value between −1 and +1. Any value close to +1 indicates strong positive relationships that are always significant statistically. This coefficient is significant or not is purely depends on the standard error that is obtained by doing bootstrapping procedure in the SmartPLS software [15]. This will allow computing the t value to estimate the significance of path coefficient linking between the two constructs. Commonly used critical values for two-tailed tests are 1.65 for the significance level 10%, 1.96 for the significance level 5% and 2.67 for the significance level 1%. Hence, if the result of the path coefficient is higher than the critical value, then the path coefficient is significant at certain probability of error.

Coefficient of Determination (R2). The next step involves is evaluating the R^2 value, which indicates the variance explained the predictive accuracy and is calculated as a squared correlation in each of the endogenous constructs. The R^2 value ranges between 0 and 1 with higher levels indicating the higher levels of predictive accuracy. As a rule of thumb, the accepted level of R^2 values of 0.75, 0.50, and 0.25 can be considered substantial, moderate, and weak [22, 24, 25].

Effect Size (f2). In addition to evaluating the R^2 values of all endogenous constructs, the change in the R^2 value when a specific predictor construct is omitted from the model can be used to evaluate whether the omitted construct has a substantive impact on the endogenous constructs. This measure is referred to as the f^2 effect size [24]. This can be calculated by using the formula:

$$f^2 = \left(R_i^2 - R_e^2\right)/\left(1 - R_i^2\right) \tag{1}$$

where R_i^2 is when the path is included in the model, R_e^2 is when the path is excluded from the model. The following guidelines are outlined for interpreting the effect sizes: Less than 0.02 – no effect, Small – 0.02, medium 0.15 and Large – 0.35.

5   Result and Discussion

The relationships between the exogenous and endogenous latent variables are assessed in the structural model, after the reliability and validity of the measurement model assessment is completed and the results of the measurement model analysis is available in [8]. The assessment methods involved in the structural model are through evaluating R^2 value, which is coefficient of determination and β values, the path coefficients of the model. For a good model, the value of R^2 of endogenous latent variable should be higher than 0.25 [23, 24]. Since the R^2 value for the proposed conceptual model is 0.607 which is higher than the suggested values, the model is considered to have substantial degree of explained variance of perceived quality by inhibiting factors (Fig. 2).

Fig. 2. Conceptual Framework path model in PLS-SEM

The result from Table 3 has demonstrated that all the paths, barring organizational structure, attained the t-values which are higher than the cut-off point for a significance level of 5%, that is 1.96. This implies that all the paths in this research model have a strong effect on the quality aspects. The highest β value is 0.767 for Exploratory Testing. This is the most significant construct influences critically to improve the effectiveness and efficiency of the software quality in the SME's.

Table 3. Path coefficients with t-values for the structural model

Hypothesis	Original sample	t-values	p-values	Support
H1	0.056	0.843	0.200	No
H2	0.128	2.505	0.006*	Yes
H3	0.572	12.948	0.000*	Yes
H4	0.562	10.718	0.000*	Yes
H5	0.442	7.877	0.000*	Yes
H6	0.361	6.635	0.000*	Yes
H7	-0.092	2.420	0.008*	Yes
H8	0.767	16.547	0.000*	Yes

*P < 0.05

6   Model Validation

The conceptual model developed for this research study must be validated to check its usefulness. This validation process is carried out by checking the stability of the model through calculating the sample size using power analysis test. Power analysis test (1- β) is to check the stability of the model's parameter with sample size used for the analysis [26]. It is confirmed that the sample size used for this research study is sufficient for generating a stable model. The test is conducted by calculating the power of the model through G*Power 3.1.9.2 software package [27, 28]. Input parameters required for the software are the significance level (α) of the test, the sample size (N) used for the research study and the effect size of the population.

Based on the above details, the input parameters used for this research study are significance at 0.05 (i.e., 95% confidence level), sample size (N) as 200, and the effect size 0.15. The generated values of power analysis for the various sample sizes are shown in Fig. 3 which indicates that the power of overall model increases

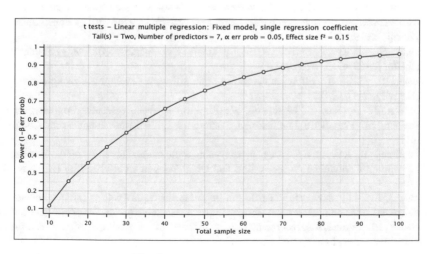

Fig. 3. Generated power analysis

correspondingly when there is an increase in the sample size. It achieved 100% power at sample size of 100 since this study uses 200 samples and it is obvious that it is more adequate for achieving substantial power.

7 Conclusion

This study highlighted the effectiveness and efficiency of implementing Exploratory Testing to improve the software quality in the SME's by various inhibiting factors. These factors are grouped and modeled into 7 categories in SmartPLS software where it was analyzed for assessing the effect on software quality. The major conclusion drawn from this study are the R^2 value of the model is more than 0.50 and classified as a good model where it has a substantial degree of variance explained for quality constructs with adoption of Exploratory Testing. The hypothesizes were tested using bootstrapping procedure and proved that from H2 to H8 all the hypothesis had positive relationship and have a significant impact to improve the software quality, barring the hypothesis H1. The sample size of 200 involved in the study was adequate and validated through power analysis. Thus, it is obvious from the result and analysis that implementing Exploratory testing in the SMEs' provide a most significant factor that increases the efficiency of bug identification process which helps to improve the software quality considerably.

Acknowledgment. This paper/research is partly supported by the Ministry of Education, under the Fundamental Research Grant Scheme (FRGS) with Ref. No FRGS/1/2018/ICT04/UTP/02/04.

References

1. Shongwe, M.M.: Knowledge management in small software development organisations: a South African perspective. South Afr. J. Inf. Manage. **19**(1), a784 (2017). https://doi.org/10.4102/sajim.v19i1.784
2. Gruner, S., van Zyl, J.: Software testing in small IT companies: a (not only) South African problem. SACJ **47**, 7–32 (2011)
3. Coelho, E., Basu, A.: Effort estimation in agile software development using story points. IJAIS **3**(7), 7–10 (2012)
4. Järvinen, J., Huomo, T., Mikkonen, T.: Running software research programs: an agile approach. In: 39th IEEE International Conference on Software Engineering Companion (2017)
5. Choudhary, B., Rakesh, S.K.: An approach using agile method for software development. In: International Conference on Innovation and Challenges in Cyber Security (ICICCS 2016) (2016)
6. Ghahrai, A.: Why Exploratory Testing is Important in Agile Projects (2015). https://www.testingexcellence.com/exploratory-testing-important-agile-projects/. Accessed 25 Nov 2017
7. Felderer, M., Ramler, R.: Risk orientation in software testing processes of small and medium enterprises: an exploratory and comparative study. Softw. Qual. J. **24**, 519–548 (2015)

8. Murugan, T., Basri, S., Dominic, P.D.D.: Analyzing the conceptual model for exploratory testing framework using PLS-SEM. In: 3rd International Conference on Computing, Mathematics and Statistics (2017)
9. Suresh Kumar, P., Samba Siva Rao, N.: Automation of software testing in agile development - an approach and challenges with distributed database systems. Global J. Res. Anal. **3**(7) (2014)
10. Moritz, E.: Case study: how analysis of customer found defects can be used by system test to improve quality. In: ICSE, pp. 123–129 (2009)
11. Afzal, W., Ghazi, A.N., Itkonen, J., Torkar, R., Andrews, A., Bhatti, K.: An experiment on the effectiveness and efficiency of exploratory testing. J. Empirical Softw. Eng. **20**, 844–878 (2014)
12. Tamara: Exploratory Testing in Agile Software Development Teams (2015). http://www.meritsolutions.com/software-development/exploratory-testing-in-agile-software-development-teams/. Accessed 17 Nov 2017
13. Dunlop, C.: 3 Reasons Agile Team Loves Exploratory Testing (2017). https://www.tricentis.com/blog/2017/02/14/agile-exploratory-testing-love/. Accessed 23 Nov 2017
14. Thangiah, M., Basri, S.: A preliminary analysis of various testing techniques in Agile development - a systematic literature review. In: ICCOINS 2016 (2016)
15. Ringle, C.M., Wende, S., Becker, J.-M.: "SmartPLS 3." Boenningstedt: SmartPLS GmbH (2015). http://www.smartpls.com
16. Matthews, L.M., Zablah, A.R., Hair, J.F., Marshall, G.W.: Increased engagement or reduced exhaustion: which accounts for the effect of job resources on salesperson job outcomes? J. Market. Theory Pract. **24**(3), 249–264 (2016)
17. Rutherford, B.N., Park, J., Han, S.-L.: Increasing job performance and decreasing salesperson propensity to leave: an examination of an Asian sales force. J. Pers. Sell. Sales Manage. **31**(2), 171–184 (2011)
18. Hair, J.F., Hult, G.T.M., Ringle, C.M., Sarstedt, M.: A Premier on Partial Least Squares Structural Equation Modeling (PLS-SEM). Sage, Thousand Oaks (2014)
19. Bollen, K.A.: Structural Equations with Latent Variables. Wiley, New York (1989)
20. Hair Jr., J.F., Matthews, L.M., Matthews, R.L., Sarstedt, M.: PLS-SEM or CB-SEM: updated guidelines on which method to use. Int. J. Multivariate Data Anal. **1**(2), 107–123 (2017)
21. Hair, J.F., Hult, G.T.M., Ringle, C.M., Sarstedt, M.: A Primer on Partial Least Squares Structural Equation Modeling (PLS-SEM), 2nd edn. SAGE, Thousand Oaks (2017)
22. Hair, J.F., Ringle, C.M., Sarstedt, M.: PLS-SEM: indeed a silver bullet. J. Market. Theory Pract. **19**, 139–151 (2011)
23. Cohen, J., Cohen, P., West, S.G., Aiken, L.S.: Applied Multiple Regression/Correlation Analysis for the Behavioral Sciences, 3rd edn. Lawrence Erlbaum Associates, Mahwah (2003)
24. Hair, J.F., Sarstedt, M., Ringle, C.M., Mena, J.A.: An assessment of the use of partial least squares structural equation modeling in marketing research. J. Acad. Mark. Sci. **40**(3), 414–433 (2012)
25. Henseler, J., Ringle, C.M., Sinkovics, R.R.: The use of partial least squares path modeling in international marketing. Adv. Int. Market. **20**, 277–319 (2009)
26. Chin, W.W.: The partial least squares approach to structural equation modeling. In: Marcoulides, G.A. (ed.) Modern Methods for Business Research, pp. 295–336. Erlbaum, Mahwah (1998)
27. Erdfelder, E., Faul, F., Buchner, A., Lang, A.-G.: Statistical power analyses using G*Power 3.1: tests for correlation and regression analyses. Behav. Res. Methods **41**(4), 1149–1160 (2009)

28. Faul, F., Erdfelder, E., Lang, A.-G., Buchner, A.: G*Power 3: a flexible statistical power analysis program for the social, behavioral, and biomedical sciences. Behav. Res. Methods **39**(2), 175–191 (2007)
29. Lohmöller, J.-B.: Latent Variable Path Modeling With Partial Least Squares. Physica, Heidelberg (1989)
30. Wold, H.O.A.: Partial least squares. In: Kotz, S., Johnson, N.L. (eds.) Encyclopedia of Statistical Sciences, vol. 6, pp. 581–591. Wiley, New York (1985)
31. Jöreskog, K.G., Wold, H.O.A.: The ML and PLS techniques for modeling with latent variables: historical and comparative aspects. In: Wold, H.O.A., Jöreskog, K.G. (eds.) Systems Under Indirect Observation, Part I, pp. 263–270. North-Holland, Amsterdam (1982)
32. Austin, A., Williams, L.: One technique is not enough: a comparison of vulnerability discovery techniques. In: Proceedings of the 5th International Symposium on Empirical Software Engineering and Measurement, ESEM 2011, pp. 97–106 (2011). https://doi.org/10.1109/esem.2011.18
33. Prakash, V., Gopalakrishnan: Testing efficiency exploited: scripted versus Exploratory testing (2011)
34. Almomani, M.A.T., Basri, S., Mahmood, A.K., Baashar, Y.M.: An empirical analysis of software practices in Malaysian Small and Medium Enterprises. In: 2016 3rd International Conference on Computer and Information Sciences (ICCOINS). IEEE (2016)
35. Almomani, M.A.T., Basri, S., Mahmood, A.K., Amos Orenyi Bajeh, A.O.: Software development practices and problems in Malaysian small and medium software enterprises: a pilot study. In: 2015 5th International Conference on IT Convergence and Security (ICITCS). IEEE (2015)
36. Rahman, N.A., Yaacob, Z., Radzi, R.M.: The challenges among Malaysian SME: a theoretical perspective. World J. Soc. Sci. **6**(3), 124–13 (2016)
37. Mishra, D., Mishra, A.: Software process improvement in SME's: a comparitive view. In: ComSIS, vol. 6, no. 1 (2009)
38. Basri, .S., O'Connor, R.: Evaluation on knowledge management process in very small software companies: a survey (2012)
39. Basri, S., O'Connor, R.: Understanding the perception of very small software companies towards the adoption of process standards. In: Riel, et al. (eds.) Systems, Software and Services Process Improvement. CCIS, vol. 99, pp. 153–164. Springer (2010)
40. Akimoto, S., Yaegashi, R., Takagi, T.: Test case selection technique for regression testing using differential control flow graphs. In: 16th International Conference on Software Engineering, Artificial Intelligence, Networking and Parallel/Distributed Computing (SNPD) (2015)
41. Wang, Y.: What are the biggest obstacles to growth of SMEs in developing countries?–An empirical evidence from an enterprise survey. Borsa Istanbul Rev. **16**(3), 167e176 (2016)
42. Musa, H., Chinniah, M.: Malaysian SMEs development: future and challenges on going green. In: 6th International Research Symposium in Service Management, IRSSM-6 2015, 11–15 August 2015, UiTM Sarawak, Kuching, Malaysia (2016)
43. Itkonen, J. Mäntylä,, M.V., Lassenius, C.: How do testers do it? An exploratory study on manual testing practices. In: Third International Symposium on Empirical Software Engineering and Measurement (2009)
44. Itkonen, J., Mäntylä, M.V.: Are test cases needed? Replicated comparison between exploratory and test-case-based software testing. J. Empirical Soft. Eng. **19**(2), 303–342 (2014)
45. Hellmann, T.D., Maurer, F.: Rule-based exploratory testing of graphical user interfaces. In: Proceedings of Agile Conference (2011)

46. Sharif, B., Khan, S.A., Bhatti, M.W.: Measuring the impact of changing requirement on software project cost: an emperical investigation. Int. J. Comput. Sci. Issues **9**(3), 170 (2012)
47. Vasconcelos, A., Silva, C., Wanderley, E.: Requirements engineering in agile projects: a systematic mapping based in evidences of industry. In: XVIII Ibero-American Conference on Software Engineering, pp. 460–473 (2015)
48. Higgins, T.: Poor Requirements – What Impact Do They Have? (2012). http://www. blueprint.com/poor_requirements_what_impact_do_they_have/. Accessed 7 Sept 2017
49. van der Broek, R., Bonsangue, M.M., Chaudron, M., van Merode, H.: Integrating testing into Agile software development processes. In: 2nd International Conference on Model-Driven Engineering and Software Development (MODELSWARD), pp. 561–574 (2014)
50. Hellman, T.D., Sharma, A., Ferreira, J., Maurer, F.: Agile testing: past, present, and future. In: Agile Conference, Dallas, TX, pp. 55–63 (2012)

Shapp: Workload Management System for Massive Distributed Calculations

Tomasz Gałecki and Wiktor B. Daszczuk[(✉)]

Institute of Computer Science, Warsaw University of Technology,
Nowowiejska Street 15/19, 00-665 Warsaw, Poland
tgalecki@mion.elka.pw.edu.pl, wbd@ii.pw.edu.pl

Abstract. The paper presents an overview of existing workload management systems and diagnoses difficulties in their use. Such systems require a complex configuration, and the process of ordering tasks to be carried out is burdened with many restrictions. A new solution is presented which supports the use of massive distributed computations. The work presents the process of designing, implementing and testing the workload management Shapp library, based on the HTCondor system. It implements a convenient application interface in the form of a dynamically linked library, which extends the capabilities of existing applications with a convenient mechanism allowing the use of massive distributed processing. Recursive computations in tree-like structure are possible using Shapp library.

Keywords: Distributed computing system · Massive calculations ·
Workload management · HTCondor

1 Introduction

Nowadays, calculations in the IT, engineering, economy etc. are experiencing huge solution spaces. The increasing application of machine learning and simulation is also contributing to this trend. In these fields, the processing of the widest possible space of solutions is particularly critical. Increasing computing power and falling storage prices support the possibility of massive computations.

The good example of such calculations are simulations of PRT (Personal Rapid Transit) operation with various sets of environment and technical parameters [1]. Without going into the specificity of the problem, for larger models (e.g., the Corby benchmark [1]), a single simulation takes just over 2 h and completely utilizes one processor core. For the complete simulation of the 100-element space on a single 8-core processor, the researcher would have to spend more than a day of CPU time to process the set. Moreover, adding another dimension to the examined state space, increases the computation time complexity to 1250 h (about 52 days). One of the solutions is to use many machines and manually order them to analyze consecutive sets. However, such a solution leads to incomplete utilization of the machines.

© Springer Nature Switzerland AG 2019
R. Silhavy (Ed.): CSOC 2019, AISC 984, pp. 318–329, 2019.
https://doi.org/10.1007/978-3-030-19807-7_31

In addition, collecting the results (extracting them form text files) is cumbersome and prone to error. Another option is to use scripting languages for starting the calculations. Such a script must deal simultaneously with:

- matchmaking tasks for machines,
- maintaining the computers (controlling the state of the calculations),
- reporting on the performance of consecutive tasks,
- supplying the input data,
- distribution of executable files,
- collecting the results,
- solving of environmental problems (suspension of operating systems, restarting of failed tasks).

The problem of distribution of input data and acquisition of results in both solutions (manual or using scripts) can be easily implemented using a multi-access database. However, this is not always the optimal solution (e.g., for binary data). Thus we need a system for massive calculations performed in a network of computers, called *workload management*.

Several such systems are available. Most of them offer a wide spectrum of functionalities, however, generally they are specialized for given applications. Some examples of such systems are listed in Sect. 2. Due to their specialization in solving one type of problem, it is not possible to use them in arbitrary applications. It is necessary to use middleware that supports mass processing in the distributed environment.

The calculation program itself must meet certain requirements. The majority of the command line-invoked programs, which run autonomously on the input data and return results as a set of output parameters (or files), fit the requirements of massive calculations, even if the input data vary for each program call. The requirements of the simulation program Feniks PRT simulator [2] are specific:

- long processing time of the computing unit (single simulation for a specific set of parameters),
- no parallelization possibility inside a single computing unit (instead, multiple parallel invocation of the simulator are possible),

To fulfil these requirements, due to lack of framework that supports workflow management from inside the program, the special library was elaborated, which simplifies the run of many computational tasks invoked from the central simulation unit. Such mode of operation is called *farming*. Our library *Shapp* (Sharable Application) responds the specific requirements described in Sect. 3.1.

In the paper, the details of our workload management solution are presented. In Sect. 2, available mass computation frameworks are described, with a special attention to the HTCondor system [3], which is the base of the implemented solution. Section 3 presents the details of our Shapp library. Section 4 describes functional tests. In Sect. 5, a summary is provided along with a list of other possible development directions for the library.

2   Related Work on Workload Management

The distributed workload management systems are the implementations of the idea of mass distributed computing. Computational units are individual programs that are repeatedly executed for different sets of input data [3]. Examples of the use of distributed massive processing systems include:

- *BOINC* [4] - a system based on the use of computing power provided by Internet users. Example subprojects are: *ATLAS@Home* [5] - simulation of the ATLAS project (in LHC), *Asteroids@home* [6] - broadening the knowledge of asteroids, Big and Ugly Rendering Project (*BURP*) [7] - a non-commercial and open project focusing on the rendering of three-dimensional scenes in the cloud.
- *Comcute* [8] - the system dealing, among others, with the genetic testing of a radical settlement, techniques and processing of images.

The majority of systems use already existing stable middleware. This middleware is a basis for creation of distributed computation solutions. Typical middleware is responsible for:

- administration of resources (computers in the network),
- management of a task queue,
- support for task prioritization,
- matchmaking of tasks to be carried out, which fulfill specific requirements,
- providing an Application Programming Interface (API) that simplifies the management of the entire system.

These types of tasks appear in every distributed computing system. For example, the BIONIC project [4], which is a framework program for computations on machines made available by Internet users, is based on the HTCondor system [9]. Individual systems are distinguished by their configuration (in case they are based on the same software).

2.1   Types of Distributed Processing Systems

2.1.1   Structural Classification

- *Computer cluster.* A computer cluster is a group of computers connected to each other so as to allow each system to be transparent. Computer clusters typically carry out high-performance tasks (Sect. 2.1.2), like *distcc* [10] - automatized to run distributed compilation, or Linux Virtual Server [11] - software for building general purpose clusters. It is based on several systems with greater specialization, e.g.: *UltraMonkey* [12] - a process-based process system for supporting a computing pool over the internet/intranet, *Keepalived* [13] - a distributed process-based system simplifying the use of Linux Virtual Server to ensure greater reliability and high availability of the calculation pool.

- *Network processing.* Network processing uses highly distributed, heterogeneous computer units. From the point of view of the user, this idea uses open protocols and general purpose interfaces in order to build a virtual and powerful computing environment from a network of connected different nodes. The BOINC project [4] mentioned above is an example of this. Network processing most often accomplishes tasks from the high-throughput processing class (Sect. 2.1.2). Middleware implementations are similar, they differ in additional functionalities that facilitate the use of given systems for specific types of tasks. Common examples are: *Eucalyptus* [14] - a paid tool for the integration and use of Amazon Web Services, *HTCondor* [9] - a free tool with support for heterogeneous network systems.

2.1.2   Functional Classification

There are 3 types of distributed computing systems due to the type of processing: high-throughput computing [15], high-performance computing [16] and multi-task computing [17].

- *High-throughput computing.* This processing mode focuses on supporting computing units that require chaining. The class operates on a macro scale - the processing time of individual computing units is calculated in hours or even days. Examples of the use of this type of systems are: development of new plastics [18] or protein development [19].
- *High-performance computing.* In high-performance processing, computing units can be completely paralleled (pool computing). This class operates on a micro scale - the processing time of individual computing units (tasks) is counted in milliseconds. It applics to the computing clusters. Examples of the use of this type of systems are: development of results of collisions in Large Hadron Collider [20] or distributed compilation [10].
- *Multi-task computing.* Multi-tasking is an intermediate solution between high-throughput and high-performance processing. Tasks can be both sequential and parallel. The process concentrates on cooperation with the file system. Examples of the use of this type of systems are: high-performance, distributed databases [21] or search engines [17].

2.2   The Choice of the Basic Environment

Among the existing solutions, it is difficult to find one that would fit the problem of simulation as massive computations. The main reasons are:

- *Type of processing.* The first aspect that strongly limits the choice of available solutions is the type of processing required. Simulations of long-time tasks that in cannot be parallelized (like Feniks simulations [2]) meet the class of high-performance processing systems.
- *Execution environment.* An important requirement is the ability to use the currently available computing resources. The available computers are diverse, controlled by various operating systems (Linux, Windows). Only network processing supports such heterogeneous environment.

- *Correct handling of .NET environment on Linux and Windows systems*. The Feniks simulator is developed under the .NET environment, which is not fully supported on Linux systems, but it is possible to use the support of the Mono environment [22].
- *License and cost*. One of the main requirements of the system was its opening to further development. It was assumed that the used environment should be free, at least for academic use.

Five potential systems were selected.

- *Oracle Grid Engine* [23] - closed and paid software; supports homogeneous cluster solutions.
- *Globus Toolkit* [24] - supports only nodes under the control of Linux class operating systems. In addition, open source support ended in January 2018.
- *TORQUE* [25] - supports only nodes running Linux class operating systems; the code was closed in June 2018.
- *SLURM* [26] - open source and free software supporting homogeneous cluster solutions; it only works under Linux.
- *HTCondor* [9, 27] - open source and free software supporting network processing with support of *scavenging* (described below) on heterogeneous execution nodes in the high-throughput processing mode.

Given the previous requirements for a distributed computing system, HTCondor has proved to be the most accurate choice. Additional support for scavenging (in addition to simply using all available computer resources) helps in the utilization of free computing resources that may still have their functions as personal computers.

CPU scavenging. Scavenging [27–29] is one of the functionalities offered by the HTCondor distributed computing system. It is based on the utilization of the computing power of the processor at times when the user does not use the computer. When the system detects that user input devices are inactive by the predefined time, the service proceeds to perform calculations. This functionality is optional. It can be replaced by, for example, reserving a number of cores for a user.

2.3   HTCondor Weak Points

After selecting the most suitable distributed processing system (HTCondor), we tried to apply it for simulations without additional support. The way of executing computing units is very difficult to apply to perform multiple simulations with differing parameter sets. This would require an additional script to create files defining batches that would differ only in the input data.

There is also a different way to request the execution of the calculation units - application interface in Python 2. HTCondor does not provide support for any other programming language than Python, in particular, there is no solution supporting the C# language.

2.4   Bridging the Gap

The biggest problem with the use of a distributed processing system is the impossibility of creation of the descendant batches in the simulator itself. It should be organized similarly to the *fork()* procedure known from Posix class systems.

Such a solution would allow to create simple batches (in the problem - the simulator application together with the simulation model and a set of simulation parameters) in the body of the calculation unit.

3   Solution – the Shapp Library

3.1   Functional Requirements

Below is the list the functional requirements for the new Shapp framework, both concerning distributed computations themselves and additional ones that are not necessary, but may be added in the future.

1. An application equipped with the Shapp library should be able to distinguish whether its execution is a base (initial) or descendant-computational unit.
2. The library should automatically send the newly created result from the descendant calculation unit to its parent unit.

   – The library should be able to create many descendant units (via chaining, pool calculations or recursive invocations). The nested calls might be aware of the expected operating system, required operational memory, needs for processor cores etc.
   – The library should asynchronously inform about changes in the descendant status (start, end, suspension, deletion) in order to make a decision about further processing or displaying information to the user.

3. It should enable direct communication of a parent unit with its descendants (the rules of communication are described in [30]).
4. The library, as part of the creation of a descendant unit, should return its descriptor to the parent. This could allow the parent unit to control the operation of its descendants, and to communicate with them.
5. Instance of the library at the time of its closing (finishing the encapsulating program) should not require the removal of any descendant units (they can continue to run without their supervisor). However, if such behavior is desired, it must be implemented by the user by calling the descendant-killing operation, provided by the library.
6. The library should allow invoking a computational unit which is a simple batch application (or operating system script), without the Shapp library attached.
7. The library should provide for the use of shared file systems. In this case, there are no unnecessary file transfers to run the descendant units (executables, input and output files).

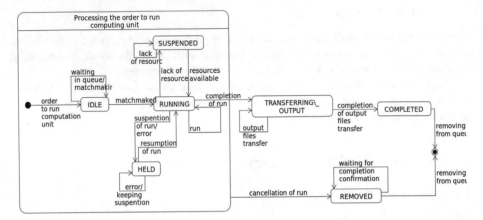

Fig. 1. State diagram of tasks in the HTCondor system. The naming of states has been borrowed directly from the system documentation.

3.2 The Operation of HTCondor

Figure 1 shows the state diagram of the HTCondor system. The diagram has been prepared by the authors based on the observations of the HTCondor in operation, as well as extracts of information from the documentation and research publications.

When the task is executed, the HTCondor system analyzes the catalog tree of the started task. Any newly created files (with arbitrary extensions) are considered as result files to be sent back to the node containing the parent unit. This mechanism can be limited, and the user can simply define the names of result files. Then the remaining files will be lost. The system thus configured became the basis for the operation of Shapp library – the executable file, input data and configuration files are simply transferred to the target node as a set of files in specified locations.

3.3 Shapp Interface

During the development of the Shapp library, several important implementation decisions were made that significantly affected its final shape:

- The solution design is in the form of a Dynamic-Linked Library (DLL).
- The library interface based on composition. In order to use the library interface, one should create an interface class object and invoke methods for it [31].
- Communication with the HTCondor system is carried out through Python scripts.
- Additional library parameters are stored in the environment variables.
- Information about changes in the system state is carried out on an event basis (*events* mechanism in .NET).

4   Functional Test

The aim of the test was to measure specific data on the overheads associated with the use of distributed processing. The overall throughput was determined by HTCondor capabilities. In this scenario, the Shapp library was used only as a convenient interface to HTCondor via C# language with cloning ability. Scheduling, executables, input/output file transfer and system administration is handled directly by HTCondor. The Shapp library provided a smart configuration of the HTCondor's submitting procedure in a way to perform the whole task in a comfortable manner via a single application.

The test consisted in running simultaneous 100 PRT network simulations of the Corby model [1] using the Feniks simulator [2] (approximate duration of such simulation is about 20 min):

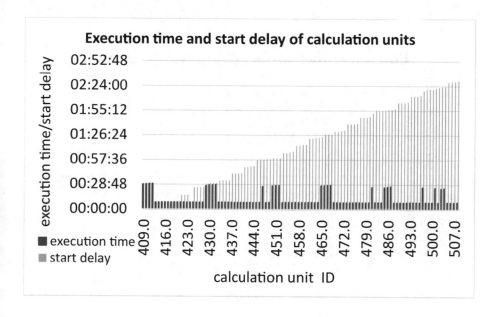

Fig. 2.   Diagram of the execution time and start delay of individual calculation units.

A set of 2 computers was used in ICS, WUT laboratory with the following specification: processor: i5-4670 at 3.40 GHz, operational memory: 32 GB DDR3, operating system: Windows 10 Education.

A *slot* is the smallest calculation unit available in HTCondor. One slot corresponds to one processor core at a time. Each task carried out in the HTCondor can use many slots, and the default is one slot. The above mentioned specification translates into 4 slots with 8 GB operating memory each. In total, 8 slots participated in the calculations.

Figure 2 presents a column chart containing 2 series of data:

- *Execution time*: dominant values can be noticed: 8 min and 29 min. The source of 29 min execution time was identified as a different calculation made on one of the computers in parallel.
- *Start delay*: a linear increase can be observed.

The collected results are consistent with the expectation. Figure 3 shows the number computing units run in parallel. On its basis, several conclusions can be drawn:

- The dominant number of units executed at the same time is 8, which coincides with the number of available slots.
- The time overhead coming from the HTCondor scheduler is small. For the most of the time, all the system has been fully utilized. At the worst intervals (after several dozen seconds), a maximum of 3 slots did not perform tasks.

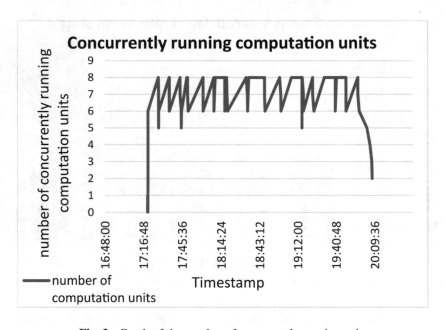

Fig. 3. Graph of the number of concurrently running units.

It follows that the HTCondor system efficiently manages the available nodes and maintains their high utilization.

5 Summary

We present a workload management system for massive distributed computations in C# programs. Our solution, using own Shapp library based on HTCondor system, overcomes disadvantages reported in Sect. 2.3. Especially, it supports tree-based computations, in which every computing unit may be a root of the calculation subtree.

The created library supports the use of the massive distributed computing system just from the application code, which creates descendant computational tasks. In addition, apart from the basic mechanism for creating new copies of other accounting units, the library offers additional functionalities:

- creating workspaces for each computing unit (then the input and output files are placed in a dedicated directory that can be different for each specific unit)
- monitoring the status of processed descendant processing units,
- enabling communication between the parent unit and the descendant units,
- cloning of the computing unit itself.

Thanks to the extension of the functionality of the HTCondor system, many classes of problems, which were then successfully implemented in the large-scale distributed system, gained new possibilities. In addition, the use of these functionalities is very simple.

All functionalities (Sect. 3.1) required for the proper functioning of the library have been implemented. More extensive support for inter-unit communication is under development.

Only small number of the massive distributed processing scenarios have been applied and tested. Further test are planned:

- Analysis of the impact of a computational task type on the utilization of slots.
- The scale of computations. The issue to check is still its efficiency at larger sets of computational units, for example organized into a tree.

In its current state, the Shapp library supports recursive processing that has not been used in the experiment (only pool processing has been used). However, as part of the further development of the library, such research should also be carried out.

References

1. Daszczuk, W.B., Mieścicki, J., Grabski, W.: Distributed algorithm for empty vehicles management in personal rapid transit (PRT) network. J. Adv. Transp. **50**(4), 608–629 (2016). https://doi.org/10.1002/atr.1365
2. Daszczuk, W.B.: Discrete event simulation of personal rapid transit (PRT) systems. Autobusy-TEST **17**(3), 1302–1310 (2016). arXiv:1705.05237
3. Sfiligoi, I.: glideinWMS—a generic pilot-based workload management system. J. Phys. Conf. Ser. **119**(6), 062044 (2008). https://doi.org/10.1088/1742-6596/119/6/062044
4. Anderson, D.P.: BOINC: a system for public-resource computing and storage. In: Buyya, R. (ed.) Fifth IEEE/ACM International Workshop on Grid Computing, Pittsburgh, PA, 8 November 2004, pp. 4–10. IEEE (2004). https://doi.org/10.1109/grid.2004.14

5. Aad, G.: The ATLAS experiment at the CERN large hadron collider. J. Instrum. **3**, 407 (2008). https://iopscience.iop.org/article/10.1088/1748-0221/3/08/S08003/meta

6. Méndez, B.J.H.: SpaceScience@Home: authentic research projects that use citizen scientists. In: Garmany, C., Gibbs, M.G., Moody, J.W. (eds.) EPO and a Changing World: Creating Linkages and Expanding Partnerships, Chicago, IL 5–7 September 2007, pp. 219–226. ASP Press, San Francisco (2008). http://adsabs.harvard.edu/full/2008ASPC..389..219M

7. Patoli, M.Z., Gkion, M., Al-Barakati, A., Zhang, W., Newbury, P., White, M.: An open source Grid based render farm for Blender 3D. In: 2009 IEEE/PES Power Systems Conference and Exposition, Seattle, WA, 15–18 March 2009, pp. 1–6. IEEE (2009). https://doi.org/10.1109/psce.2009.4839978

8. Czarnul, P., Kuchta, J., Matuszek, M.: Parallel computations in the volunteer–based comcute system. In: Wyrzykowski, R., Dongarra, J., Karczewski, K., Waśniewski, J. (eds.) International Conference on Parallel Processing and Applied Mathematics, PPAM 2013, Warsaw, Poland, 8–11 September 2013. LNCS, vol. 8384, pp. 261–271. Springer, Heidelberg (2014). https://doi.org/10.1007/978-3-642-55224-3_25

9. HTCondor. https://research.cs.wisc.edu/htcondor/

10. Pool, M.: distcc, a fast free distributed compiler. In: The Linux Conference, Las Vegas, NV, June 2004, pp. 1879–1885 (2004). https://fossies.org/linux/distcc/doc/web/distcc-lca-2004.pdf

11. Zhang, W.: Linux virtual server for scalable network services. In: Ottawa Linux Symposium, Ottawa, Canada, 22 July 2000, pp. 1–10 (2000). www.linuxvirtualserver.org/ols/lvs.pdf

12. Owsiany, M.: High availability in Linux System (in Polish: Wysoka dostępność w systemie Linux) (2003). http://marcin.owsiany.pl/studia/inf-4_rok/swn/referat.pdf

13. Cassen, A.: Keepalived. http://www.keepalived.org/

14. Nurmi, D., Wolski, R., Grzegorczyk, C., Obertelli, G., Soman, S., Youseff, L., Zagorodnov, D.: The eucalyptus open-source cloud-computing system. In: Cappello, F., Wang, C.-L., Buyya, R. (eds.) 2009 9th IEEE/ACM International Symposium on Cluster Computing and the Grid, Shanghai, China, 18–21 May 2009, pp. 124–131. IEEE (2009). https://doi.org/10.1109/ccgrid.2009.93

15. Raman, R., Livny, M., Solomon, M.: Matchmaking: distributed resource management for high throughput computing. In: Seventh International Symposium on High Performance Distributed Computing, Chicago, IL, 31 July 1998, pp. 140–146. IEEE (1998). https://doi.org/10.1109/hpdc.1998.709966

16. Santos, A., Almeida, F., Blanco, V.: Lightweight web services for high performance computing. In: Oquendo, F. (ed.) European Conference on Software Architecture, Aranjuez, Spain, 24–26 September 2007. LNCS, vol. 4758, pp. 225–236. Springer, Heidelberg (2007). https://doi.org/10.1007/978-3-540-75132-8_18

17. Raicu, I., Foster, I.T., Zhao, Y.: Many-task computing for grids and supercomputers. In: Workshop on Many-Task Computing on Grids and Supercomputers, Austin, TX, 17 November 2008, pp. 1–11. IEEE (2008). https://doi.org/10.1109/mtags.2008.4777912

18. Satyanarayana, K.C., Gani, R., Abildskov, J.: Polymer property modeling using grid technology for design of structured products. Fluid Phase Equilib. **261**(1–2), 58–63 (2007). https://doi.org/10.1016/j.fluid.2007.07.058

19. Zakrzewska, K., Bouvier, B., Michon, A., Blanchet, C., Lavery, R.: Protein–DNA binding specificity: a grid-enabled computational approach applied to single and multiple protein assemblies. Phys. Chem. Chem. Phys. **11**(45), 10712 (2009). https://doi.org/10.1039/b910888m

20. Bird, I.: Computing for the Large Hadron Collider. Annu. Rev. Nucl. Part. Sci. **61**(1), 99–118 (2011). https://doi.org/10.1146/annurev-nucl-102010-130059

21. Raicu, I., Foster, I., Zhao, Y., Szalay, A., Little, P., Moretti, C.M., Chaudhary, A., Thain, D.: Towards data intensive many-task computing. In: Kosar, T. (ed.) Data Intensive Distributed Computing: Challenges and Solutions for Large-Scale Information Management, pp. 28–73. IGI Global (2012). https://doi.org/10.4018/978-1-61520-971-2.ch002

22. Nishimura, H., Timossi, C.: Mono for cross-platform control system environment. In: 6th International Workshop on Personal Computers and Particle Accelerator Controls, Newport News, VA, 24–27 September 2006 (2006). https://escholarship.org/uc/item/3hn297s0

23. Kolici, V., Herrero, A., Xhafa, F.: On the performance of oracle grid engine queuing system for computing intensive applications. J. Inf. Process. Syst. 10(4), 491–502 (2014). https://doi.org/10.3745/JIPS.01.0004

24. Foster, I., Kesselman, C.: Globus: a metacomputing infrastructure toolkit. Int. J. Supercomput. Appl. High Perform. Comput. 11(2), 115–128 (1997). https://doi.org/10.1177/109434209701100205

25. Krieger, M.T., Torreno, O., Trelles, O., Kranzlmüller, D.: Building an open source cloud environment with auto-scaling resources for executing bioinformatics and biomedical workflows. Futur. Gener. Comput. Syst. 67, 329–340 (2017). https://doi.org/10.1016/j.future.2016.02.008

26. Yoo, A.B., Jette, M.A., Grondona, M.: SLURM: simple linux utility for resource management. In: Feitelson, D., Rudolph, L., Schwiegelshohn, U. (eds.) Job Scheduling Strategies for Parallel Processing, pp. 44–60. Springer, Heidelberg (2003). https://doi.org/10.1007/10968987_3

27. Thain, D., Tannenbaum, T., Livny, M.: Distributed computing in practice: the Condor experience. Concurr. Comput. Pract. Exp. 17(2–4), 323–356 (2005). https://doi.org/10.1002/cpe.938

28. Asagba, P., Ogheneovo, E.: Qualities of grid computing that can last for ages. J. Appl. Sci. Environ. Manag. 12(4) (2010). https://doi.org/10.4314/jasem.v12i4.55218

29. Georgatos, F., Gkamas, V., Ilias, A., Kouretis, G., Varvarigos, E.: A grid enabled CPU scavenging architecture and a case study of its use in the greek school network. J. Grid Comput. 8(1), 61–75 (2010). https://doi.org/10.1007/s10723-009-9143-2

30. Galecki, T.: The environment of support of a massive distributed computing (in Polish: Srodowisko wsparcia masowego przetwarzania rozproszonego), BSc thesis, Warsaw University of Technology, Institute of Computer Science, 50p. (2019). http://repo.bg.pw.edu.pl/index.php/pl/r#/info/bachelor/WUTcac04f4e732f434590a18a4b4d6fcf68/?r=diploma&tab=&lang=pl

31. Ossher, H., Kaplan, M., Harrison, W., Katz, A., Kruskal, V.: Subject-oriented composition rules. ACM SIGPLAN Not. 30(10), 235–250 (1995). https://doi.org/10.1145/217839.217864

A Fog Node Architecture for Real-Time Processing of Urban IoT Data Streams

Elarbi Badidi$^{(\boxtimes)}$

College of Information Technology,
United Arab Emirates University, Al Ain, UAE
ebadidi@uaeu.ac.ae

Abstract. Sensors and IoT devices in smart cities generate vast volumes of data that need to be harnessed to help smart city applications make informed decisions on the fly. However, time-sensitive applications cannot tolerate sending data streams to the cloud for processing because of the unacceptable latency and network bandwidth requirements. Cities require the ability to efficiently stream data and process data streams in real-time at the edge. This paper describes a fog node-based architecture that aims to deal with the streaming and processing of data generated by the various devices and equipment of the city to enable the creation of value-added services. The core components of the fog node are a powerful distributed messaging system and a powerful stream processing engine that can adequately scale with the data volumes.

Keywords: Edge computing · Fog computing · Data streams processing · Urban data · IoT gateway

1 Introduction

Over the last decade, the sensing technology has witnessed phenomenal development and growth. The technological advances have permitted the development of sensors that are small enough to be implanted into various kind of devices to monitor, for instance, a variety of parameters such as temperature, light, humidity, moisture, motion, pressure, sound, vibration, and other physical aspects of the external environment. Besides, advances have been made in the development of tiny sensors that can be embedded in clothing, human tissue, and other materials.

In parallel with this phenomenal growth of the sensing technology, communication technologies and wireless connectivity have improved significantly to the point that wireless connectivity is becoming an integral part of modern electronic equipment. Sensors and actuators, embedded in connected devices, can send and receive data over the network. The data from sensors is flowing in a constant stream from the device to the network leading to massive amounts of data, known as big data, which need to be harnessed.

Data is becoming the fuel of the digital economy. It is more valuable than ever. That's why organizations and corporates are investing tremendous amounts of money in capturing and storing vast amounts of data and getting insights from collected data. The challenge is using data streams that are still in motion and extracting valuable information from them even before the data is stored for post-event processing and analysis.

© Springer Nature Switzerland AG 2019
R. Silhavy (Ed.): CSOC 2019, AISC 984, pp. 330–341, 2019.
https://doi.org/10.1007/978-3-030-19807-7_32

In a smart city context, the growing adoption and deployment of IoT, cloud computing, advanced analytics, and artificial intelligence promised to take remote monitoring in smart cities to another level. Indeed, leveraging IoT-based data streams means increased visibility into events taking place across the city and continuous improvement of services offered to citizens. Deployed IoT devices and sensors generate large volumes of data while monitoring road traffic conditions continuously, night-time activity and traffic, energy and water consumption, weather conditions, structural integrity of bridges and historical monuments, and parking spaces availability. With in-depth insights into events in real time, cities can improve services and reduce costs, often without any disruption to citizens. Monitoring services can:

- capture meters and sensors readings
- monitor various city assets to detect abnormal issues
- apprehend critical alert notifications
- analyze sensor data and alert histories
- learn normal conditions and detect anomalies
- Leverage data from connected products

With the above benefits of deploying IoT solutions, the ability to store and move massive amounts of sensed data is becoming challenging. The volume of data continuously being generated at the edge, in different parts of the city, is growing so fast that the networks are unable to process it. Instead of transmitting data to a remote data center or the cloud for processing and storage, end devices should pass the data to an edge computing device that aggregates, processes or analyzes that data. The ultimate goal of this operation is to minimize costs and latency while controlling network bandwidth. A significant benefit of edge computing is the reduction of data that must be transmitted and stored in the cloud.

In this paper, we propose an architecture of a fog node, which will be at the edge of the network to extend the capabilities of IoT gateways by allowing the streaming of large amounts of data and its processing in near-real time. The architecture includes several components that are chained to build a pipeline. Data streams originating from IoT devices and sensors are conveyed to nearby IoT gateways using a variety of communication protocols (MQTT, CoAP, Zigbee, WiMAX, etc.). These IoT gateways can perform some preliminary data aggregation and then publish aggregated data to topics in an Apache Kafka[1] cluster. Data (events) in the Kafka topics are made available for consumption by an Apache Storm[2] topology.

The remainder of the paper is organized as follows: Sect. 2 provides background information on the new paradigms of edge computing and fog computing. Section 3 presents related work on real-time processing of IoT data streams mainly at the edge. Section 4 describes our proposed fog node-based architecture that will serve for the real-time processing of IoT data streams. Section 5 describes the components of the fog node for data streams processing. Section 6 illustrates an implementation scenario. Finally, Sect. 7 concludes the paper.

---

[1] http://kafka.apache.org.

[2] https://storm.apache.org.

2 Edge and Fog Computing

Edge computing essentially represents the process of decentralizing processing capabilities and services by moving them close to data sources. This process can have a very significant impact on the latency and can reduce the volumes of data transmitted to cloud servers.

As sensors and IoT smart devices generate vast amounts of data, the well-established data management systems and practices will no longer be sufficient to take full advantage of IoT. The basic idea behind edge computing is to place computing resources and storage at the edge of the network, close to the data generation location [9, 16].

Since cloud computing is not always efficient for data processing when data is produced at the edge of the network and applications are time-sensitive, much effort has gone into designing and deploying cloudlets [8, 9], micro-data centers [2, 3], and fog computing solutions [4] at the edge of the network. According to IDC's forecast for 2019, at least 40% of the data generated by IoT devices would be stored, processed and analyzed at the edge of the network [5].

In smart cities where time-sensitive events occur very often, edge computing will play a significant role in implementing smart services. For autonomous cars to become a reality, they must react in real-time to external factors. For example, if an autonomous car drives on the road and suddenly pedestrians or animals walk in front of it, the car must stop immediately. It does not have time to send data to the cloud server and wait for instructions; it must have the capabilities to process the signal instantly and react accordingly.

Smart cities are ideal for the use of edge computing. Indeed, sensors and actuators can receive commands based on decisions made locally, without having to wait for decisions made in another distant place. Cities can use edge computing for video surveillance applications and getting up-to-date data concerning the conditions of roads, intersections, and buildings to take remedial actions before accidents occur. They also can use it for controlling lighting, energy and power management, water consumption, and many more. The processing of urban IoT data streams can be pushed from the cloud to the edge, which will reduce the network traffic congestion and shorten the end-to-end latency. Edge Computing can encompass operations such as data collection, parsing, aggregation, and forwarding as well as rich and advanced analytics that involve machine learning and event processing and actions at the edge.

Cisco Systems defined Fog computing as: "Fog Computing is a highly virtualized platform that provides compute, storage, and networking services between end devices and traditional Cloud Computing Data Centers, typically, but not exclusively located at the edge of the network." [4]. Most of the earlier literature treated fog and edge computing as being synonymous and used the words fog and edge interchangeably. Fog Computing and Edge Computing are both concerned with processing and filtering IoT data before it arrives in a data center or cloud server. Edge computing is a subdivision of fog computing. The main difference between them lies in the place where data processing takes place. With Fog Computing, data processing typically occurs in a fog node near the local area network. Edge Computing processes data within edge devices and uses the communication capabilities of edge gateways to send the data to the data center or the cloud. Edge devices are often battery-powered and run full operating systems such as Linux and Android. They process raw data they receive from sensors and IoT devices and send commands to the actuators. Edge gateways also run full operating systems and have unlimited power, more CPU power, memory, and storage. They aggregate the data and act as intermediaries between the fog node and the edge devices by transmitting selected raw or pre-processed datasets to the fog node and receiving commands, such as configurations or data queries. Fog node components store and analyze collected data using machine learning algorithms and analytics. Fog computing is characterized by its openness, which is essential for the success of this paradigm for IoT platforms and applications. Proprietary solutions are typically developed in edge computing.

3   Related Work

Over the last few years, there have been several research efforts that investigated the implementation of real-time processing of IoT data streams at the edge of the network. Furthermore, some research works studied harnessing IoT and edge and fog computing in smart cities.

de Assunção et al. [6] surveyed the different solutions for stream processing and methods to manage resource elasticity. They classified the stream processing engines into three generation: (i) Early stream processing solutions (Niag-araCQ, STREAM, Aurora, etc.), (ii) Current stream processing solutions (Apache Storm, Apache Flink, Apache Samza, Apache S4, Twitter Heron, Spark streaming), and (iii) other solutions (ESC, TimeStream, Google's MillWheel, etc.). They discussed existing efforts to manage resource elasticity in stream processing engines and the emerging distributed architectures for stream processing.

Yang [14] investigated the issues of data streams processing and analytics in a fog computing environment by describing some typical fog streaming applications and proposing a fog and cloud data streaming multi-layer architecture. The author also discussed the design of a fog data streaming system by considering four dimensions: (i) system (stream processing engine, streaming task partitioning), (ii) data (data stream acquisition, stream mining and analytics), (iii) human factor (pricing and incentivization, privacy, quality of control), and (iv) optimization (dynamic optimization, resource allocation).

Chowdhery et al. [1] proposed an edge-assisted architecture to deal with the challenges faced by smart cities such as traffic congestion, public safety and security, sanitation, and high energy use. The features of the proposed architecture include (i) computation-aware information acquisition, where edge computing will allow to implement adaptive information acquisition, (ii) content and processing aware networking, enabled by edge devices, and (iii) useful information availability by making data available at the edge of the network. The proposed architecture focused on the communication of data and its distribution and processing at the edge. However, it does not describe in detail the data streams processing at the edge.

Kamburugamuve et al. [12] proposed the IoT cloud, a platform that enables connecting IoT devices to cloud services to perform data real-time processing and control. IoTCloud consists of three layers: Gateway layer, publish-subscribe messaging layer, and cloud-based big data processing layer. IoTCloud shares some common architectural components with our proposed architecture. However, IoTCloud did not consider using edge and fog computing to deal with the issues of latency and network bandwidth restrictions.

Chardonnens et al. [15] investigated the integration of high-velocity data streams from Twitter and Bitly to implement trend detection using Apache storm. The use case showed that storm supports scalable and complex analyses on high-velocity data streams. However, finding the right storm topology or scaling out a running topology required manual intervention.

Sarkar et al. [13] mathematically formulated the characteristics of fog computing and conducted a comparative analysis of fog computing against conventional cloud computing. They characterized the fog computing network with regards to service latency, power consumption, CO_2 emissions, and cost. Then, they evaluated its performance in an environment with a high number of interconnected devices requesting real-time service. They found out that fog computing outperforms cloud computing in the case of a high number of latency-sensitive IoT applications. Enhanced performance is expressed in terms of QoS and eco-friendliness.

Laska et al. [7] proposed a real-time stream processing pipeline to integrate spatiotemporal data streams from IoT devices. The Apache stream processing engine and the Apache Kafka messaging system are at the heart of the data pipeline. The data integration layer uses the GeoMQTT protocol. They implemented map matching algorithms and evaluated their latency on a local cluster. The authors stated that one of the algorithms they implemented, using a sliding window approach, gave more accurate results than comparable algorithms based on the Hidden Markov model (HMM).

Dautov et al. [10] proposed an edge computing-based architecture for implementing an Intelligent Surveillance system (ISS) at an urban scale. The solution aimed at pushing the processing tasks and data management close to where the data was generated to enhance the critical issues of performance and security levels of surveillance systems. The prototype results showed the benefits of the approach when compared with the traditional model of offloading tasks to the cloud.

Hossain et al. [11] implemented an edge computing framework to process IoT data in a smart city context. They stated that the approach is effective with regards to latency and provides situational awareness for the decision makers.

In this work, we propose a fog node-based architecture for the processing of IoT data streams in a smart city environment. The primary goal –that we share with some of the above efforts– of the architecture is to promote the streaming and processing of urban IoT data and its management and processing in a scalable fashion.

4   Architecture for IoT Data Streams Processing at the Edge

To design our proposed architecture, we assume that IoT devices publish data to a nearby IoT gateway able to aggregate and summarize data to reduce the amount of data conveyed to a fog node having more computing power and storage. Given that several processing tools and applications might be used in the fog node, it is essential to have a messaging system capable of persisting received data and implementing a publish/subscribe model as well as a pull model that processing tools and applications can use to access the data streams. The messaging system, as well as the processing platform, need to be able to scale up with the vast amount of data transmitted by the IoT gateways. This section gives an overview of our proposed architecture, depicted in Fig. 1, and describes its components.

Our proposed architecture consists of five tiers. The first tier comprises networked devices typically sensors and actuators while the second one includes IoT gateways responsible of analog-to-digital conversion and aggregation of sensed data. The third tier includes fog nodes (or edge IT systems) responsible for preprocessing of data received from the second tier before it moves to the cloud for more profound insights. The cloud computing tier is in charge of processing, analyzing, and storing non-time-sensitive data on cloud servers. Finally, the application tier provides monitoring and alerting dashboards.

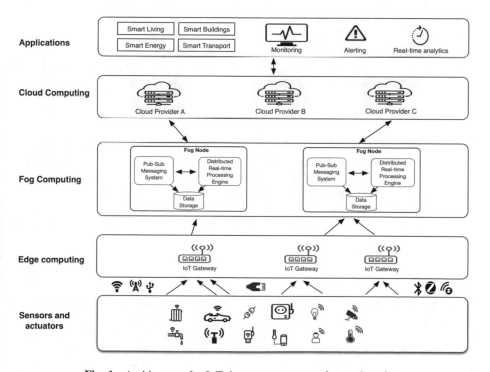

Fig. 1. Architecture for IoT data streams processing at the edge.

4.1   IoT Infrastructure

This tier consists of sensors and smart IoT devices. Smart devices permit to automate the operations of a city by collecting data on its physical assets (buildings, facilities, equipment, vehicles, etc.). Collected data allows monitoring the behavior and status of these assets and optimizing the resources and processes of the city. A device detects some input, like light, motion, speed, vibration, pressure, water level, heat, or any other environmental phenomenon, from its surrounding environment and responds to it. The device reading, which is then converted into a human-readable form, is sent over a network to a gateway for further processing. Devices and actuators, without a direct connection to the Internet, connect to edge devices or edge gateways using Ethernet or WiFi connections of a Local Area Network or using other protocols such as Bluetooth and ZigBee.

4.2   Edge Computing Tier

IoT gateways are a critical component in any IoT implementation. This component is in charge of aggregating data, translating sensor protocols, and preprocessing data before conveying it to other tiers (or to the cloud) for more processing. Aggregation and preprocessing of sensed data are required to cope with the high volume of data coming from sensors and devices. In contrast with traditional network gateways, which typically perform protocols translations, today's IoT gateways are full-fledged computing systems capable of performing advanced tasks such as protocol and data bridge between devices using heterogeneous communication protocols and data formats, malware protection, and storage and analytics. For example, the Dell Edge Gateway 5000 Series can aggregate data, perform local analytics, and convey only meaningful information to the next tier.

4.3   Fog Computing Tier

Digitized and aggregated data, by IoT gateways, for which immediate feedback is not expected may require further processing before moving to the cloud. Such processing is performed in fog nodes located at the network edge closer to the sensors. These fog nodes, which can perform analytics at the edge, permit to lessen the burden on core IT infrastructures because the massive amounts of IoT data can easily swamp data center resources and eat up network bandwidth. Section 5 describes the components of a fog node.

4.4   Cloud Computing Tier

Data that does not require immediate feedback but requires further processing is transferred to physical data centers or cloud systems, where more powerful computer systems can securely store, analyze, and manage data. Data processing at the cloud allows for more in-depth analysis and allows to combine sensor data with data from other sources for deeper insights. However, waiting for the data to reach this tier can delay getting the results.

5   Fog Node Components

The components of a fog node include the distributed messaging system, the platform for data stream processing and edge analytics, and the data storage.

5.1   Distributed Messaging System

In recent years, several implementations of the publish-subscribe message broker have been developed. However, the most popular are ActiveMQ, RabbitMQ, and Kafka. Apache ActiveMQ[3] is written in Java with JMS, WebSocket, and REST interfaces. It supports protocols like MQTT, AMQP, and OpenWire that can be used by applications in various programming languages. RabbitMQ[4] is a messaging system that is lightweight and easy to deploy on premise and in the cloud. It implements AMQP, MQTT, and many other protocols. RabbitMQ runs on several operating systems and cloud environments and provides developers with libraries for interfacing with the broker for all popular programming languages. RabbitMQ can be deployed in various configurations to meet the scalability and availability requirements.

Apache Kafka is renowned as the most advanced message broker that can handle data streams efficiently and in a scalable manner. It is an open-source distributed messaging system that is easy to manage and use. Kafka is ideal for real-time scenarios such as telemetry from sensors and connected devices, social analytics, click-stream analysis, financial alerts, and network monitoring. Organizations can infuse real-time events from devices, sensors, applications, and websites. In general, Kafka integrates with a scalable stream processing engine such as Apache Storm, Apache Flink or Apache Spark.

5.2   Distributed Real-Time Processing Engine

After enabling the efficient storage and data access in the distributed messaging system, the data streams processing engine would enable city stakeholders to efficiently transform and analyze vast amounts of IoT data streams at the edge. The output of this component would typically be stored in the messaging system to allow applications, such as monitoring and visualization applications, to use it to take appropriate actions and generate valuable dashboards. An extensible set of well-established open source and commercial solutions for data processing can be used as the data streams processing engine. These solutions include Apache Storm, Apache Flink, Apache Samza, Amazon IoT, Google Cloud Dataflow, and Amazon Elastic MapReduce.

---

[3] http://activemq.apache.org.

[4] https://www.rabbitmq.com.

6   A Scenario of Implementation

Figure 2 depicts the stream processing components in a Fog Node. These components include a Kafka cluster, a Storm topology, and a Zookeeper server, which provides configuration management for them.

Kafka usually runs on a cluster of one or more brokers. It immutably stores messages from multiple sources (producers) in queues (topics), which are organized into multiple partitions. The messages in a partition are indexed and saved with a timestamp. Other processes (consumers) can query messages stored in Kafka partitions. Kafka partitions are replicated across brokers in the cluster to ensure fault tolerance. Kafka has four APIs:

- Producer API - enables applications and services to publish data streams into Kafka topics.
- Consumer API - enables applications and services to subscribe to Kafka topics of interest and retrieve data streams for processing.
- Streams API - enables applications to convert the input streams, stored in the topics, into other output streams that other applications can consume.
- Connector API - enables applications to connect Kafka clusters with external sources such as key-value stores or relational data-bases.

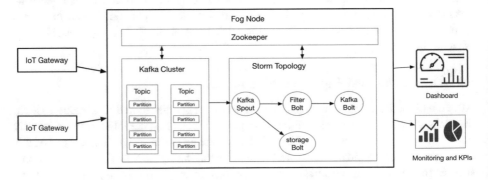

Fig. 2. Fog Node component for real-time stream processing

Apache Storm is a distributed open-source framework that facilitates real-time data processing. Apache Storm is written in the Clojure language, which is the first functional Lisp programming language. It works for processing data in real-time, just as Hadoop works for batch processing of data. Apache Storm relies heavily on Apache Zookeeper for its cluster state management operations, such as message acknowledgments, processing reports, and other similar messages. A Storm application is designed as a workflow, called a topology, in the form of a directed acyclic graph (DAG) with spouts and bolts as vertices of the graph. The edges of the graph are named streams and direct data from one vertex to another. Apache Storm can handle more than one million tuples per second per node, which is highly scalable and offers processing work guarantees. Apache Storm provides several components for working with Apache Kafka. KafkaSpout and KafkaBolt can be used to read and write data from Kafka respectively.

Zookeeper is a distributed coordination service for managing a large number of hosts. Coordinating and managing a service in a distributed environment is a complex process. Zookeeper solves this problem with its simple architecture and API. It allows developers to focus on the core business logic without worrying about the distributed nature of the application.

We are experimenting with this implementation scenario using ThingsBoard IoT gateway. ThingsBoard is an open-source IoT platform that enables rapid development and management of IoT projects. ThingsBoard provides a Kafka plugin, which is responsible for sending messages to the Kafka cluster triggered by specific rules. We deployed a single instance of ThingsBoard with PostgreSQL database in a Docker container on a MacBook Pro. ThingsBoard web interface allows describing the assets, such as buildings, and their IoT devices. It provides a rules chains interface that allows specifying the actions to take for input events such as telemetry measurements. We use a generator that emulates the temperature readings of a thermometer. The readings are then forwarded to the Kafka plugin. Figure 3 depicts the rule chain for temperature telemetry.

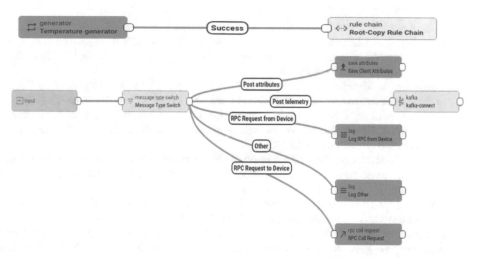

Fig. 3. Specification of a rule chain for temperature telemetry in ThingsBoard IoT Gateway.

7   Conclusion

Effective governance and management of smart cities rely heavily on the deployment of a large number of smart sensors and devices to monitor road traffic, energy and water consumption, and safety on the streets and different places of the city, and many more. The vast amounts of data generated by these sensors and devices need to be harnessed to enable city stakeholders to make informed decisions on the fly. For this to happen, it's essential to process IoT data streams in near real-time and use data analytics tools to gain insights from the events happening in the city. Data streams are typically sent to cloud servers for storage and processing. The abundance and elasticity of cloud resources make this scenario feasible in many use cases. However, time-sensitive applications, such as processing videos recorded by surveillance cameras, cannot tolerate sending data streams to the cloud for processing due to network bandwidth requirements and high latency.

In this paper, we have described a fog node-based architecture, which aims at dealing with the streaming and processing of data generated by the various devices and equipment of the city at the edge of the metropolitan network. Processing data streams at the edge reduces network traffic and improves the latency of time-sensitive applications. The architecture is based on a distributed messaging system and a powerful stream processing engine that can accommodate data volumes. An implementation scenario of the architecture was presented. It uses the Kafka messaging system, considered the most advanced messaging system, capable of handling large amounts of data and implementing data replication and persistence, as well as the Storm streaming engine that can process vast volumes of data. A prototype of the system is being developed using the ThingsBoard IoT gateway to ingest data streams into the Kafka system.

References

1. Chowdhery, A., Levorato, M., Burago, I., Baidya, S.: Urban IoT edge analytics. Fog computing in the internet of things: intelligence at the edge, pp. 101–120 (2017). https://doi.org/10.1007/978-3-319-57639-8_6
2. Greenberg, A., Hamilton, J., Maltz, D.A., Patel, P.: The cost of a cloud: Research problems in data center networks. ACM SIGCOMM Comput. Commun. Rev. **39**(1), 68–73 (2008)
3. Cuervo, E. et al.: MAUI: making smartphones last longer with code offload. In: Proceedings of 8th International Conference on Mobile Systems, Applications, and Services, San Francisco, CA, USA, pp. 49–62 (2010)
4. Bonomi, F., Milito, R., Zhu, J., Addepalli, S.: Fog computing and its role in the Internet of things. In: Proceedings of 1st Edition MCC Workshop Mobile Cloud Computing, Helsinki, Finland, pp. 13–16 (2012)
5. IDC.com.: IDC FutureScape: Worldwide Internet of Things 2017 Predictions. https://www.idc.com/research/viewtoc.jsp?containerId=US40755816. Accessed 20 Feb 2019
6. de Assunção, M.D., da Silva Veith, A., Buyya, R.: Distributed data stream processing and edge computing_ A survey on resource elasticity and future directions. J. Netw. Comput. Appl. **103**, 1–17 (2018)
7. Laska, M., Herle, S., Klamma, R., Blankenbach, J.: A scalable architecture for real-time stream processing of spatiotemporal IoT stream data—performance analysis on the example of map matching. IJGI **7**(7), 238 (2018)
8. Satyanarayanan, M., Bahl, P., Caceres, R., Davies, N.: The case for VM-based cloudlets in mobile computing. IEEE Pervasive Comput. **8**(4), 14–23 (2009)
9. Satyanarayanan, M., Simoens, P., Xiao, Y., Pillai, P., Chen, Z., Ha, K., Hu, W., Amos, B.: Edge analytics in the Internet of Things. IEEE Pervasive Comput. **14**(2), 24–31 (2015)
10. Dautov, R., Distefano, S., Bruneo, D., Longo, F., Merlino, G., Puliafito, A., Buyya, R.: Metropolitan intelligent surveillance systems for urban areas by harnessing IoT and edge computing paradigms. Softw. Pract. Exper. **48**(8), 1475–1492 (2018)

11. Hossain, S.A., Rahman, M.A., Hossain, M.A.: Edge computing framework for enabling situation awareness in IoT based smart city. J. Parallel Distrib. Comput. **122**, 226–237 (2018)
12. Kamburugamuve, S., Christiansen, L., Fox, G.: A framework for real time processing of sensor data in the cloud. J. Sens. (2015). https://doi.org/10.1155/2015/468047
13. Sarkar, S., Chatterjee, S., Misra, S.: Assessment of the suitability of fog computing in the context of Internet of Things. IEEE Trans. Cloud Comput. **6**(1), 46–59 (2015)
14. Yang, S.: IoT stream processing and analytics in the fog. IEEE Commun. Mag. **55**(8), 21–27 (2017)
15. Chardonnens, T., Cudre-Mauroux, P., Grund, M., Perroud, B.: Big data analytics on high velocity streams: a case study. In: Proceedings of the 2013 IEEE International Conference on Big Data, Big Data 2013, pp. 784–787 (2013). https://doi.org/10.1109/bigdata.2013.6691653
16. Shi, W., Cao, J., Zhang, Q., Li, Y., Xu, L.: Edge computing: vision and challenges. IEEE Internet Things J. **3**(5), 637–646 (2016)

Optimizing Performance of Aggregate Query Processing with Histogram Data Structure

Liang Yong[(⊠)] and Mu Zhaonan

Network and Information Center,
Guizhou University of Commerce, Guiyang 550014, China
124265276@qq.com, 54762999@qq.com

Abstract. In today's big data era, the capability of analyze massive data efficient and return the results within an short time limit is critical to decision making, thus many big data system proposed and various distributed and parallel processing techniques are heavily investigated. Among previous research, most of them are working on precise query processing, while approximate query processing (AQP) techniques which make interactive data exploration more efficiently and allows users to tradeoff between query accuracy and response time have not been investigate comprehensively. In this paper, we study the characteristics of aggregate query, a typical type of analytical query, and proposed an approximate query processing approach to optimize the execution of massive data based aggregate query with a histogram data structure. We implemented this approach into big data system Hive and compare it with Hive and AQP-enabled big data system BlinkDB, the experimental results verified that our approach is significantly fast than these existing systems in most scenarios.

Keywords: Massive data · Approximate query processing · Histogram · Aggregate query · Performance optimization

1 Introduction

In the big data era, corporations and businesses and netizens are increasingly depending on analytical query processing over enormous amounts of data (spanning terabytes or even petabytes in size) to make intelligent business and personal decisions. The time it takes to make these decisions is often to be critical. Since approximate query processing (AQP) can efficiently reduce the involved data and workload, it has becoming an effective technology to optimize massive data processing.

Aggregate query is a method of deriving group and subgroup data by analysis of a set of individual data entries. It is a type of frequently used analytical query which is very useful in decision support application, meanwhile, it often to be costly from the perspective of query processing. Through in-depth analysis of the massive data processing system, this paper proposed a histogram based approximate query processing technique to optimize the performance of analytic aggregate query. This approach has been implemented into big data system Hive [1, 2] cluster for efficient parallel query processing over massive data set. Since it features to obtain a higher degree of approximation query

© Springer Nature Switzerland AG 2019
R. Silhavy (Ed.): CSOC 2019, AISC 984, pp. 342–350, 2019.
https://doi.org/10.1007/978-3-030-19807-7_33

results only with a slight expense of accuracy, which is very useful to interactive data exploration and efficient analytical query processing over massive data. The experimental results show that our approach is significantly fast than existing techniques such as Hive and BlinkDB [3, 4], which can quickly return approximate query results over huge amounts of data.

The rest of this paper is organized as following. Section 2 briefly reviewed the related work of related big data systems. Section 3 present the needed preliminary knowledge and our histogram based aggregation approximate query processing approach. The experimental results are presented in Sect. 4. Finally, we summarize our work and draw the conclusion in Sect. 5.

2 Related Work

Optimization for massive data analytical applications have been a very critical research topic, and many researchers have devoted to make better massive data analytical system. In recent decade, many research outputs have been applied into popular massive big data analysis system such as Hive [1, 2], Dremel [5, 6], Impala [7, 8], BlinkDB [3, 4, 9], Spark [10–12], SnappyData [13], etc. However, most of aforementioned research are optimized for precise query processing, only a relative small portion of work [14–19] dedicate to approximate query processing which make interactive massive data exploration more efficiently and allows users to tradeoff between query accuracy and response time. Among them, few of work proposed to approximate aggregate query processing [20–22]. In this paper, since we implement our approach into Hive and compare it with both Hive and BlinkDB, thus we outline their architecture in following paragraph.

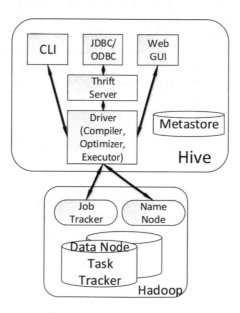

Fig. 1. Architecture of Hive

Hive is a data warehouse based on Hadoop. It supports a SQL-like language HiveQL, as well as Mapreduce scripts written by user as the plug-in to solve complicated jobs which built-in mappers and reducers cannot processing. The architecture of Hive is shown in Fig. 1. Thrift Server provides a simple client API to support the execution of HiveQL. Metastore is like the catalog of database system, it store all the table and data related information of Hive. Driver maintains the whole life cycle of compilation, optimization and execution of HiveQL. Compiler is used for compiling the HiveQL sentence to a DAG which consists of Mapreduce jobs as the execution plan and then submitting the query plan to the execution engine according to certain logical relations. Hadoop is the actual execution engine of Hive.

BlinkDB is a parallel approximate query processing engine for interactive SQL queries and large-scale data. BlinkDB allows users to trade-off accuracy for response time, enabling interactive queries over massive data by running queries on data samples and presenting results annotated with meaningful error bars. For example, reliably detecting a malfunctioning server in a distributed collection of system logs does not require knowing every request the server processed. Based on this insight, BlinkDB allows one to tradeoff between query accuracy and response time, enabling interactive queries over massive data by running queries on data samples. BlinkDB allows users to store their data in SQL-based aggregate queries, as well as response time or error bound constraint.

In AQP area, there are two basic types of approximate query processing methods: on-line query processing and pre-computing method. Online query processing do not required prior data processing, by querying representative data to approximate the query results at runtime. The pre-computing method preprocess the data in advance, generated to represent the original data and the data set is much less, using the data collection in the query to obtain similar results.

BlinkDB adopt the pre-computed method into Hive and analyze the data stored in Hadoop Distributed File System with built-in rich SQL query method of Hive. It has following advantage: BlinkDB allow user use SQL statements and restrict the deviation's bound or the time of responding. BlinkDB dynamically choose the best samples to satisfy the error constraint and responsive time constraint during the runtime; only need a few seconds instead of many minutes to query multiple large volume of data by BlinkDB. However, the sample cannot be pretreated quickly in BlinkDB to get the optimal set of samples for the memory size.

3   Approximate Aggregate Query Processing with Histogram

3.1   Preliminary

During approximate aggregate query processing over massive data, users always want to obtain feedback with a customized error bound as quickly as possible rather than spending minutes or even hours for accurate response. For example, suppose you have a table named *Customer* with five columns: *ID*, *Address*, *Telephone*, *City*, and *Email*. A query is as follows: SELECT COUNT (*) FROM *Customer* WHERE *City* = 'China' GROUP BY OSERROR 10 CONFIDENCE 95%. Return result will be accompanied by an error rate ± 10 and 95% confidence intervals.

The histogram is a variable number of different levels of relative frequency plotted with the rectangular block diagram (the area of each rectangle corresponds to the frequency). Histogram can parse the data regularity and clearly visualize the distribution of the data characteristics. During histogram construction, how to do reasonable data grouping is one of the key issues. In our work, in order to make the implementation to be relative simple, we construct the histogram to make every group of data with equal width. In decision making oriented analytical aggregate queries context, typically aggregate functions include COUNT, SUM, AVG, MAX, MIN, MEDIAN, etc. Benefit from the characteristics of histogram, it will be easy to gain the statistical insights of data and answer the query quickly.

3.2   Our Methodology

As a distributed data warehouse system leveraged on Hadoop system, Hive has been frequently used for SQL based massive data analytics. In this work, we proposed a histogram based approximate aggregate query processing approach. In our approach, a histogram is constructed by stratified sampling of the original data. When a query issued, the sample set who can help to quickly returned results will involve in query processing. Assume the column set in WHERE GROUP BY, and HAVING are known advance, the histogram will obtained by stratified sampling these columns before query processing in the preprocessing stage.

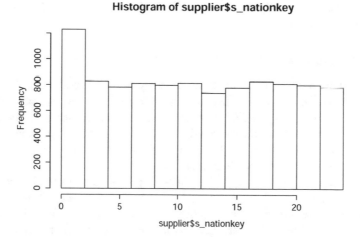

Fig. 2. Column set distribution

Figure 2 showed the histogram of *s_nationkey*, which is a column of table supplier. In this histogram, according to the percentage of each barrel to total data, different percentage of the sample for each bucket, it does not change the distribution of state. It can represent all of the data using a random sample of small and quickly returning similar results. For example, denote the original data as *R*, and the sample represent as

S. Assure *R.a* is {0, 1} and *S* is a 20% sample. The results of "select count (*) from *R* where *R.a* = 0" is identical to "select 5*count (*) from S where S.a = 0", while the number of result sets is 5*2 = 10.

Suppose a query *Q* with a WHERE or GROUP BY statement, let $\varphi = \{c1, c2, \ldots, ck\}$ to be the original columns set of table *T*, the response time constraints denote as *t* and the error rate bound as *e*, sample set $\text{SFam}(\varphi) = \{S(\varphi, Ki)|0 \le i \le m\}$ can be obtained from stratified sampling of each subset.

In table *T*, Let $D(\varphi)$ be the set of unique values *x* on the columns in φ, $T_x = \{r : r \in T \text{ and } r \text{ takes values } x \text{ on columns } \varphi\}$. We will choose a sample $S \subseteq T$ with $|S| = n$ rows. For each group T_x there is a corresponding sample group $S_x \subseteq S$ that is a subset of T_x, which will be used instead of T_x to obtain an aggregate value. The aggregation calculation for each of the S_x will be subject to error that will depend on its size. The standard method is sampling uniformity with equal to the probability of extraction of *n* rows from the table *T*. However, for the calculation of aggregate functions, uniform sampling is not always the best solution.

According to the distribution of each group in the histogram, random sampling are employed to each group with specific probability. For each $x \in x_0, \ldots x_{|D(\varphi)|-1}$, assigned a *counter*, forming a $|D(\varphi)|$-vector of counts N_n^*, Compute N_n^* as: $N(n') = \left(\min\left(\left\lceil \frac{n'}{|D(\varphi)|} \right\rceil, |Tx_0| \right), \min\left(\left\lceil \frac{n'}{|D(\varphi)|} \right\rceil, |Tx_1| \right), \ldots \right)$, the optimal *counter-vector* for a total sample size n'. Then choose $N_n^* = N (\max \{n' : \|N(n')\| \le n\})$. For each x, sample N_{nx}^* rows uniformly at random without replacement from T_x, forming the sample S_x. When $|T_x| = N_{nx}^*$, then the sample consists of all rows in T_x, then this group will be no sampling error.

3.3   Confidence Bound

In order to allow users to understand the accuracy of query results and determine the correct query results in a probability interval, it requires a combination of the probability distribution of query results to give confidence intervals, then we can estimate the confidence interval based on the overall distribution of the data. To be noted that, in our approach, we assume probability distribution of the sampled data who constitute the histogram is identical to the original data.

For the unknown parameters θ, in addition to get its point estimation, we not only want to obtain an estimated range, but also to know the confidence of the parameter θ. Such range is usually given in the form of interval, the confidence level is given by the probability. Such an estimate is called interval estimates or confidence intervals. In the following paragraph, we will present the concept of confidence interval.

Let θ defined as an unknown parameter X in general, α $(0 < \alpha < 1)$ is a number given in advance, $\hat{\theta}_1 = \hat{\theta}_1(X_1, X_2, \ldots X_n)$, $\hat{\theta}_2 = \hat{\theta}_2(X_1, X_2, \ldots X_n)$ is two estimators of θ, if $P\left\{\hat{\theta}_1 < \theta < \hat{\theta}_2\right\} = 1 - \alpha$ called random interval $\left(\hat{\theta}_1, \hat{\theta}_2\right)$ for the unknown parameter θ who hold a confidence level of *1-α* confidence interval. Confidence is also often referred as the confidence level or confidence coefficient. Typically α taken 0.05, 0.01, 0.10, depending on the specific needs.

The procedure of interval estimation of the general requirements is as follows: Firstly, access the random variable (pivot variable) $T = T(X_1, X_2, \ldots X_n; \theta)$ in the catalog and obtain its distribution. Secondly, for a given level of confidence $1-\alpha$, and two constants C1, C2 determined by the distribution of T, then the $P\{C_1 < T(X_1, X_2, \ldots X_n; \theta) < C_2\} = 1 - \alpha$. Finally, the event about that $\{C_1 < T(X_1, X_2, \ldots X_n; \theta) < C_2\}$ is represented as the $\{T_1(X_1, X_2, \ldots X_n) < \theta < T_2(X_1, X_2, \ldots X_n)$ then $P\{T_1(X_1, X_2, \ldots X_n) < \theta < T_2(X_1, X_2, \ldots X_n)\} = 1 - \alpha$, a confidence level of is $1-\alpha$ and confidence interval is (T_1, T_2).

4 Experimental Results

The experimental evaluation is conduct over a 4 nodes cluster, the configuration of each node is as following: 8 GB RAM, 500 GB hard drive. The dataset is generated by follow TPC-H specification and several queries originate from TPC-H is used to evaluate the performance of our histogram based technique and existing systems, e.g., Hive and BlinkDB.

4.1 Performance of Precise Query Processing

In the first experiments, we examine performance tradeoffs of Hive, BlinkDB and our histogram-based approach over precise query. In this experiment, table *lineitem* containing about 600 million records. The data stored in a Hive cluster without sampling or compression. The queries involved in experiments is Q1: SELECT SUM (l_tax) from *lineitem*. After execute the query by Hive, BlinkDB and our histogram-based query engine (an optimized Hive engine), the results are depicted in Fig. 3. It's clearly that our Histogram based approach to be 2.x – 4.x faster than original Hive engine.

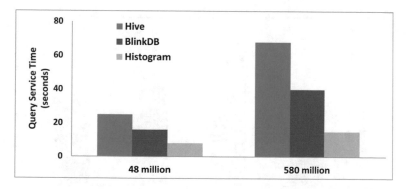

Fig. 3. Response time of accurate query processing

4.2   Performance of Histogram Based Approximate Query Processing

In these experiments, we evaluate the performance of aggregate queries processing by compare our histogram based method with Hive and BlinkDB over TPC-H dataset. During the experiments, the effect of type of query and different error rate are evaluated. Firstly, the experiments are conducted in the same query with same error rate conditions, and under different data sizes. Both BlinkDB and histogram-based approximate query with 5% relative error are run on 48 million rows and 580 million rows. Figure 4 shows the query service time (seconds), our approach only consume near half time of BlinkDB.

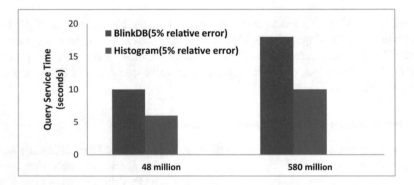

Fig. 4. Response time of approximate query processing in various data size

Secondly, we compare our approach with BlinkDB with more TPC-H queries. The results are presented in Fig. 5. In most case, our histogram based method are achieve better performance than BlinkDB.

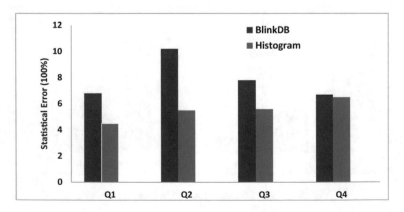

Fig. 5. Response time of approximate query processing in various data size

Thirdly, we conduct experiments to evaluate BlinkDB and our approach under different error rate scenarios, while data size is 48 million. The results are described in Fig. 6. As the results showed in above experiments, in each error rate context, our histogram based Hive engine still to be better than BlinkDB.

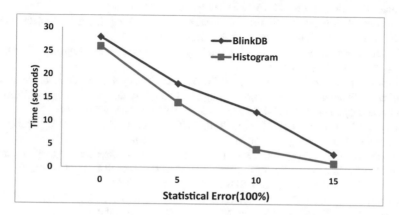

Fig. 6. Response time of different error rate scenarios

5 Conclusions

In our work, we present a histogram based approximate approach for massive data based aggregate query processing. This method utilize the statistical insights of histogram to accelerate the aggregation, e.g., average (), sum (), count (), etc., and can enable user make trade-off between accuracy and response time. We implemented it into the Hive cluster and evaluate it with well-known Hive and BlinkDB system. The experimental study over TPC-H dataset verified that our histogram based approach is significantly fast than these existing system in most scenarios.

Acknowledgements. This paper is supported by Guizhou University Science and Technology Talent Support Program (No.KY [2016] 086).

References

1. Thusoo, A., Sarma, J.S., Jain, N., Shao, Z., Chakka, P., Zhang, N., Anthony, S., Liu, H., Murthy, R.: Hive - a petabyte scale data warehouse using Hadoop. In: ICDE, pp. 996–1005 (2010)
2. Huai, Y., Chauhan, A., Gates, A., Hagleitner, G., Hanson, E.N., O'Malley, O., Pandey, J., Yuan, Y., Lee, R., Zhang, X.: Major technical advancements in apache hive. In: SIGMOD Conference 2014, pp. 1235–1246 (2014)
3. Agarwal, S., Mozafari, B., Panda, A., Milner, H., Madden, S., Stoica, I.: BlinkDB: queries with bounded errors and bounded response times on very large data. In: EuroSys 2013, pp. 29–42 (2013)

4. Agarwal, S., Panda, A., Mozafari, B., Iyer, A.P., Madden, S., Stoica, I.: Blink and it's done: interactive queries on very large data. PVLDB 5(12), 1902–1905 (2012)
5. Melnik, S., Gubarev, A., Long, J.J., Romer, G., Shivakumar, S., Tolton, M., Vassilakis, T.: Dremel: interactive analysis of web-scale datasets. PVLDB 3(1), 330–339 (2010)
6. Afrati, F.N., Delorey, D., Pasumansky, M., Ullman, J.D.: Storing and querying tree-structured records in dremel. PVLDB 7(12), 1131–1142 (2014)
7. Kornacker, M., Behm, A., Bittorf, V., Bobrovytsky, T., Ching, C., et al.: Impala: a modern, open-source SQL engine for hadoop. In: CIDR 2015 (2015)
8. Wanderman-Milne, S., Li, N.: Runtime code generation in Cloudera Impala. IEEE Data Eng. Bull. 37(1), 31–37 (2014)
9. Agarwal, S., Milner, H., Kleiner, A., Talwalkar, A., Jordan, M.I., Madden, S., Mozafari, B., Stoica, I.: Knowing when you're wrong: building fast and reliable approximate query processing systems. In: SIGMOD Conference 2014, pp. 481–492 (2014)
10. Zaharia, M., Chowdhury, M., Franklin, M.J., Shenker, S., Stoica, I.: Spark: cluster computing with working sets. In: HotCloud 2010 (2010)
11. Zaharia, M., Chowdhury, M., Das, T., Dave, A., Ma, J., McCauly, M., Franklin, M.J., Shenker, S., Stoica, I.: Resilient distributed datasets: a fault-tolerant abstraction for in-memory cluster computing. In: NSDI 2012, pp. 15–28 (2012)
12. Armbrust, M., Xin, R.S., Lian, C., Huai, Y., Liu, D., Bradley, J.K., Meng, X., Kaftan, T., Franklin, M.J., Ghodsi, A., Zaharia, M.: Spark SQL: relational data processing in spark. In: SIGMOD Conference 2015, pp. 1383–1394 (2015)
13. Mozafari, B., Ramnarayan, J., Menon, S., Mahajan, Y., Chakraborty, S., Bhanawat, H., Bachhav, K.: SnappyData: a unified cluster for streaming, transactions and interactice analytics. In: CIDR 2017 (2017)
14. Li, K., Li, G.: Approximate query processing: what is new and where to go? - a survey on approximate query processing. Data Sci. Eng. 3(4), 379–397 (2018)
15. Han, X., Wang, B., Li, J., Gao, H.: Efficiently processing deterministic approximate aggregation query on massive data. Knowl. Inf. Syst. 57(2), 437–473 (2018)
16. Park, Y., Mozafari, B., Sorenson, J., Wang, J.: VerdictDB: universalizing approximate query processing. In: SIGMOD Conference 2018, pp. 1461–1476 (2018)
17. Peng, J., Zhang, D., Wang, J., Pei, J.: AQP++: connecting approximate query processing with aggregate precomputation for interactive analytics. In: SIGMOD Conference 2018, pp. 1477–1492 (2018)
18. Galakatos, A., Crotty, A., Zgraggen, E., Binnig, C., Kraska, T.: Revisiting reuse for approximate query processing. PVLDB 10(10), 1142–1153 (2017)
19. Chaudhuri, S., Ding, B., Kandula, S.: Approximate query processing: no silver bullet. In: SIGMOD Conference 2017, pp. 511–519 (2017)
20. Kaiping, F., Hua, Z., Chaoying, F., Heng, C.: In: Application of Histogram Method on Cost Estimate in Query Optimization. Computer & Digital Engineering (2010)
21. Acharya, S., Gibbons, P.B., Poosala, V.: Congressional samples for approximate answering of group-by queries. In: ACM SIGMOD, May 2000
22. Cormode, G.: Sketch techniques for massive data. In: Synopses for Massive Data: Samples, Histograms, Wavelets and Sketches (2011)

Software Algorithm of EM-OLAP Tool: Design of OLAP Database for Econometric Application

Jan Tyrychtr[✉], Martin Pelikán, and Ivan Vrana

Faculty of Economics and Management, Department of Information Engineering,
Czech University of Life Sciences in Prague, Prague, Czechia
tyrychtr@pef.czu.cz

Abstract. The in-field econometric analysis with specific OLAP technology is of benefit to non-econometric analysts for efficient analysis. Integration of the transformation of the econometric model (TEM) method with the software tool is useful for designing a multidimensional database. This research presents the design prototype of such software and its integration with an in-field TEM method to implement database specific for OLAP by a star schema. The prototype of the EM-OLAP Tool software was designed by a simple click-and-play menu using a graphical user interface and optimized to adapt the database design requirements. An algorithm for transformation of econometric models was developed to design a multidimensional schema of database according the TEM method.

Keywords: Algorithm · Econometric software · OLAP · Multidimensional model · Decision support

1 Introduction

Econometrics uses a variety of methods from mathematics, statistics, and economics. The basis for econometric modeling is the regression analysis used to estimate the depended variable based on the knowledge of independent variables. Currently, attention is focused primarily on financial economics in this area. Financial econometrics works with a large set of data. For example, stock price, discount rate estimation, cash flow estimation, dividend policy, etc. Many of these indicators require analyzing a large amount of historical data and data that can be generated at monthly, weekly, daily or even minute intervals.

Generally, the OLAP (on-line analytical processing) technology [1–3, 5] is commonly used to analyze a large amount of data, which allows many indicators to be calculated from data stored in a multidimensional form. OLAP uses a multidimensional view of data where dimensions represent data categories (e.g. product, customers, employees, time) and fact tables that store measurement calculations as indicators relevant to analysis (profit, number of sales, EBITDA, etc.).

To save a large amount of data for econometric analysis, we introduced the TEM (transformation of the econometric model) method [9] for the econometric model

© Springer Nature Switzerland AG 2019
R. Silhavy (Ed.): CSOC 2019, AISC 984, pp. 351–359, 2019.
https://doi.org/10.1007/978-3-030-19807-7_34

transformation of multidimensional schema through the use of mathematical notation. In this paper, we introduce an algorithm of TEM method for the software tool to transform econometric equations to multidimensional paradigm of star schema. This research also presents the design prototype of EM-OLAP (Econometric Model OLAP) software for designing the database by a star schema.

2 Methods

To create the algorithm and the prototype design of the EM-OLAP tool, we based on the TEM method, introduced in [9]. The formal rules of the TEM method consider a set Y and set X, where:

$Y = \{y_s\} \cup \{y_{st}\}$ is a finite set of endogenous variables,
$X = \{x_r\} \cup \{x_{rt}\}$ is a finite set of exogenous variables and
$Rel \subseteq (X \times Y) \cup (Y \times Y)$ is a set of structural relations in the econometric model.

The star schema is any set with 5 elements (*Ent*, *Key*, *Att*, *Ass*, *getKey*), where:

Ent is a non-empty finite set of entities in the schema,
Key is a finite non-empty set of keys in the schema,
Att is a finite non-empty set of attributes in the schema,
Fact $\subseteq$ *Ent* is a finite set of facts in the schema,
Dim $\subseteq$ *Ent* is a finite set of dimensions in the schema, and
Measure $\subseteq$ *Fact* is a finite set of measures in the schema.

Each entity $e \in Ent$ is described by the collection of keys and attributes $\forall e \in Ent : \exists(\{k \in Key\} \cup \{a \in Att\})$.
getKey is a function that returns the *Key* entities in the star schema: $getKey(e) : Ent \rightarrow Key_e \subseteq Key$.
Ass $\subseteq (Dim \times Fact)$ is a finite set of relationships of the entities.

To describe the proposed algorithm, the pseudocode is used together with the flowchart [6]. We use a pseudocode to show how to implement new software algorithm of TEM method. Pseudocode [4, 7, 8] represents an informal high-level description of the program. It omits details like variable declarations and specific syntax of programming language.

We divide the algorithm into several steps. In the first step the algorithm describes entering input data - econometric equations. The next step is to describe the algorithm of the TEM method, which is split into three phases: creating a basic schema, creating dimensions and creating association between dimensions and facts. This corresponds to the steps of the TEM method.

3 Results

In this section we propose new software algorithm of EM-OLAP Tool intended for designing of OLAP database for an econometric application.

3.1   The General Algorithm

In the first phase of the EM-OLAP tool algorithm, the user enters the econometric model (the equation with a variable name). Subsequently, the user is prompted to select the type of econometric variables, whether it is an endogenous or exogenous one. After completing this step, the TEM method is applied. The whole general algorithm is described by Pseudocode 1 and shown in Fig. 1.

```
Start EM-OLAP Tool;
block Adding of equation {
Get equation;
Select type of variables;
}
block TEM Method {
   Create basic star schema environment;
   Create dimensions;
   Create relations;
}
Display schema;
```

[Pseudocode 1: Steps of General Algorithm of EM-OLAP Tool]

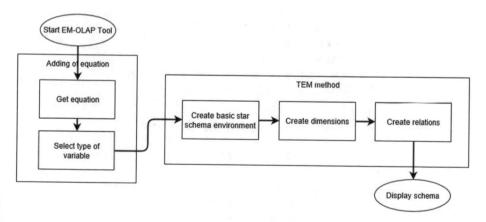

Fig. 1.   General algorithm of EM-OLAP Tool

3.2   The Algorithm for Entering Equation

This algorithm begins by obtaining equations of the econometric model (*Get equation*). The user enters the equation at an input (*User writing equation*). Subsequently, the number of variables from the given equation (*Get count of variables*) is calculated, and a call is displayed for selecting the different types of variables (*Select variable type*). It is important for distinguishing the endogenous and exogenous variables in the model. If the user has chosen the type of variable, then the algorithm continues the *for-loop*.

First, initialization of loop is performed by setting integer $i = 0$. The loop $i + 1$ is repeated until i is less or equal to the number of variables in the equation. If a user has selected the variable as endogenous, the variable is added to the endogenous variable set (*Y array*). If the variable is not selected as endogenous, then it is verified that the user has selected it as exogenous. If so, the variable is added to the array of exogenous variables.

Finally, it increments the counter before running the loop starts from the beginning with another variable. It also performs in case that the variable is neither endogenous nor exogenous.

The whole algorithm is described by Pseudocode 2 and it is shown in Fig. 2.

```
Get equation;
/User writing equation/;
block Adding of equation {
Get count of variables;
before_check:
Select variable type;
if(User selected variables) {
  for(Set i=0;// initialize
  i <= count of variables; i++) {
  if(User selected endogenous)
    variable y[i] ∈ Y;
  else if(User selected exogenous)
    variable x[i] ∈ X;
    }
}
else
  loop before_check;
}
block TEM Method {
  Create basic star schema;
  Create dimensions;
  Create relations;
}
Display schema;
```

[Pseudocode 2: Steps of Adding of Equation]

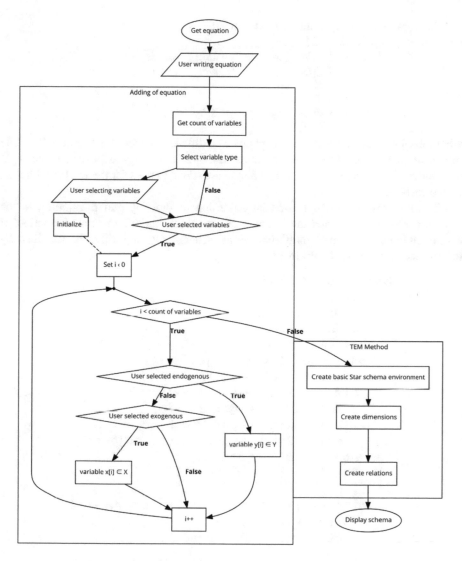

Fig. 2. Algorithm for entering equation

3.3   The TEM Algorithm

An important part of the EM-OLAP tool is an algorithm based on the TEM method. This algorithm transforms the econometric variables into a multidimensional paradigm. The algorithm uses three two-dimensional arrays (matrices) for storing the properties of dimensional objects, fact table, and associations in the diagram.

The array $Dim(n, m)$ is the N-by-5 matrix:

$$\begin{bmatrix} title_1 & x_1 & y_1 & width_1 & height_1 \\ title_2 & x_1 & y_2 & width_2 & height_2 \\ \vdots & \vdots & \vdots & \vdots & \vdots \\ title_N & x_N & y_N & width_N & height_N \end{bmatrix}$$

where each row of the matrix corresponds to a different object in a diagram. The *title* represents the name of the variable or its description. *x* and *y* are axes for rectangle positioning in the diagram. Width and height are added to axes for the width and height of the rectangle.

In this paper, we do not consider algorithm of drawing and positioning in the diagram. This issue is not the essence of our algorithm for the transformation of econometric equations into multidimensional schema. Our algorithm is described by the Pseudocode 3 and shown in Fig. 3.

```
block TEM Method {
   Start;
   Set Dim(n,m),
   Fact(n,m),
   Ass(n,m);
   for(i ← 0; i < X.length; i++) {
      Dim(i,title)  ← X(i);
      Dim(i,Xaxis)  ← get X axis;
      Dim(i,Yaxis)  ← get Y axis;
      Dim(i,width)  ← get width object;
      Dim(i,height) ← get height object;
      Ass(i,Dim)  ← getKey(i);
      Ass(i,Fact) ← getKey(i);
      }
   for(n ← 0; n < Equation.length; n++) {
      Measures(n)  ← Equation(n);
      Fact(n,Measure) ← Measures(n);
      }

}

Draw Dim, Fact and Ass;
Display schema;
```

[Pseudocode 3: Steps of TEM Algorithm]

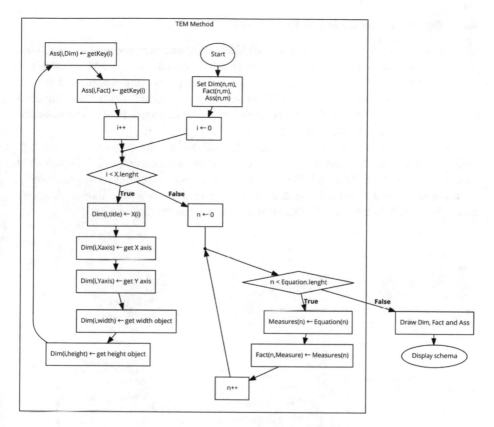

Fig. 3. Algorithm for application TEM method

After *Dim*, *Fact* and *Ass* variables were set, then the algorithm continues in the *for-loop*:

First, initialization of the loop is performed by setting $i = 0$. The loop is repeated until i is less than the length of array X with exogenous variables. Each exogenous variable is stored in the *Dim* array, including other information such as the title, the x and y axes, the width and height of the object. Simultaneously, a key (primary and foreign) is created and stored in the *Ass* array. It is important for creating a link between the fact table and dimensions. After this loop, another loop is performed, which assigns the individual equations into the *Measure* array and then the individual measures are stored in the *Fact* array.

In the last phase, a dimension, a fact table, and an association between the fact table and dimensions are drawn and displayed to the user.

3.4   The Prototype Design of EM-OLAP Tool

To enable an easy design of the multidimensional schema from econometric model, the previously described algorithm was implemented in special prototype of software tool called "EM-OLAP Tool". It will allow automated design of multidimensional schema based on a user-defined source data about econometric models - equations, variables and its descriptions. The software prototype was implemented in the Visual Studio 2017 with C# programming language.

An expert user can enter the econometric equation into the textbox. Subsequently, individual variables are loaded from the econometric equations into the tab-page. The user must manually indicate whether this is endogenous or exogenous variable. Finally, a multidimensional scheme is created for the selected econometric model. A simple example of how to use the EM-OLAP tool is shown in the Fig. 4.

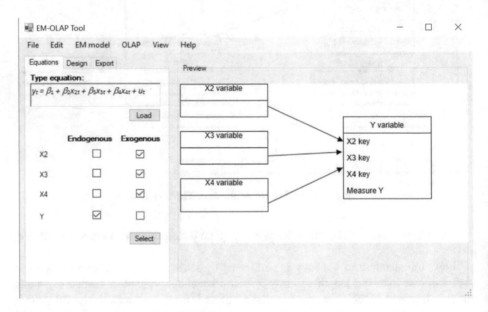

Fig. 4. Prototype design of EM-OLAP Tool

4   Conclusion

This paper provides readers with details of algorithm for development of software for econometric OLAP database design. The algorithm is based on the TEM method. Each step of this method was included. An algorithm can be used for development of software to facilitate development of a database.

In this paper, we introduced a prototype of this software. It is the first prototype, which should be further improved especially with regard to another econometric factors as e.g. linear or nonlinear models, time variable, export and data source definition capability etc., which we did not consider in this paper.

It would be appropriate to simulate evaluate the functionality of the proposed algorithm in our future research.

Acknowledgements. This work was conducted within the project Ambient intelligence in decision-making problems in uncertainty conditions (2019B0008) funded through the IGA foundation of the Faculty of Economics and Management, Czech University of Life Sciences in Prague.

References

1. Abelló, A., Romero, O.: On-line analytical processing. In: Liu, L., Özsu, M.T. (eds.) Encyclopedia of database systems, pp. 1949–1954. Springer, Heidelberg (2009)
2. Berson, A., Smith, S.J.: Data Warehousing, Data Mining, and OLAP. McGraw-Hill Inc., New York (1997)
3. Celko, J.: Joe Celko's Analytics and OLAP in SQL. Elsevier, San Francisco (2010)
4. Erciyes, K.: Guide to Graph Algorithms: Sequential, Parallel and Distributed. Springer (2018)
5. Han, J., Pei, J., Kamber, M.: Data Mining: Concepts and Techniques. Elsevier, Waltham (2011)
6. Chapin, N.: Flowchart, pp. 714–716. Wiley, Chichester (2003)
7. Kopriva, D.A.: Implementing Spectral Methods for Partial Differential Equations: Algorithms for Scientists and Engineers. Springer, Dordrecht (2009)
8. Robertson, L.A.: Pseudocode, Encyclopedia of Information Systems. Elsevier (2003)
9. Tyrychtr, J., Pelikán, M., Štiková, H., Vrana, I.: EM-OLAP framework. Bus. Inf. Syst. Eng. **60**(6), 543–562 (2018)

Discovery of Important Location from Massive Trajectory Data Based on Mediation Matrix

Xu Zhang$^{(\boxtimes)}$ and Yongsen Hu

College of Computer Science and Technology, Chongqing University
of Posts and Telecommunications, Chongqing 400065, China
zhangx@cqupt.edu.cn

Abstract. Analyzing large volume of trajectory data plays an important role in understanding user behaviors and providing personalized recommendations. However, existing work faces many challenges in important location discovery processing speed and accuracy. This paper proposes a general computing framework to improve the accuracy of occupational and residential location detection in cellular network. An important location discovery module and an index structure is included, which improves the efficiency and accuracy. A mining algorithm MMA (Matrix base Mining Algorithm) is proposed, which improves the accuracy of user important location. Experimental evaluation shows that the proposed algorithm has higher accuracy and efficiency in real environment.

Keywords: Low quality trajectory · Trajectory clustering · Important location · Mediation matrix

1 Introduction

In the era of rapid development of Internet, LBS technology and mobile communication, data recording the location and trajectory information of mobile objects are generated all the time. These location information data based on time series generated by vehicles, crowds, animals and natural phenomena belong to trajectory data. It is significant important to study occupational, residential and travel activities of human beings based on massive trajectory data.

The understanding of user's important location is based on user's movement. Long times ago, user movement can only be recorded by manual methods such as questionnaires. These time-consuming and time-consuming data acquisition methods make the amount of data small and imperfect. With the continuous development and popularization of smart phones, which play an important role in people's daily life, massive trajectories of people movement have been recorded. When people use smart phones to answer or dial calls, send and receive messages, and surf the Internet, they will connect with their nearest base station, generate and record location information log records based on time series, that is called mobile phone trajectory data [1]. To some extent, these trajectories record users' daily activities and travel. Mobile trajectory data has the characteristics of large volume in both population and data records, which makes it possible to extract and learn user movements.

© Springer Nature Switzerland AG 2019
R. Silhavy (Ed.): CSOC 2019, AISC 984, pp. 360–369, 2019.
https://doi.org/10.1007/978-3-030-19807-7_35

However, "large volume" are involved with "low quality". Although mobile trajectory data record users' daily stay and travel behavior, data quality brings many inconveniences and misunderstanding of human movement. For important location discovery, the low quality of mobile phone trajectory data is mainly reflected in the following two aspects:

Poor Accurate of Acquired Raw Trajectory Data. Raw trajectory data itself is the record set of the user's smart devices connecting to the base station. When user connects to the base station, it means that the user appears in the area around the base station, but the specific location of the user cannot be determined. On the other hand, there are various types of base stations, including micro-stations, macro-stations, repeaters and radio frequency remote stations. The coverage of all types of base stations ranges from several hundred meters to several kilometers, which brings greater challenges to user location discovery.

Unbalanced Data Distribution. The distribution of base stations varies significantly in different regions of the city, for example, the density of base stations in the central area is much higher than that in the suburbs. There are more data recorded in the downtown area and fewer data recorded in the countryside.

Aiming at the problem of low accuracy of user important location recognition, this paper proposes a general computing framework and a mining algorithm. When processing massive trajectory data, spatio-temporal correlation is considered with the concept of active points sequences, which can significantly improve the accuracy of important location discovery. A mediation matrix correction method is proposed to improve the accuracy of important location further.

The paper is organized as: Sect. 2 introduces the related work. Section 3 defines several important concepts and important location mining frameworks. Section 4 verifies the performance of the proposed algorithm through experiments. Section 5 give a conclusion of this work

2 Related Work

The location information data generated in the process of human social activities is the main data source and basis for studying human occupational, residential and travel behavior. With the development of information technology, the pattern and scope of human activities have undergone tremendous changes. However, due to geographical and social factors, occupational, residential and travel activities of human beings tend to show certain regularity [3–5]. People periodic around their places of residence and work. Finding periodic behaviors [6] is essential to understanding human movements. Therefore, it is found that the important location represented by residence and work place has high research significance and application value [6–10].

There are many location points in the spatio-temporal trajectory of moving objects, but not all location points are of research significance. Different trajectory points have different authenticity and importance. The combination of certain location points in the trajectory reflects people's stay and travel behavior in a period of time, such as shopping mall, work place, home, etc. In the paper [11, 12], it explores the refueling behavior of taxi drivers in the urban area by using the residence point analysis technology, excavates the points of interest of the crowd and the refueling events of taxi drivers, so as to reduce the waiting time of taxi drivers and optimize the layout of gas stations. In paper [13], it combines with the residence point analysis technology to improve the efficiency of taxi operation and save cost through different trajectory strategies. Important location mining methods based on moving trajectory data are mainly divided into two categories: grid statistics and clustering.

The method based on grid statistics is the earliest proposed method. Paper [9, 13] first rasterized the research area, then matched the base station position to the corresponding grid. By counting the number of users appearing in each grid, the grid with the largest number of times was identified as the grid with important locations. Use the center point of the grid or the average value of the location where the grid appears as an important location for the user. The method based on grid statistics is simple and feasible, but the accuracy and accuracy of the results are not high.

Clustering analysis is the latest research direction. Paper [7, 15] directly cluster the base station location points connected by users and regard the cluster center as the important location of users. Zhang [16] proposed a general solution framework to improve the availability of trajectory data. The framework includes a state-based filtering module, which improves the availability of data and an important location mining module. Based on this framework, two distributed mining algorithms are designed: GPMA (grid-based parallel mining algorithm) and SPMA (station-based parallel mining algorithm) are grid-based parallel mining algorithms. Although the cluster-based analysis method considers the duration and number of users in each cluster, it does not consider the connection time and duration of each base station in the cluster.

3   Computation Framework

In this paper, we proposed a general framework which includes index module, clustering module and correction module, as shown in Fig. 1. First, we import user trajectory data into HDFS for storage to ensure the reliability of data. Secondly, we need to construct a mediation matrix to complete the process because the implementation of the modification and optimization of the important location of users based on mediation matrix. Mediation matrix is a microcosm of the overall user's life activity position, which can well reflect the main activity position of the crowd in life. It will provide a strong basis for us to identify the important location of users. Finally, after building the mediation matrix, with the help of index query, when submitting a user's core location query, it will quickly return to the important location of the final user.

3.1   Definitions

Definition (Important Location). User's important location refers to the position where users stay for a long time and frequently. Residence and workplace are two of the most typical important locations. This article focuses only on these two important locations.

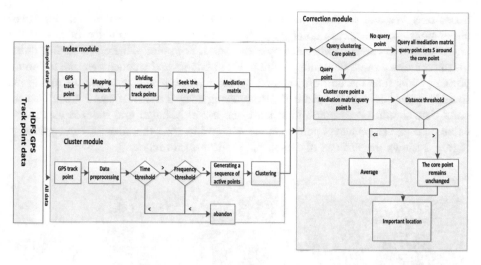

Fig. 1. Framework of proposed method

Definition (Mediation Matrix). Mediation matrix stores the main activity location of the crowd in the grid, which is a description of the geographical clustering characteristics of the user population. It consists of four attributes: grid number, longitude, latitude and the number of residence points.

Definition (Trajectory Point). The trajectory point is the original data point in the user's trajectory.

Definition (Activity Point). Active points refer to the trajectory points of users recorded in trajectory data in a certain spatial or temporal region.

Definition (Dwell Point). Residence point is the clustering core point of a set of trajectory points recorded in trajectory data in a certain spatial or temporal region.

3.2   Mediation Matrix

Index technology is an important method to improve computational efficiency. In this paper, the index method is used to map the extracted crowd activity location to the hierarchical index grid, so as to modify and optimize the important user location extracted subsequently.

People are good at living in groups, and their range of activities and ways of traveling also show group phenomena. With the rapid development of modern society, people's work and home locations are more centralized, showing the group nature of people. Therefore, it is particularly necessary to use the characteristics of people's groups to optimize and modify individuals, which will also more accurately highlight the characteristics of individuals.

Structure. We can assume that the spatial area is a rectangle. There is a given two-dimensional spatial access point data set D, and the position point p of any of the trajectories in D can be represented by coordinates. For point p, there is a method index (p) that returns a mesh containing point p. From this it can be concluded that the point $p(p_x, p_y)$ falls into the unit $c[p_x/r, p_y/r]$, where r is the grid size. After we divide the spatial region into regular grid cells using a single-layer mesh of size r * r, all access point data should be assigned to at least one mesh. Manage grid indexes using key-value pairs and store access points in the same grid in areas that map the same grid key. Figure 2 shows the process of establishing a group grid index.

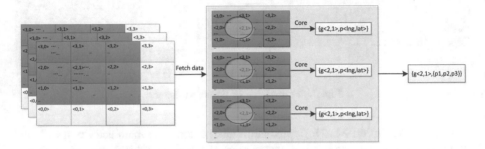

Fig. 2. Grid indexing process

3.3   Clustering Analysis

Clustering module is used to further narrow the search scope of the areas that have been found to contain important locations and to find the areas that really contain important locations. Because users may have multiple important locations of the same type at the same time, for example, users may have multiple workplaces at the same time. In addition, the user's important location will change. Therefore, clustering algorithm needs to be able to automatically find out the user's multiple important locations and changes in important locations. In this paper, the improved density clustering algorithm is used, and the clustering results are analyzed in depth to find out each important location and its effective time.

Active Point Sequence. The purpose of activity point generation is to analyze the main residence of the user according to their mobile phone signals. Firstly, the module searches the area range according to the location of the user, that is, the base station coverage searches the user's activity range, and generates the residence sequence of the corresponding activity points by recording all the active residence points. In the process of generating active point sequence, the base station coverage is used to represent the range of user's active area, and the active point sequence about the base station can be generated. Considering the characteristics of mobile signaling data, i.e. base station jump, elegance and low accuracy, this paper adopts the method of generating multi-base station active point sequence, so as to identify the main active area of users more accurately (Table 1).

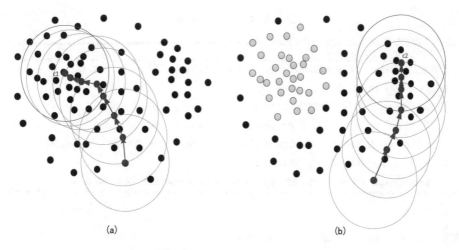

(a)   (b)

Fig. 3. Clustering procedure

Table 1. Active point sequence generation algorithms

Algorithm 1:Sequence Generation of Active Points
Input: User trajectory residence point Tra(userID,p_0,p_1,...,p_n)
Output: Active Point Sequence
1. cacheList=null ; //Cache Active Point Sequences
2. dist ; //Trajectory Point-to-Point Distance Threshold
3. for each point1 <- Tra do
4. cache=null;
5. for cach point2 <- Tra do
6. Distance(point1, point2) < dist;
7. Add point2 to cache
8. end for
9. Add cache to cacheList
10. end for
11. output<cacheList>

Cluster. In the algorithm, the main activity location of users is calculated, and all possible clustering results satisfying the conditions are calculated. Then, according to clustering analysis, the main activity location of users is finally determined. In this algorithm, the sequence of travel activity points of a single user is taken as the input. The specific process of the algorithm is described as follows. Figure 3 shows two clusters and core points c1 and c2 of the same user dataset (Table 2).

Table 2. Clustering algorithm

Algorithm 2:cluster analysis

Input: cacheList, user activity point sequence Tra (userID, p_0, p_1, p_2... p_n)

Output: important location

1. All activities in the cacheList are taken as central points as input.

2. If more than one cluster is obtained after clustering, the time overlap of the corresponding states between clusters is checked. If there are overlaps, the clusters with fewer user connections will be deleted.

3. Each cluster is analyzed step by step to get the important location of users.

3.4   Important Location Optimization

In order to improve the accuracy of user's important location discovery, this paper uses the group activity characteristics of the crowd to correct and optimize the single user. Important location correction algorithm is shown in Table 3.

Table 3. Important location optimization algorithms

Algorithm 3:Important location optimization

Input: User Clustering Core Point P0, Mediation Matrix

Output: User Location Identification Points

1. Define a distance threshold *dist* ;

2. Define final result *rs*;

3. Data = pointMap. get (p0)// Get the index data of the grid in which the core point is located

4. If (data = null) {// Core Points Correspond to no data in the grid, see Fig. 4(b)

5. Get the data in the grid near the core point grid and store it in the data

6. Get the mean of all the data within the *dist* centered on the core and store it in the result.

7. else{// Core Points grid has raw data in the corresponding grid, see Fig. 4(a)

8. Get the mean of all the data within the *dist* centered on the core and store it in the result.

9. }

10. Output *rs*

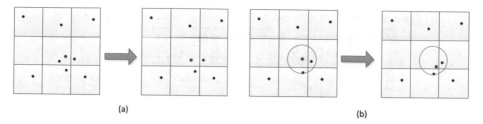

Fig. 4. Important location optimization

4 Experimental Evaluation

4.1 Environment

The experiment is implemented on IBM Big Insight Platform. The algorithm is developed in Scala and run in Spark 1.4. Seven nodes have the same configuration, one master node and six data nodes. The node system is Centos 6.4. Twelve executors are used. The memory of each executor is 8 GB, and the memory of driver is 5 GB.

4.2 Dataset

In this paper, we have desensitized mobile phone signaling data in Chongqing, including encrypted user-ID, longitude, latitude, time and other information. The data set is from April 1, 2018 to April 30, 2018 (a total of 30 days). The data source contains 3 billion recorded data and about 150 GB user trajectory data.

4.3 Evaluation

In this paper, the accuracy of the algorithm is evaluated from three aspects: Precision, Recall and F1-measure. Mean error is used to evaluate the accuracy. Assuming that the number of important locations found by the algorithm is P, the number of important locations correctly identified is R, and the actual number of important locations is T, so there are, Precision = R/P, Recall = R/T, F1-measure = 2(Precision × Recall)/(Precision + Recall).Assuming that the user's actual important location is l, the location found by the algorithm is f, the average error is

$$mean\_error = \sum_{i=0}^{i<Q} dis(l_i, f_i)/P \tag{1}$$

Where, the distance between the identification and the two locations is identified dis(l, f).

Table 4 shows the performance comparison of the algorithm in terms of accuracy of volunteer data under different grid sizes. As it is shown in the table, with the decrease of mesh size, the running time of the algorithm increases with the use of mediation matrix. Smaller mesh needs more data partitioning operations, which cause a slight increase in the running time of the algorithm.

Table 4. Performance comparison of algorithms with different mesh sizes

Mesh size (m)	Average running time (min)	
	Mediation matrix (MMA)	Unmediated matrix
100	92	84
200	98	84
300	80	90
400	85	108
500	88	130

Table 5. Comparisons of algorithms performance

Algorithm	Accuracy rate (%)	Recall rate (%)	F_1 value	Average error (m)
SPMA [16]	88.1	91.9	0.899	300
MMA	90.1	92	0.91	187

In the experiment, the mesh size of the two algorithms is defined as 300 m, and the coverage of the base station is assumed to be 300 m. As it is shown in Table 5, the accuracy of MMA method is better than SPMA, and the average error is significantly improved in real dataset.

5   Conclusion

A general computing framework for important location discovery is proposed in this paper. Compared with the existing work, this paper improves the availability of low-quality trajectory data through the sequence of active points, and constructs a mediation matrix using the characteristics of human group activities, which further improves the accuracy of user important location mining. The experimental results show that the F_1 value of the proposed algorithm is higher and the discovery location output high accuracy in real environment.

Acknowledgement. The work is supported by the National Nature Science Foundation of China (41571401).

References

1. Wu, R., Luo, G., Shao, J., Tian, L., Peng, C.: Location prediction on trajectory data: a review. Big Data Min. Anal. **1**(2), 108–127 (2018)
2. Xu, J.-J., Zheng, K., Chi, M.-M., Zhu, Y.-Y., Yu, X.-H., Zhou, X.-F.: Trajectory big data: data, applications and techniques. J. Commun. **36**(12), 97–105 (2015)
3. Gonzalez, M.C., Hidalgo, C.A., Barabasi, A.L.: Understanding individual human mobility patterns. Nature **453**(7196), 779–782 (2008)
4. Song, C., Qu, Z., Blumm, N., Barabasi, A.L.: Limits of predictability in human mobility. Science **327**(5968), 1018–1021 (2010)

5. Song, C., Koren, T., Wang, P., Barabasi, A.L.: Modelling the scaling properties of human mobility. Nat. Phys. **6**(10), 818–823 (2010)
6. Li, Z., Ding, B., Han, J., et al.: Mining periodic behaviors for moving objects. In: ACM SIGKDD International Conference on Knowledge Discovery and Data Mining, pp. 1099–1108. ACM (2010)
7. Ashbrook, D., Starner, T.: Using GPS to learn significant locations and predict movement across multiple users. Pers. Ubiquit. Comput. **7**(5), 275–286 (2003)
8. Cho, E., Myers, S.A., Leskovec, J.: Friendship and mobility: user movement in location-based social networks. In: ACM SIGKDD International Conference on Knowledge Discovery and Data Mining, San Diego, CA, USA, pp. 1082–1090. DBLP, August 2011
9. Bao, J., Zheng, Y., Mokbel, M.F.: Location-based and preference-aware recommendation using sparse geo-social networking data. In: International Conference on Advances in Geographic Information Systems, pp. 199–208. ACM (2012)
10. Yuan N.J., Wang, Y., Zhang, F., Xie, X.: Reconstructing individual mobility from smart card transactions: a space alignment approach. In: Proceedings of the 13th International Conference on Data Mining (ICDM 2013), pp. 877–886. IEEE (2013)
11. Zhang, F., Wilkie, D., Zheng, Y., Xie, X.: Sensing the pulse of urban refueling behavior. ACM Trans. Intell. Syst. Technol. **6**(3), 13–22 (2013)
12. Zhang, F., Yuan, N.J., Wilkie, D., Xie, X.: Sensing the pulse of urban refueling behavior: a perspective from taxi mobility. ACM Trans. Intell. Syst. Technol. **6**(3), 1–23 (2015)
13. Zhang, D., Sun, L., Li, B., Chen, C., Pan, G., Li, S., Wu, Z.: Understanding taxi service strategies from taxi GPS traces. IEEE Trans. Intell. Transp. Syst. **16**(1), 123–135 (2015)
14. Scellato, S., Noulas, A., Lambiotte, R., Mascolo, C.: Socio-spatial properties of online location-based socialnetworks. In: Proceedings of the 5th International Conference on Weblogs and Social Media, vol. 11, pp. 329–336. AAAI Press, Palo Alto (2011)
15. Chen, J., Hu, B., Zuo, X., et al.: Personal profile mining based on mobile phone location data. Wuhan Daxue Xuebao **39**(6), 734–738 (2014)
16. Zhang, Z.G., Jin, C.Q., Wang, X.L., Zhou, A.Y.: Discovering important locations from massive and low-quality cell phone trajectory data. J. Softw. **27**(7), 1700–1714 (2016). (in Chinese)

The Evolution of Blockchain Virtual Machine Architecture Towards an Enterprise Usage Perspective

Andrei Tara[1], Kirill Ivkushkin[2], Alexandru Butean[1(✉)],
and Hjalmar Turesson[3]

[1] Lucian Blaga University of Sibiu, Victoriei 10, 550024 Sibiu, Romania
{andrei.tara, alexandru.butean}@ulbsibiu.ro
[2] Insolar Technologies, 6300 Zug, Switzerland
kirill.ivkushkin@insolar.io
[3] York University, Keele Street, Toronto 4700, Canada
hturesson@schulich.yorku.ca

Abstract. Virtualization in the context of blockchain systems represents an essential phase in the development and migration of services from public chains to enterprise logic. Most of the ongoing blockchain uses-cases are using the existing public ledgers, but for business products and services, there is a need for custom tailored solutions to ensure flexibility and security. The Ethereum Virtual Machine has opened new ways to solve problems that require a public proof by executing logic on a decentralized ecosystem. In a natural evolutive process, virtualization logic was shaped by numerous architectures and business requirements. Beside performance and scalability, enterprise virtual machines are starting to focus on features like: formal verification, execution of business logic, cross-domain rules, upgradable smart contracts, data privacy politics and audit tools. The current paper presents an introduction, an architectural walkthrough of the most relevant virtual machines that currently exist in the enterprise blockchain ecosystem and an analytical perspective on the future of industry ready VM's.

Keywords: Blockchain · Virtualization · Architecture

1 Introduction

1.1 The Role of the Virtual Machine on Blockchain

The first implementation of a decentralized VM on blockchain was introduced in 2014 by Ethereum [1] and was upgraded ever since by numerous other approaches. In a blockchain architecture, the virtual machine (VM) has the capacity to abstract the entire network and make it work as a single supercomputer that solves numerous computational tasks. The role of the VM is to ensure the proper execution environment for smart contracts. Initially the smart contracts were automatically enforcing really simple rules between 2 parties without the need of an arbiter. In order to adapt to enterprise models, more advanced VMs are required to support business logic, interoperability and integration with all kinds of industrial applications, standards and data privacy regulations.

© Springer Nature Switzerland AG 2019
R. Silhavy (Ed.): CSOC 2019, AISC 984, pp. 370–379, 2019.
https://doi.org/10.1007/978-3-030-19807-7_36

1.2   Virtual Machine Categories

In order to categorize the VMs, there are two important perspectives: integration and execution engine.

From the integration perspective, there are two types of virtual machines:

- Platform dependent VM - embedded into the core components, delivers good performance but it is hard to maintain and verify (ex: EVM [1], NEO VM [2]);
- Pluggable VM and tools - comes as a plugin, needs a complex adaptor to communicate with all the other layers, it is easy to maintain, improve and usually supports formal verification (ex: LLVM [3], COQ [4]).

From the execution engine perspective, there are two categories that can be used to perform decentralized computations:

- Virtual machine - provides a robust and isolated execution sandbox where the smart contract execution can be tracked and controlled on a fine-grained level. Comes with a performance penalty that must be paid because the smart contract needs to be tailored for a particular virtual machine [5].
- Execution containers - a more permissive sandbox, usually based on a docker engine. Allows more dynamic smart contracts written in generic programming languages. Compared to the virtual machine approach this provides more flexibility to the user but also requires smart contracts to have additional logic to prove they are deterministic [6].

2   Architecture of the Existing Models

Since there is no detailed documentation available, for understanding the architecture of the existing popular VM models, an experimental analysis was conducted. The analysis started from the existing resources (open-source code repositories, whitepapers) and 4 months later concluded into the architectural models described in this chapter. All the diagrams, explanations and conclusions are original and were formulated after multiple sessions of code evaluation and experiments.

2.1   Ethereum

The EVM (Ethereum Virtual Machine) [7] is a quasi-turing complete 256-bit VM, and a critical component of the Ethereum network. The EVM architecture has adopted a stack-and-memory model with a 32-byte instruction word size. This model provides access to the program stack, a register space, an expandable temporary memory and a permanent storage (written into the blockchain). From a smart contract design level, it lacks support for a standard library and makes the development of contracts difficult since users must provide support for all required functions. Another shortcoming is related to the fact that 256-bit integers are not natively supported in 32 or 64-bit architectures. The result is a more complicated assembly for simple operations [8].

The architecture (see Fig. 1) has 3 main elements: (a) The machine state - contains the mechanisms required to control the execution flow, gas control component and memory/ stack management structure; (b) The world state - abstracts the virtual machine access to the outside word (account, ledger, etc.); (c) The virtual read only memory - contains a set of subroutines used to impose certain runtime rules to secure correct contextual execution.

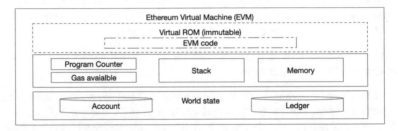

Fig. 1. Ethereum - virtual machine overall architecture.

2.2   EOS

EOS aims to be the first decentralized operating system, with an almost zero price for transactions [9]. Instead of creating a custom VM, EOS relies on the Web Assembly Virtual Machine (WAVM) [10], a stack-base model that allows EOS to use its own set of intelligent contract engine based on WebAssembly (WASM) [11], that enables users to write smart contracts using C/C++.

As displayed in Fig. 2, the WAVM [12] architecture has the following main components: (a) The IR (Intermediate Representation) bridges the parser, the serialization system, and the runtime communication module. IR is very simple to use in manipulating memory and closely resembles with the semantics of the WebAssembly binary format; (b) The parser uses a table-driven deterministic finite automaton to scan the input string for tokens. The tables are generated from a set of tokens that match literal strings and a set of tokens that match regular expressions. The parser is a standard recursive descent parser [13]; (c) The Runtime component is the primary consumer of the byte code. It provides an API for instantiating WebAssembly modules and calling functions (to instantiate modules).

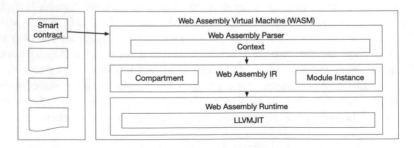

Fig. 2. EOS - WebAssembly virtual machine overall architecture

2.3   NEO

For the VM implementation, NEO introduces an optimal solution that has all the advantages of EVM without crippling the developers with language barriers. The NEO virtual machine [2] provides a general-purpose solution that resembles Java Virtual Machine (JVM) [14] and .Net Runtime [15]. In other words, it is similar to a virtual CPU that executes instructions from the contract sequence [16]. Compared to other approaches, NEO provides an important feature by decoupling the gas from the actual token, so instead of having gas as a subunit of the token, it uses two token systems: one for the gas and one for the assets.

The architecture, as presented in Fig. 3, is composed by 3 major elements: (a) The execution flow is responsible with machine execution logic and flow control; (b) The evaluation stack keeps the actual state of the execution; (c) The interoperability service works as an abstraction layer used to communicate with the outer word.

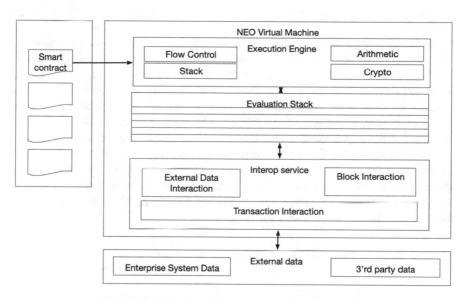

Fig. 3. NEO - virtual machine overall architecture

2.4   Zilliqa

Zilliqa was one of the first projects in the ecosystem of tokenized distributed ledgers, that introduced the concept of sharding [17] in a blockchain architecture [18]. This approach increases the overall throughput of the system but adds more complexity to the execution of smart contracts creating the problem of sharded smart contracts [19]. In this context, in order to facilitate the execution and to ensure the correctness of smart contracts, Zilliqa created Scilla [20]. Scilla is an intermediate-level smart contract language, that imposes a structure on smart contracts that will make applications less vulnerable to attacks by eliminating known vulnerabilities directly at the language-level. Furthermore, the proposed structure also allows formal verification proofs. [21].

The Zilliqa architecture (Fig. 4), uses the Scilla interpreter as an execution environment [22]. Benefiting from the Shard Abstraction layer, when a contract is called, there is a mechanism that passes the ledger state and contract rules to the interpreter.

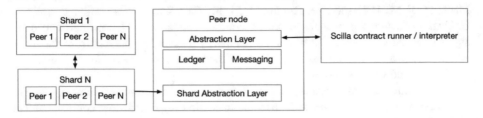

Fig. 4. Zilliqa - virtual machine overall architecture

2.5   Hyperledger Fabric

One of the most promising projects in the Hyperledger family is IBM's Fabric [23]. With a modular architecture, Fabric is intended to be a framework for the development of various enterprise industrial solutions. The Fabric chaincode executor relies on a docker container. The purpose of the container is to ensure the isolation of the smart contract execution from the peer itself, so that a smart contract (chaincode) cannot crash or even access the peer because of an error or malicious code. The chain code is instantiated in the peer node that injects the container image with the contract logic and deploys that image [24]. Once the contract execution started, all the transactions and proposals received by a peer will be dispatched (for execution) on the container [25]. Fabric architecture uses a docker container where Go smart contracts are executed. The access to the ledger and message bus is performed using an abstraction layer as displayed in the classic docker approach (Fig. 5).

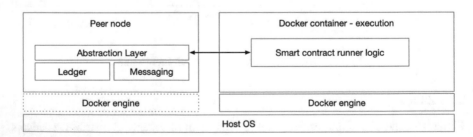

Fig. 5. Hyperledger Fabric - virtual machine overall architecture

2.6   Insolar

The Insolar platform [26] offers flexible governance allowing users to choose between an Insolar public network or a private network with domain-based rules. The smart contracts are designed with enterprise logic in mind and the current VM model relies on a Go Plugin [27] that creates a separate execution container accessible through Remote Procedure Calls (RPC). The usage of RPC provides an extension mechanism for integration of various RPC compliant containers such as any JVM-based language.

As displayed in Fig. 6, the central component of the VM is the Logicruner that receives calls from the MessageBus (abstracts the communication with the network) and stores information using the ArtifactManager (abstracts the ledger and the persistent storage). Once the contract invocation is intercepted, the Go Plugin engine deploys the binary on the execution container through an RPC client, providing an isolated context that will be accessible only from executed contracts. Some other major classic components such as Wallet or Allowance are abstracted as contracts but on runtime they are mapped to proxy stubs.

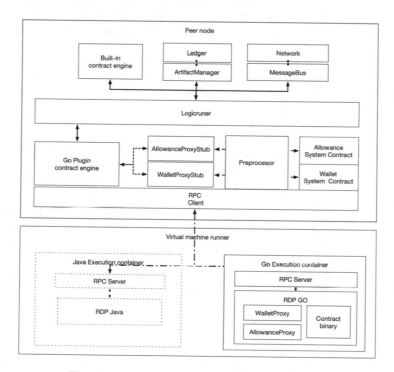

Fig. 6. Insolar - virtual machine overall architecture

3 The Future of Industry Ready VM's

Blockchain as a Service is starting to be used for a large variety of use-cases [28]. Simple value transfer applications like digital currencies or tokenized markets are already benefiting from the existing VM models. The real challenge occurs when the current models are confronted with the dynamics and rigor of the industrial processes. Any model that doesn't have dedicated components to match the variety of integration cases is somehow considered obsolete. The architecture models presented in a gradual evolution in chapter 2 (from Ethereum to Insolar), were chosen from a series of many others just because they have the potential to match the enterprise needs.

In order to prepare for a multi-stage industry adoption process, we propose a series of features and architecture details that are mandatory for current and future VMs in order to adapt to the rapidly changing landscape:

Critical a) Evaluate and integrate standardized procedures and containers, according to under development ISO/TC 307 - Blockchain and distributed ledger technologies [29];

b) Guarantee a safe compilation and a consistent output using advanced formal analysis and verification methods, frameworks and tools [30];

c) Extend security or high-load functionality with non-turing-complete languages preventing vulnerability at the language-level - change the purpose of the VM so that it can extend/amend platform behavior, e.g. introduce more consensus algorithms, security features or policies;

d) Prepare abstraction and access layers for domains and shards - model architecture and programming connectors for shard abstraction or shard access features while guaranteeing the same contextual environment for a contract;

e) Combine within a single platform different VMs by target purpose

f) Favor the integration with external services, get contextual data or control the privacy by using off-chain contracts. If a specialized logic is restricted to a specific segment, that means that the same code for a different segment will behave in a different way because there are different governance rules;

g) Upgradable smart contracts are hard to integrate into public networks, but private networks have control over decision and should use this type of contracts in correlation with business process modeling;

h) Favor the usage of high level languages to allow integration with existing technologies or new products and services from emerging fields: IoT [31], AI[32], Smart City [33];

i) Add support for natural language contracts to allow non-technical

Important   stakeholders to contribute directly to the formulation of rules

In a fast-paced environment, any initiative that aims to build a new VM model or to use/extend an existing one for blockchain integration into industrial applications, should consider the above capabilities as reasonable requirements.

4 Conclusion

This paper presented a summarized evolution of blockchain virtual machines from an enterprise usage perspective and an analysis on how current and future models should mature in order to adapt to the industry needs. The most important architectural details were described from an integration point of view to highlight the expansion from the classic models to enterprise-tailored models as follows:

- The first generation of VMs represented by EVM was a game changer in the ecosystem but the programming language constraints and lack of interoperability are limiting the business integration perspectives;
- EOS made another step towards industry adoption by using a standardized VM;
- NEO removed the programming language barriers (using Javascript, C#, Java) and introduced an interoperability service for interactions with the outer world;
- Zilliqa fragmented the network into shards for improving the performance and found a way for cross-shard smart contracts that are formally verified;
- Hyperledger Fabric adapted the classic docker container to a modular blockchain architecture to achieve adaptability to a large variety of use-cases;
- Insolar, introduced multiple abstraction layers to support business ready smart contracts developed with any RPC compliant language.

As presented in the analysis section, the capabilities of smart contracts are the main reason for the usage of virtualized environments. Their logic is expanding towards an open world of decentralized services. In terms of requirements for industry integration of blockchain, standards and security are critical, interoperability remains imperative and adaptive business logic should be prioritized. In order to ensure the progress, the focus should stay on all the emphasised characteristics because multiple virtualized environments should be able to interact seamlessly to allow the development of trusted machine-to-machine protocols.

Acknowledgements. This work is supported by the York University, Toronto, Canada through the Blockchain Lab at http://blockchain.lab.yorku.ca/.

This work is supported through the DiFiCIL project (contract no. 69/08.09.2016, ID P_37_771, web: http://dificil.grants.ulbsibiu.ro), co-funded by ERDF through the Competitiveness Operational Programme 2014–2020.

This work was possible thanks to the open source communities of Ethereum, Insolar, NEO, EOS, Hyperledger and Zilliqa that offered the code, technical papers and explanations for all required details.

References

1. Wood, G.: Ethereum: A Secure Decentralised Generalised Transaction Ledger. https://ethereum.github.io/yellowpaper/paper.pdf. Accessed 15 Nov 2018
2. NEO White Paper - A distributed network for the Smart Economy. http://docs.neo.org/en-us/whitepaper.html. Accessed 15 Dec 2018
3. Lattner, C., Adve, V.: LLVM: a compilation framework for lifelong program analysis & transformation. In: Proceeding of the 2004 International Symposium on Code Generation and Optimization, p. 75. ACM, Palo Alto-California (2004)
4. Paulin-Mohring, C.: Introduction to the Coq proof-assistant for practical software verification. In: Meyer, B., Nordio, M. (eds.) Tools for Practical Software Verification LASER 2011. Lecture Notes in Computer Science, vol. 7682, pp. 45–95. Springer, Heidelberg (2011)

5. Neo Contract White Paper. http://docs.neo.org/en-us/basic/neocontract.html. Accessed 15 Dec 2018
6. Rad, B., Bhatti, H., Ahmadi, M.: An introduction to docker and analysis of its performance. IJCSNS Int. J. Comput. Sci. Netw. Secur. **17**(3), 228–235 (2017)
7. Ray, J.: A Next-Generation Smart Contract and Decentralized Application Platform. https://github.com/ethereum/wiki/wiki/White-Paper. Accessed 15 Dec 2018
8. Working with Big Numbers Using x86 Instructions. http://x86asm.net/articles/working-with-big-numbers-using-x86-instructions/. Accessed 15 Dec 2018
9. Grigg, I.: EOS - An Introduction. https://eos.io/documents/EOS_An_Introduction.pdf. Accessed 15 Dec 2018
10. McFadden, B., Lukasiewicz, T., Dileo, J., Engler, J.: Security Chasms of WASM. https://i.blackhat.com/us-18/Thu-August-9/us-18-Lukasiewicz-WebAssembly-A-New-World-of-Native_Exploits-On-The-Web-wp.pdf. Accessed 15 Dec 2018
11. Haas, A., Rossberg, A., Schu, D., Titzer, B., Holman, M, Gohman, D., Wagner, L., Zakai, A., Bastien, J.F.: Bringing the web up to speed with webAssembly. In: Proceedings of the 38th ACM SIGPLAN Conference on Programming Language Design and Implementation, pp. 185–200. ACM, Barcelona-Spain (2017)
12. WebAssembly Virtual Machine. https://github.com/WAVM/WAVM. Accessed 15 Dec 2018
13. Johnstone, A., Scott, E.: Generalised recursive descent parsing and follow-determinism. In: Koskimies K. (eds.) Compiler Construction, CC 1998. Lecture Notes in Computer Science, vol. 1383, pp. 16–30. Springer, Heidelberg (1998)
14. Singh, T.: The Hotspot Java virtual machine: memory and architecture. Int. J. Allied Pract. Res. Rev. 60–64 (2014)
15. Kennedy, A., Syme, D.: Design and implementation of generics for the .NET Common Language Runtime. In: Proceedings of the ACM SIGPLAN 2001 conference on Programming language design and implementation. ACM, New York (2001)
16. NEO Smart Contract Introduction. http://docs.neo.org/en-us/sc/introduction.html. Accessed 15 Dec 2018
17. Kokoris-Kogias, E., Jovanovic, P., Gasser, L., Gailly, N., Syta, E., Ford, B.: OmniLedger: a secure, scale-out, decentralized ledger via sharding. In: IEEE Symposium on Security and Privacy. IEEE Xplore, San Francisco (2018)
18. The ZILLIQA Technical Whitepaper. https://docs.zilliqa.com/whitepaper.pdf. Accessed 15 Dec 2018
19. Al-Bassam, M., Sonnino, A., Bano, S., Hrycyszyn, D., Danezis, G.: Chainspace: A Sharded Smart Contracts Platform. https://arxiv.org/pdf/1708.03778.pdf. Accessed 15 Dec 2018
20. Sergey, I., Kumar, A., Hobor, A.: Scilla: a Smart Contract Intermediate-Level LAnguage. https://arxiv.org/pdf/1801.00687.pdf. Accessed 15 Dec 2018
21. The Not-So-Short ZILLIQA Technical FAQ. https://docs.zilliqa.com/techfaq.pdf. Accessed 15 Dec 2018
22. Scilla: Syntax and Semantics. https://github.com/Zilliqa/scilla/blob/master/docs/scilla-spec.pdf. Accessed 15 Dec 2018
23. Hyperledger Architecture, Volume 1. https://www.hyperledger.org/wp-content/uploads/2017/08/Hyperledger_Arch_WG_Paper_1_Consensus.pdf. Accessed 15 Dec 2018
24. Androulaki, E., Barger, A., Bortnikov, V., Cachin, C., Christidis, K., De Caro, A., Enyeart, D., Ferris, C., Laventman, G., Manevic, Y.: Hyperledger fabric: a distributed operating system for permissioned blockchains. In: EuroSys 2018 Proceedings of the Thirteenth EuroSys Conference, Article No. 30, Porto, Portugal (2018)

25. Benhamouda, F.; Halevi, S.; Halevi, T.: Supporting private data on Hyperledger fabric with secure multiparty computation. In: IEEE International Conference on Cloud Engineering (IC2E), pp. 357–363. IEEE Xplore, Florida (2018)
26. Ivkushkin, K.: Insolar Blockchain Platform, unpublished material (2019)
27. Go Plugin System over RPC. https://github.com/hashicorp/go-plugin. Accessed 15 Dec 2018
28. Zīle, K., Strazdiņa, R.: Blockchain use cases and their feasibility. Appl. Comput. Syst. **23**(1), 12–20 (2018)
29. International Organization for Standardization: ISO/TC 307 - Blockchain and distributed ledger technologies. https://www.iso.org/committee/6266604.html. Accessed 15 Dec 2018
30. Chatterjee, K., Goharshady, A.K., Velner, Y.: Quantitative analysis of smart contracts. In: Ahmed A. (eds.) Programming Languages and Systems. ESOP Lecture Notes in Computer Science, vol. 10801, pp. 739–767. Springer, Heidelberg (2018)
31. Hanada, Y., Hsiao, L., Levis, P.: Smart Contracts for Machine-to-Machine Communication: Possibilities and Limitations. https://arxiv.org/abs/1806.00555. Accessed 8 Jan 2018
32. Dinh, T.N., Thai, M.T.: AI and blockchain: a disruptive integration. Computer **51**(9), 48–53 (2018)
33. Michelin, R.A., Dorri, A., Lunardi, R.C., Steger, M., Kanhere, S.S., Jurdak, R., Zorzo, A.F.: SpeedyChain: a framework for decoupling data from blockchain for smart cities. In: MobiQuitous 2018 Proceedings of the 15th EAI International Conference on Mobile and Ubiquitous Systems: Computing, Networking and Services, pp. 145–154. ACM, New York (2018)

Modeling and Evaluating a Human Resource Management Ontology

Wiem Zaouga[1(✉)] and Latifa Rabai[1,2]

[1] SMART Laboratory, Université de Tunis, ISG Tunis, 2000 Bardo, Tunisia
`Wiemzaouga89@gmail.com`, `Latifa.Rabai@gmail.com`
[2] College of Business, University of Buraimi,
512 Al Buraimi, Sultanate of Oman

Abstract. Over the last few years, there is a growing attention toward Human Resources (HRs) as one of the most valuable asset of any organizations; and managing successfully this asset is crucial in project management. However, despite its relevance, the literature review was shown a lack of a shared, interoperable framework on Human Resource Management (HRM) that allow HRs as a team to interchange their knowledge, skills and facilitates their proper use of Tools and Techniques (T&T); further, HRM processes are not enough to perform all HR duties due to a lack of a common terminology and a complete understanding about project requirements. This paper deals with two main contributions. First, we model a semantic point of view of the main concepts related to HRM domain in the context of PMBOK 5th Guide; the outcome of this model will be an ontology promoting interoperability among HRs as well as their efficient use of T&T. After that, we propose an evaluation of the proposed ontology using both criteria-based evaluation approach to focus on the ontology content and its structure and automated consistency checking approach in order to consider ontology consistency.

Keywords: Domain ontology · Ontology evaluation ·
Ontology evaluation approaches · Human resource management

1 Introduction

Human resource (HRs) as one of the most valuable asset of any organizations play a key role in their success [1]; and managing successfully this asset is crucial in project management. Within this context, HRM is the processes that organize, manage and lead the team to achieve project objectives, using its knowledge, skills and a set of T&T defined for each process [2]. Therefore, when HRs are brought together as a team, a common shared framework to interchange their knowledge, skills and facilitates their efficient use of T&T, is required. As a formal, explicit representation of a shared conceptualization [3], ontology is the semantic means to represent and share knowledge into specific domain, provides a common vocabulary as well as a description of the meaning of concepts, as a way to promote interoperability, knowledge reuse and information integration [4].

© Springer Nature Switzerland AG 2019
R. Silhavy (Ed.): CSOC 2019, AISC 984, pp. 380–390, 2019.
https://doi.org/10.1007/978-3-030-19807-7_37

Several prominent efforts have been developed to address various issues related to HRM. For instance, Gomez Perez et al. [5] proposed ontology based on standards namely SEEMP Reference Ontology to describe the detail of a job posting and the curricula data of a job seeker. SEEMP Ontology is structured of thirteen sub-ontologies: Competence, Compensation, Driving License, Economic Activity, Education, Geography, Job Offer, Job Seeker, Labor Regulatory, Language, occupation Skill and Time. Other reference with similar idea can be found in Mochol et al. [6]. An ontology-supported web-based HR recruitment was proposed by Ma et al. [7] that exploit Formal Concept Analysis (FCA) for building HR ontology without the involvement of domain experts, this ontology performs the matching between the position requirements and applicants' competences by using ALN (description logics). In [8], Szekely developed a HR-ontology for an IT company which provide support for modeling a common vocabulary for those share information about HRs, this ontology offers possibility to query the employees implied in a project, to match the employees according to their job, skills, etc. In [9], Stormier and Röhrs presented a methodical paradigm of conceptual modeling related to HRM domain in general and in employee assignment in particular by developing domain ontology for employee assignment, also, a HR modeling language and a HR modeling tool are developed. The work presented in [10] proposes an ontology-based approach to assign HRs to software projects, this work explores the use of ontology for building a decision support system that helps project managers to select those employees who are best fit within a new software project based on semantic similarity metrics. In [11], Schmidt and Kunzmann presented a reference model for ontology-based approaches to competency-oriented human resource development that integrates management competencies and offers learning opportunities for employees. An Ontology for Skill and Competency Management was proposed by Zarandi and Fox [12] using a formal ontology that model HRs skills and competencies at particular levels of proficiency in a dynamic environment as well as measuring, inferring, and validating competencies through the use of Process Specification Language (PSL). Miranda et al. [13] developed a novel integrated model that defines the knowledge about all the key aspects related to competence management; in order to present a recruiting and training model able to integrate the professional competences of users in the context of SIRET project. A summary of the works above with their main focus related to HRM processes is presented in Table 1.

This literature review showed a variety of ontologies for managing HRs. The common aspect of all these works is that they address separately such HRM process. However, none of them consider the overall HRM area with a common, interoperable framework.

This paper achieves two principal objectives. The first one is a semantic point of view related to HRM domain in the context of PMBOK 5th Guide proposed as a HRM ontology modeling. The second objective of this work is to evaluate the proposed ontology using two main approaches namely criteria-based evaluation which focus on the ontology content and its structure and automated consistency checking approach to assess ontology consistency.

Table 1. An overview of ontologies related to HRM.

Ontologies references	HRM processes			
	Recruitment	Assignment	Development competency	HRM
Gomez Perez et al. [5]	×			
Mochol et al. [6]	×			
Ma et al. [7]	×			
Szekely, A [8]				×
Stormier and Röhrs [9]		×		
Paredes et al. [10]		×		
Schmidt and Kunzmann [11]			×	
Fazel-Zarandi and Fox [12]			×	
Miranda et al. [13]			×	

This paper is structured as follows: Sect. 2 describes the process of developing the ontology as well as detail the proposed ontology. Evaluation of the ontology is discussed in Sect. 3. Finally, Sect. 4 presents the conclusion of this work and some future directions.

2   The Proposed HR-Ontology

We adopt METHONTOLOGY as a method for developing ontologies which provides techniques and activities used over the ontology life cycle [14]. The ontology development process proposed by METHONTOLOGY consists of the following steps (see Fig. 1).

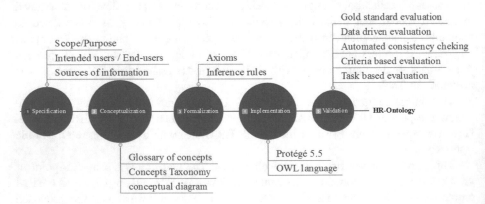

Fig. 1. The development process of our proposed HR-Ontology.

2.1   Ontology Specification

As shown in Fig. 1, we are interested in the scope, users and sources of information. To specify the scope, purposes of the target ontology, a set of Competency questions (CQs) [15] will be developed and seem as queries that could be answered by the ontology [16]. These CQs include, but are not limited to, "What Tools and Techniques are appropriate for each HR process and its related activity?, "What skills are important to consider when assigning staff?", Which are the needed artifacts to produce in each activity? etc. With regard to HR-Ontology scope, we focus on developing domain ontology that covers HRM processes with their related T&T, artifacts and the required competencies. To specify users, we consider the project manager and project management team as they are the main decision makers in HRM domain. Further, the source of information is provided by PMBOK Guide which is the international recognized consensus on the field of Project management [2, 4].

2.2   Ontology Conceptualization

The conceptualization is the main step in the ontology development in which the concepts, attributes, and their relationships in the HRM Knowledge area are collected based on PMBOK 5th Guide. In this context, 140 words are suggested. All these terms, including the relevant classes, subclasses, instances and relations between them. Out of the 140 words, 11 terms are concepts that create ontology classes, 32 subclasses, and 80 instances. One term "is a" is about taxonomic relations and 15 terms are about association. For further portraying this step, the relationships among concepts were also described in Fig. 2 which presents the conceptual model of our HR-Ontology.

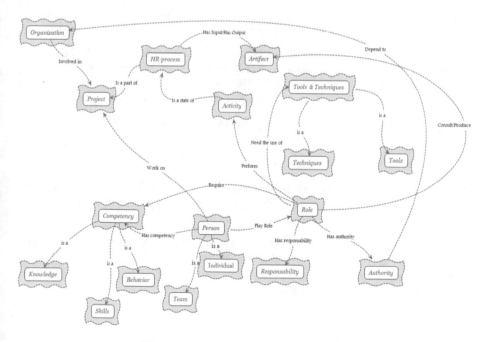

Fig. 2. The conceptual model for HRM.

2.3   Ontology Formalization

During this step, the step2 outcome is transformed into a formalized model, using axioms as depicted in Fig. 3, some terminological (Tbox) and assertion (Abox) axioms are defined to describe the domain knowledge (subsumption relationships among concepts) and concepts to which individuals belong (relationships between individuals and concepts), respectively [16]. These axioms are represented by description languages (DLs) such as SHOIN logic DL.

//— **Axioms used to describe classes** —//

Activity $\subseteq$ T($\exists$ ID. Integer) $\cap$ ($\exists$Name. String) $\cap$ ($\exists$ Descrpition. String) $\cap$ ($\geq$ 1 IsStateOf. HR Process)

Role $\subseteq$ T($\exists$ ID. Integer) $\cap$ ($\exists$Name. String) $\cap$ ($\exists$ Descrpition. String) $\cap$ ($\geq$ 1 (Consult $\cap$ Produce). Artifact) $\cap$ ($\geq$ 1 perform. Activity) $\cap$ ($\geq$ 1 Has Responsability. Responsability) $\cap$ (= 1 Has Authority. Authority) $\cap$ ($\geq$ 1 Need the use. Tools&*Techniques*) $\cap$ ($\geq$ 1 Require. Competency)

//—**Axioms used to express relationships among classes** —//

Skills $\subseteq$ Competency / Knowledge $\subseteq$ Competency/ Behavior $\subseteq$ Competency / Skills $\cap$ Knowldege $\cap$ Behavior $\subseteq$ $\perp$

Document $\subseteq$ Artifact / Template $\subseteq$ Artifact/ Document $\cap$ Template $\subseteq$ $\perp$

Tool $\cap$ Technique $\subseteq$ $\perp$

Plan HR management $\cap$ Acquire project team $\cap$ Develop project team $\cap$ Manage project team $\subseteq$ $\perp$

//—**Individual assertions**—//

Tools= {Organization chart & Position description, Organizational theory, Resource histogram, Multi criteria decision analysis, Personal assessment tool, Observation & Conversation}

Interpersonal_skills= {Leadership, Influencing, Team building, Motivation, Communication, Decision Making, Negotiation, Trust building, Conflict management, Coaching, Political and cultural awareness}

Document= {Project management plan, Project HR plan, Activity resource requirements, Staffing management plan, Resource calendars, Project staff assignment, Team performance assessments, Issue log, work performance reports, Project document, change requests}
(…..)

Fig. 3. Some examples of axioms defined for HR-Ontology.

2.4   Ontology Implementation

This step codify the formalized representation obtained using a specific ontology language. In our case, the proposed ontology was coded by means of Protégé 5 editor tool using OWL as ontology language.

- The first step is to create the concepts hierarchy, present the relationships between these concepts and its characteristics (functional, disjoint, etc.). The second step is to associate for each concept (.i.e. Develop_Team_Activity subclass) its properties and adding its instances (see Fig. 4).

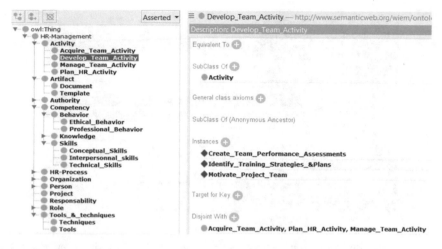

Fig. 4. Protégé screenshots of classes' hierarchy of the proposed ontology.

- Further, axioms are required for defining class restrictions. Figure 5 describes Project_Manager where object property (i.e. Has_Authority_Level), universal restriction (Only) and existential restriction (Some) are defined.

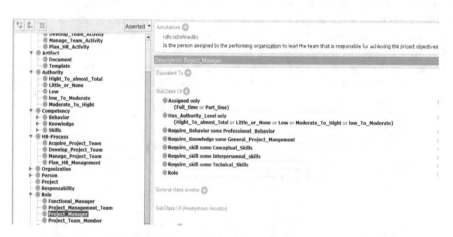

Fig. 5. Protégé screenshots of class description for Project_Manager.

- The Role class has four sub-classes; instances of each role are related to the instances of Artifact class through two properties "consult" and "produce", and related to Responsibility using "Has_Responsability", etc. Fig. 6 describes the project_management_team subclass in order to perform "Develop_Team _Activity".

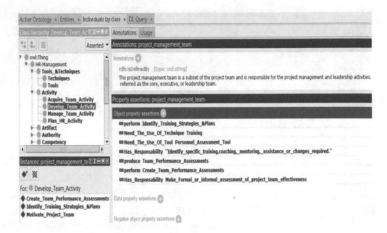

Fig. 6. Relation between Project_management_team with Artifact, Activity, Responsibility and Tools_&_Techniques classes.

3   Ontology Evaluation

Ontology evaluation is a key step on the ontologies development for measuring the quality of ontology according to certain criteria [17]. It's a "technical judgment of the ontology content with respect to a specific frame of reference" and classified in the context of *verification* (is for building ontology correctly) and *validation* (is for building the correct ontology); using metrics, methods, etc. [18]. From this view, the ontology evaluation is as depicted in Fig. 7, it measures the distance between the approximate conceptualization (graph of HR-Ontology) and the real world using metrics, methods, etc.

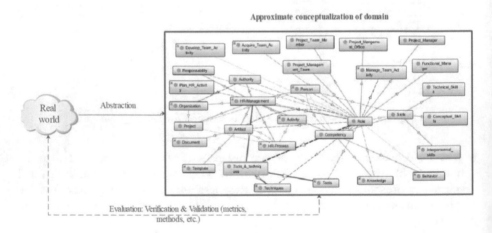

Fig. 7. Ontology evaluation.

3.1 Ontology Evaluation Approaches

Various approaches have been considered in the literature; such as gold standard, corpus (data) driven, automated consistency checking, criteria-based and finally task-based approaches. However, it's not evident to select the suitable approach, since some may not fit well to the ontology scope and its domain. A summary of these approaches with regard to strong and weak points is presented here (Fig. 8).

As shown in Fig. 8, there is no single preferred approach for evaluating ontology; the choice depends on the objective of evaluation, the domain in which the ontology will be applied in, and on what criteria we are trying to meet. In order to make sure that the adopted evaluation approach matches with our objective, we consider the strengths (+ green icon) as well as the limitations (− blue icon) of each one. In this work, the gold standard approach was not considered since there is no existing benchmark ontology in the HRM domain. Although the data driven approach is powerful to evaluate the coverage of HR-Ontology with PMBOK 5th corpus, it is not suitable to evaluate the correctness, clarity, and applicability of ontology. Task-based evaluation is an effective approach to evaluate ontology performance to achieve its intended goals in a specific task but it does not evaluate the structure, architecture, and design layer of the ontology.

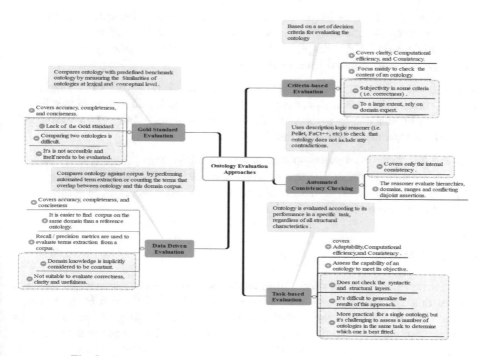

Fig. 8. Summary of ontology evaluation approaches [17, 19, and 20].

Hence, we adopted both criteria-based evaluation and automated consistency checking to assess HR-Ontology. In this perspective, the aim of the HR-Ontology is to store all the knowledge derived from HRM domain and enable searching the suitable T&T, the needed artifact and the required competencies for specific role according to each HRM process. That's why Criteria-based evaluation is relevant in our case to verify the content of ontology in term of consistency, clarity, completeness, and expandability and cover the structural layer of the ontology. The validation of these criteria was conducted using a description logic reasoner.

3.2   Evaluating the Proposed Ontology

In this section, we describe the evaluation of the HR-Ontology by examining how the proposed ontology meets consistency, clarity, completeness, and expandability criteria. Their validation was conducted by an open source reasoner FaCT++ using DL query tab for running query with Descriptive Logic language. Evaluation results are as follow:

Consistency: Automated consistency checking has been positive, objective, and no inconsistencies were found in term of hierarchies, domains, ranges, and disjoint assertions. Thus, HR-Ontology did not include any contradictory information.

Completeness: This criterion is evaluated by checking that the core concepts and attributes which are manually extracted from PMBOK are included in HR-Ontology. Thus, all intended HR-concepts and their relations were explicitly stated in our ontology or can be inferred from existing axioms.

Expandability: New definitions could be added to HR-Ontology without changing the current content. As discussed, our objective is to model a domain ontology (see Fig. 3) which can be scalable to a specific HRM process without altering the existing well defined concepts and relations within HR-Ontology.

Clarity: The intended meaning and definitions of HRM concepts are clearly specified without ambiguity; since all these concepts and their definitions are gathered from PMBOK and were listed in the Protégé using <rdfs: isDefinedBy>, synonyms class <rdfs: comment>, an instance of <rdfs: property>.

4   Conclusion

Two core contributions are driven from this work. First, the development process was proved that HR-ontology fills the knowledge gap of HRM domain by providing a formal and shared vocabulary about concepts related to HRM processes as well as enables searching the suitable T&T, the needed artifact and the required competencies for specific role according to each HRM process. Further, our proposed ontology can be exploited for various purposes especially to develop decision support systems for better managing HRs. Second, the evaluation process was shown that HR-Ontology is clear, consistent, extendable, and complete based on its main purposes. However, the limitation in this process is that the adopted approaches were largely subjective; instead, expandability, clarity, completeness criteria are hard to measure. As a future work, we can assess completeness (coverage) criterion according to data driven approach using quantitative measure such as number of overlapping terms, precision, recall, F-measure, and vector space similarity in order to quantify it objectively.

References

1. Jack-Ide, I.O., Uys, L.R., Middleton, L.E.: Mental health care policy environment in Rivers State: experiences of mental health nurses providing mental health care services in neuro-psychiatric hospital, Port Harcourt, Nigeria. Int. J. Ment. Health Syst. **7**(1), 8 (2013)
2. Project Management Institute: "A Guide to the Project Management Body of Knowledge, 5th edn." Project Management Institute, Inc., Pennsylvania (2013)
3. Gruber, T.R.: Toward principles for the design of ontologies used for knowledge sharing? Int. J. Hum Comput Stud. **43**(5–6), 907–928 (1995)
4. Sheeba, T., Krishnan, R., Bernard, M.J.: An ontology in project management knowledge domain. Int. J. Comput. Appl. **56**(5) (2012)
5. Gómez-Pérez, A., Ramírez, J., Villazón-Terrazas, B.: An ontology for modelling human resources management based on standards. In: International Conference on Knowledge-Based and Intelligent Information and Engineering Systems, pp. 534–541. Springer, Heidelberg (2007)
6. Mochol, M., Oldakowski, R., Heese, R.: Ontology based recruitment process. GI Jahrestagung (2), pp. 198–202 (2004)
7. Ma, H., Hartmann, S., Vechsamutvaree, P.: Towards FCA-facilitated ontology-supported recruitment systems. Enterp. Model. Inf. Syst. Archit. **13**, 182–189 (2018)
8. Szekely, A.: An approach to ontology development in human resources management. In: The 5th International Conference on Virtual Learning ICVL, Târgu-Mures, Romania (2010)
9. Strohmeier, S., Röhrs, F.: Conceptual modeling in human resource management: a design research approach. AIS Trans. Hum.-Comput. Interact. **9**(1), 34–58 (2017)
10. Paredes-Valverde, M.A., del Pilar Salas-Zárate, M., Colomo-Palacios, R., Gómez-Berbís, J. M., Valencia-García, R.: An ontology-based approach with which to assign human resources to software projects. Sci. Comput. Program. **156**, 90–103 (2018)
11. Schmidt, A., Kunzmann, C.: Sustainable competency-oriented human resource development with ontology-based competency catalogs. In: Expanding the Knowledge Economy: Issues, Applications, Case Studies. Proceedings of E-Challenges (2007)
12. Fazel-Zarandi, M., Fox, M.S.: An ontology for skill and competency management. In: FOIS, pp. 89–102 (2012)
13. Miranda, S., Orciuoli, F., Loia, V., Sampson, D.: An ontology-based model for competence management. Data Knowl. Eng. **107**, 51–66 (2017)
14. Fernández-López, M.: Overview of methodologies for building ontologies (1999)
15. Uschold, M.: Building ontologies: towards a unified methodology. In: Proceedings of 16th Annual Conference of the British Computer Society Specialists Group on Expert Systems (1996)
16. Rezgui, K., Mhiri, H., Ghédira, K.: Towards a common and semantic representation of e-portfolios. Data Technol. Appl. **52**(4), 520–538 (2018)

17. Guo, B.H., Goh, Y.M.: Ontology for design of active fall protection systems. Autom. Constr. **82**, 138–153 (2017)
18. Hlomani, H., Stacey, D.: Approaches, methods, metrics, measures, and subjectivity in ontology evaluation: a survey. Semant. Web J. **1**(5), 1–11 (2014)
19. Brank, J., Grobelnik, M., Mladenić, D.: A survey of ontology evaluation techniques (2005)
20. Quinn, S., Bond, R., Nugent, C.: A two-staged approach to developing and evaluating an ontology for delivering personalized education to diabetic patients. Inform. Health Soc. Care **43**(3), 264–279 (2018)

Online Social Networks Analysis for Digitalization Evaluation

Andrey M. Fedorov, Igor O. Datyev$^{(\boxtimes)}$, Andrey L. Shchur, and Andrey G. Oleynik

Institute for Informatics and Mathematical Modeling-Subdivision of the Federal Research Centre "Kola Science Centre of the Russian Academy of Sciences", 184209 Apatity, Russia
{fedorov, datyev, shchur, oleynik}@iimm.ru

Abstract. The majority of countries is committed to developing the digital economy. The article discusses the potential use of online social networks to assess the level of digitalization. A brief overview of international indices used for evaluating digitalization, as well as Russian studies of regional digitalization are provided. In particular, using the developed software, the number of users in different cities and regions of one of the most popular online social networks in the post-Soviet space VKontakte is analyzed. The obtained results are compared with official statistics on the number of residents in regions and settlements. The problems that a researcher faces when using social networks as a source of statistical data are discussed. Conclusions are drawn about the possibility of applying data obtained from the analysis of online social networks as components of the indices when evaluating digitalization.

Keywords: Digital economy · Digitalization evaluation indices · Online social network VKontakte

1 Introduction

The majority of the countries in the world seek ways to develop their digital economy. The term "digitalization" partially represents the measure of this process. But this term itself is relative, so the digitalization can be assessed differently. Let's try to understand better the meaning of these two interrelated terms.

There are several definitions of the term "digital economy". The World Bank has suggested perhaps the most capacious one: "A digital economy is a system of economic, social, and cultural relations based on the use of digital information and communication technologies" [1].

In another study [2], the author identifies the technological and "analog" components of the digital economy. The most important elements of the first stand out are mobile technologies, business analytics, cloud computing and social networks. As in many other works [3, 4], it emphasizes the importance of not only technological aspects, but also of "analogue additions", such as favorable business climate, sound human capital, and good management, which are the foundation of economy growth.

© Springer Nature Switzerland AG 2019
R. Silhavy (Ed.): CSOC 2019, AISC 984, pp. 391–402, 2019.
https://doi.org/10.1007/978-3-030-19807-7_38

The digitalization is the penetration of the digital economy into our daily lives. Sometimes the term digitalization refers to the process of introducing ICT or the transition to the digital economy as a whole. So the term is used both as a process and as a phenomenon.

Disputes about what is primary in the digital economy: technology or human capital, have not abated until now. One thing is for sure: today, with the growth of the global population and the mobilization of resources, digital economy is not limited to digital commerce and services. It affects almost every aspect of life, be it healthcare, education, culture, and so on.

One of the focused studies on digitalization touches the problems of decision making in the regional governmental administrations [5]. That study suggests the development of a distributed situational centers system, which will not only provide situational awareness for the decision makers, but also help them to develop effective solutions based on a comprehensive computer analysis of problem situations. The informational and analytical support provided by such intelligent situational centers network is applied in the light of problematics in the Russian Arctic region [6] and provides an interesting insight on the ways that can change the society functioning thanks to digitalization.

The main objective of this study is to analyze the potential of the online social network services (OSN) in assessing the level of digitalization. Can OSNs be used to measure digitalization? What problems may occur when using OSN as a data source?

The relevance of the study is justified by the following reasons:

1. Official government surveys (by Rosstat [7] in Russia) on digitalization are discrete and sparse, sometimes taking a year to gather. The analysis of OSN can give almost instant statistic numbers.
2. Rosstat makes statistics based on selective surveys; by analyzing online social networks it is possible to gather a lot more data, providing a higher accuracy of the survey.
3. A possibility to compare OSN data with official Rosstat numbers and analyze it to solve problems of verification and pattern search.

2   Digitalization Evaluation Indices

To assess the growth and implementation of the digital economy, various indices and metrics are developed. Here is a brief overview of some indices focused on measuring digitalization.

2.1   International Indices

One of the most commonly mentioned characteristics for assessing the digitalization of countries around the world is the *ICT (Information and communications technology) Development Index* or *IDI*, developed by the International Telecommunication Union, ITU. IDI is based on state statistics from 193 participating countries. It uses 11 core indicators grouped into three categories, sometimes called stages or ICT levels.

First level represents the spread of network technologies and their availability (the number of fixed phones per 100 people, the number of mobile phones and smartphones per 100 people, the proportion of international traffic (bps) per Internet user, the percentage ratio of households with computers, the percentage ratio of households with access to the Internet).

Second level characterizes the intensity of ICT use in society (the percentage of the population using the Internet, the number of users of fixed broadband Internet access per 100 people, the number of users of mobile Internet access per 100 people).

Third level is an addition to the second and is responsible for the factor of human skills, i.e. for the ability of the population to use digital technologies: literacy of the population (percentage of the population over 15 years old, able to read, write and write simple sentences and produce simple arithmetic), the general level of secondary education coverage, the general level of secondary education coverage. The ITU ICT Index report for 2018 [8, 9] has shown interesting worldwide numbers for the first stage of ICT Index: 107 mobile phones and smartphones per 100 people, more than half of the world's population uses the Internet.

The *Networked Readiness Index (NRI)* is created by the World Economic Forum and measures the propensity of countries to exploit the opportunities offered by information and communications technology. It measures, the performance of 139 economies in leveraging information and communications technologies to boost competitiveness, innovation and well-being. [10]. The framework translates into the NRI, a composite indicator made up of four main categories: (1) Environment subindex (Political and regulatory environment, Business and innovation environment), (2) Readiness subindex (Infrastructure, Affordability, Skills) (3) Usage subindex (Individual usage, Business usage, Government usage), (4) Impact subindex (Economic impacts, Social impacts) [11].

The first *International Digital Economy and Society Index (I-DESI)*, published in 2016, provided a snapshot to compare statistics from 15 non-EU countries with the performance of EU Member States. The 2018 I-DESI has been able to utilise datasets over a four year time period from 2013 to 2016 to provide trend analysis. But problems have still been encountered in trying to find perfect surrogates to replicate all the indicators used in the EU28 DESI [12].

The economy digitalization index *e-Intensity*, proposed by the Boston Consulting Group (BCG), is calculated as the average of three sub-indices: infrastructure development, online spending, and user activity. The Infrastructure Development subindex reflects the degree of infrastructure development and the availability and quality of Internet access (fixed and mobile). The Online Expenses subindex includes the costs of online retail and online advertising. The User Activity subindex is calculated as the average of the subindexes of a lower level: activity of companies, activity of consumers and activity of public institutions. In 2016, there was a change in the methodology for calculating the BCG e-Intensity index: the focus was shifted towards mobile technology. The country ranking methodology was also changed: the absolute values of the index were replaced by relative ones [13].

In addition to the above, there are a number of different international indices for measuring the development of the digital economy, that use similar calculating methods. Such indices include: E-government development index (EGDI) [14] calculated by the UN Department of Economic and Social Development (UNDESA), published by the UN Electronic Participation Index (EPART) [15], Global Connectivity Index (GCI) published by Huawei [16], Global Innovation Index – GII [17], the Digitization Index (DiGiX) [10], the World Digital Competitiveness Ranking (WDCR) [18], the Digital Society Index (DSI) [19]. Due to the limited scope in this paper, we will not delve into their detailed description.

2.2   Russian Studies on Digitalization Evaluation

There is also a separate group of digital economy indices, dedicated to its development in Russia. They include studies of the Skolkovo Business School (Digital Life of Russian Megalopolises, Digital Russia Index) and the Digital Literacy Index developed by the Regional Public Center for Internet Technologies (ROCIT).

The basis of the research methodology Digital life of Russian megalopolises holds the idea of the balance between supply and demand for digital services. The final score is shown as an index of digital life in Russian major cities (15 cities with a population of more than 1 million people).

The results of two years of monitoring (2014–2015) allow us to conclude that in 2015 major Russian cities moved to the secondary digitalization stage, when the proportion of the population using the Internet regularly overcame the threshold of 70%, and the habit of using digital services in everyday life moved into the category of "dominant" [20].

The *Digital Russia Index* is calculated on the basis of public references in open sources, taking into account the credibility, citation and tonality of events [21]. The study is based on the evaluation of publications mentioning the development of the digital economy in different regions. Experts studied media articles, legislative acts, notes on websites of various departments and universities and other open sources, and in each case evaluated the "degree of positive impact" of state initiatives on digitalization. Initiative evaluations on a scale from 0 to 100 were included in the overall evaluation of individual indicators of the digital economy. In general, there were seven such indicators: regulatory and administrative indicators; personnel and training programs; research competencies and technological background; information infrastructure; information security; economic indicators; social effects.

The highest level of digitalization is observed in the central and western parts of Russia, while the lowest – in the southwestern part. The authors of the study identified four groups of federal districts on the Digital Russia Index: the leading position was given to Ural Federal District, the catching group – Central, Volga and North-Western federal districts, the second three were Northern, Far-Eastern and Southern FDs, and the closing one was North Caucasian FD. The Ural Federal District holds the lead because of the educational centers for personnel training, research and production bases and so called territories of the advanced development.

The All-Russian study *Digital Literacy Index* monitors digital competencies of Russians and has been updated annually since 2015. For its calculation, they are used as indicators characterizing the "primary" (coverage of the fixed and mobile Internet) and "secondary" digitalization - the level of digital competences in a broad sense (the level of competence in conducting financial transactions via the Internet, the level of interaction culture in OSNs, etc.). The index structure is based on 20 key parameters and includes three levels: digital consumption, digital competencies, digital security [22].

It is also necessary to mention a study [13], where the authors calculated the e-Intensity index (with minor modifications) for the Russian regions. In that study the authors noted that the degree of digitalization of Russian regions is still heterogeneous and breaks into 4 main groups: leaders (Moscow, St. Petersburg), developing regions (most regions of the central and southern Russia, regions of Siberia and the Far East), developing regions with low population (northern regions of Russia and sparsely populated regions of the Far East), and lagging regions (some parts of the North Caucasus).

Thus, according to the latest international assessments, the factor of providing people with communications with the ability to access the Internet gradually begins to lose its primary role, switching focus to the skill of using digital technologies.

In many indices, the use of online social networks by the population is clearly taken as a sub-index or indirectly. This data is obtained, as a rule, on the basis of population surveys or even online user surveys.

Since the presence of a person in online social network can be uniquely attributed to his possession of a certain level of digital literacy, the number of users in OSN on local and regional levels can serve as a source for detailed statistics on the society digitization in that area - even taking into account any accompanied errors and assumptions.

3 Online Social Networks as a Tool for Digitalization Evaluation

Thanks to their prevalence and popularity, online social networks are a good tool for conducting sociological (and various other) studies. Despite the fact that classical methods of such studies (for example, questionnaires or a telephone survey) in many cases provide more accurate statistical information, the use of OSNs can give a wider coverage and a larger sample size.

In one of our previous papers we looked at the methods and possibilities of using social networks for the information monitoring of community's territoriality based on online social network [23]. In that work we examined different ways of using online social networks to obtain various statistical data [24] and to establish the relationship between virtual and real entities [25]. When conducting sociological research through OSNs, it is necessary to take into account the existence of objective problems, the description of which is presented later in the text.

3.1 The Problematics of Using Online Social Networks for Obtaining Statistical Data

First of all, it should be noted that not every person is registered in online social networks due to various circumstances. Also, modern information technologies, such as Google's Android mobile operating system, may include mandatory registration enabling the use of OSNs (Google Plus), but the user may never use anything there other than the registration itself. Sociological studies can be carried out in various well-known OSNs, and many users have accounts in them. But there are users of only one or few particular online social networks, who might not be registered in the OSN that is targeted for data extraction and thus they won't be represented in the gathered statistics.

In addition, not all age groups are represented in OSNs, as in some of them the rules prohibit the registration of persons under a certain age. But since the possibility of strict online age verification is not common and actively used, users usually evade this prohibition easily. This leads to another problem which is related to the unreliability of the data presented in user profiles. However, some methods have been developed that allow to determine the actual age of the respondent according to various indirect characteristics. On the other hand, there are categories of people who, for various reasons, intentionally decide to not use OSNs at all.

Every online social network has a problem of inactive accounts that were used but became abandoned. Administrations of some OSNs try to solve this problem by defining a period of inactivity and inviting the user to determine in advance what to do with his account if such period happens. Another problem of using OSN as a socio-logical research tool is the presence of accounts, not associated with a person (which is implied), but with a whole group of people - for example, organizations, associations, communities of interest. In addition, some register accounts in OSNs on behalf of animals or even non-living objects. There are also quite a few users who have several active accounts in one OSN. Some of them are used exactly for research purposes in form of bots driven by software algorithms.

But even with all the mentioned problems, the capabilities of modern OSNs allow us to conduct interesting and widescale studies of our society, for which the phe-nomenon of online social networks itself becomes an integral part of existence.

In some OSNs (for example, Twitter, YouTube) a user profile is often limited to a set of basic attributes that are insufficient for solving many problems that involve personalization of the results. In such cases personalization is carried out on the basis of actions that user performs: markings of the liked materials, comments on them, redi-rects to friends, creation of playlists, etc. The analysis of a user's activity allows to quite accurately form or correct his/her profile.

The user data security policy, usually applied in OSNs, has a rather negative effect on the considered research. Even access to data that would be logical to assume open, because it is provided exactly for the functioning in an OSN, can become a daunting task. For Russia this issue is governed by the Federal Law on the Processing of Personal Data [26]. Since May 2018, a new normative document (GDPR, General Data Protection Regulation) has been introduced into European legislation that regulates the collection and processing of personal data belonging to citizens of Europe, wherever they are territorially located [27]. In terms of conducting a scientific research of society through the prism of online social networks, Russian legislation still has a sufficient number of opportunities that do not contradict the law.

In addition, when trying to analyze OSN data, there are often technical difficulties, such as poor structurization, updates of the user model and functionality, access restriction and blocking used in order to prevent unauthorized user data collection and limit infrastructure load.

Among other things it is necessary to note the difficulty to verify present day statistics obtained from the OSN analysis, due to the lag in the statistical data gener-ation by Rosstat that often is delayed by one and sometimes even three years.

3.2   VKontakte Online Social Network as a Data Source

The online social network VKontakte offers its users the opportunity to work with the whole variety of system functions not only through the web interface, but also through API functions. Both options require to login as a VK user, which imposes that user's restrictions on access to communication and information. Such restrictions are formed by users themselves, who determine what data will be available for various categories of users: friends, friends of friends or all others. In this case, the use of API functions only expands the capabilities of users, allowing them to automate actions that require manual interaction with the elements of the web interface. This automatization ability has been applied in this study.

The OSN VKontakte was chosen to confirm the hypothesis about the possibility of using social network information to assess the level of digitalization of regions due to its high popularity in the ex-Soviet space [28, 29]. According to various estimates, the number of Vkontakte users ranges from 81 to 228 million people.

In addition to the open data from the OSN VKontakte, this study uses data from the statistical reference books of Rosstat [7]. For the purpose of this work, the most interesting is the official statistic reflecting the population of individual cities and regions of Russia, as well as direct and indirect data on the level of penetration of digital technologies into the economic and social spheres. Of course, compared with the official statistics, data from the OSN looks inaccurate and not quite serious. However, when used correctly, the openness, sheer bulk and speed of obtaining from the OSN make this data somewhat of a worthy addition to the official statistics, since it allows to more fully and accurately explore society in both digital and real life.

4   Experimental Results and Analysis

With the help of the built-in API-function providing help in advanced search of the user profiles, data on their distribution among cities of Russia was obtained. It should be noted that it was done without any direct access to user profiles themselves.

As a result, a spreadsheet was created containing information on the number of VKontakte users who named a city in their profile as a place of residence/stay. Then the obtained spreadsheet was supplemented with the official Rosstat statistics on the number of population in the respective cities (the Population column of Table 1).

Table 1. A fragment of the final "accounts and population" data spreadsheet

CityID	City	Accounts	Population
353	Borovichi	50182	50896
354	Anapa	122310	75865
360	Ishimbay	52753	65422
376	Polyarnye Zori	11957	14421
385	Severouralsk	21078	26288

Similar to Table 1, a spreadsheet was formed containing information on the number of VKontakte accounts and the population in Russian regions.

Then, for the sake of better visual analysis, the data of the obtained spreadsheets was mapped. One of the final data presentations is shown in Fig. 1. Each city in it is represented by a pair of histogram columns, one of which shows the number of VKontakte accounts and the other one – the officially counted population.

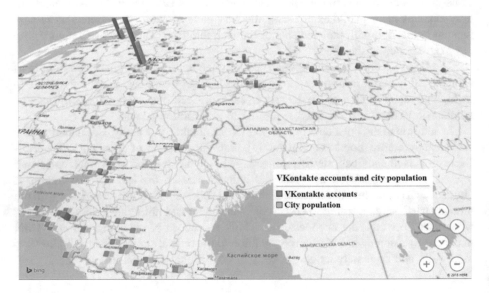

Fig. 1. Cities of Russia: accounts and population

The second variant was the presentation of each region by a pair of histogram columns, one of which displays the number of VKontakte accounts, and the second – the population of the region.

Both these variants for presenting data allowed us to identify heterogeneity and other features of territorial accounts distribution at both city and regional levels. In addition, the presentation of the spoken OSN analysis at two different levels (city and region), allowed us to correlate the results we obtained with the digitalization estimates made by other researchers.

The result of the conducted experiment clearly shows the undisputed leadership of Moscow and St. Petersburg. There's also a noticeable lagging of some regions of the North Caucasus. These examples correlate with the results of other works, such as [13, 21].

One of the main issues in the analysis of data is the problem of identifying patterns. It should be mentioned that the Pearson correlation coefficient between the population of cities according to Rosstat and the number of accounts in cities according to the results of our experiments ranges from 0.85 (with the exclusion of the cities with over one million people population) to 0.96 (in the case of processing the entire data sheet, including the aforementioned large cities). This indicates a strong relationship between the number of people and the number of accounts.

This connection is different for different locations. For cities with a large number of population and, apparently, regions with a high level of digitalization, there is a significant excess of the number of accounts over the population (approximately two times). There are three main reasons for this: (1) people from the regions ascribe themselves to the regional capital; (2) in big cities there are very strong migration and emigration flows; (3) high economic activity of the population, leading to an increase in the number of fake accounts and the number of bots. As a rule, the larger the city, the more pronounced are the above reasons and the excess of the number of VKontakte accounts over the population.

For smaller cities and towns an opposite situation is often can be observed: the excess of the population over the number of VKontakte accounts. The main reasons for this are: (1) people from the regions ascribe themselves to the capital of the region; (2) low level of Internet accessibility; (3) low economic activity of the population, usually resulting in no increase of fake accounts and bots; (4) prevalence of the elder population, not using OSNs; (5) inconsistency of the official statistics on the number of people with the real-life situation. As a rule, the smaller the population size, the more pronounced are the reasons for the excess of the population over the number of VKontakte accounts.

The key problem in correlating the official population and the number of OSN users is the counting error, which is based on the user content unreliability: multiple accounts of a same person in the OSN, manual selection of a geolocation, many bots and inactive accounts.

We should not also forget the obligatory error in calculations of the city and rural area population in Russia. Rosstat data, which is used in these cases, uses the national census of 2010 as a basis and takes into account only the officially registered population, i.e. does not include the category of temporary residents which can be rather significant in large cities.

In this aspect, the analysis of data from OSNs can be used to correct official statistics on the resident population and visitors.

5 Conclusion

Even without any specific comparison with the official population statistics the number of registered OSN users in a given settlement already allows us to establish the degree of its involvement in digital life and, therefore, digital economy.

Official population figures are not equal to the number of accounts in online social networks. However, the correlation coefficient between these two values indicates their close relationship. In different regions and cities, the fluctuation between the population size and the number of accounts is also different, which indicates the heterogeneity of digitalization of the Russian regions. The reason for such heterogeneity can be due to a combination of reasons: climatic conditions, population density, size of a city/region, availability of communication services (price, quality, availability in remote areas), migration flows, unreliable statistical data, different involvement in the virtual world due to the uneven development in different regions of the national economy, etc.

A portion of the reasons listed above arise because of the problems of using OSNs as a data source. These problems include the unreliability of user-generated content, different user representations in various online social networks (demographic, geographic reasons), technical problems of obtaining open data from OSNs for analysis, as well as legal and ethical aspects.

Analysis of online social networks requires the development and application of information technology to verificate data obtained from such networks. More accurate values, as well as new knowledge can be obtained by using algorithms for semantic contextual analysis and machine learning. A deeper analysis using these methods will make it possible to correct official statistics or even become one of the sources for these statistics. In addition, to obtain more accurate information, the data must be provided from several online social networks. Another important issue from the point of view of scientific research is the identification and data exclusion of bots and inactive accounts. Lastly, a significant aspect of OSN analysis should be dedicated to the temporal component of the research, providing the ability to solve the tasks of retrospective analysis and forecast the direction of evolution and the values of various indicators of the virtual world.

As it can be seen, there is a sufficient number of problems in the way of using OSN as a data source. However, the scientific community is already working on solutions for most of them, so that gives hope for their solution in the near future.

But even today the data gathered from OSNs can be used to assess the level of digitalization of cities, regions and even countries – if not as a separate digitalization index, then at least as a sub-index. The main advantage of OSN usage is the ability to monitor and create statistical slices on the fly, obtaining the most up-to-date information based on the analysis of open data that is posted in online social networks.

Acknowledgements. The research described in this paper is partially supported by state research of the Ministry of Education of the Russian Federation № 0226-2019-0036 and by the Russian Foundation for Basic Research for the support (project №18-29-03022).

References

1. Strelkova, I.A.: Digital economy: new opportunities and threats for the development of the world economy. Econ. Taxes. Law **2**, 18–26 (2018)
2. Panshin, B.: Digital economy: features and development trends. Sci. Innovations. **157**, 17–20 (2016)
3. Kapranova, L.D.: Digital economy in Russia: state and development prospects. Econ. Taxes Law **2**, 58–69 (2018)
4. The economic contribution of broadband, digitization and ICT regulation. https://www.itu.int/en/ITU-D/Regulatory-Market/Documents/FINAL_1d_18-00513_Broadband-and-Digital-Transformation-E.pdf. Accessed 30 Jan 2019

5. Oleynik, A., Fridman, A., Masloboev, A.: Informational and analytical support of the network of intelligent situational centers in Russian Arctic. In: Proceedings of the International Research Workshop on Information Technologies and Mathematical Modeling for Efficient Development of Arctic Zone (IT&MathAZ 2018), Yekaterinburg, Russia, 19–21 April 2018, vol. 2109, pp. 57–64. CEUR Workshop Proceedings (2018)
6. Oleynik, A., Lomov, P., Shemyakin, A., Avdeev, A.: Solutions for system analysis and information support of the various activities in the arctic. Czech Polar Rep. **2**(7), 280–289 (2017)
7. Russian Federation Federal State Statistics Service. http://www.gks.ru/wps/wcm/connect/rosstat_main/rosstat/en/main. Accessed 30 Jan 2019
8. Measuring the Information Society Report, 2017 (ITU). https://www.itu.int/en/ITU-D/Statistics/Documents/publications/misr2017/MISR2017_Volume1.pdf. Accessed 20 Dec 2018
9. Measuring the Information Society Report, 2018 (ITU). https://www.itu.int/en/ITU-D/Statistics/Documents/publications/misr2018/MISR-2018-Vol-1-E.pdf. Accessed 20 Dec 2018
10. Katz, R., Koutroumpis, P., Callorda, F.: Using a digitization index to measure the economic and social impact of digital agendas. Info. **16**(1), 32–44 (2014). https://doi.org/10.1108/info-10-2013-0051
11. Global Information Technology Report 2016. World economic forum. http://reports.weforum.org/global-information-technology-report-2016. Accessed 20 Dec 2018
12. Foley, P., Co.: International Digital Economy and Society Index. https://doi.org/10.2759/745483. https://www.ospi.es/export/sites/ospi/documents/documentos/Study_International_DESI_2018.pdf. Accessed 20 Dec 2018
13. Banke, B., Butenko, V.: Russia online? Catch up can not be left behind. June 2016. The Boston Consulting Group. (in Russian). http://image-src.bcg.com/Images/BCG-Russia-Online_tcm27-152058.pdf. Accessed 20 Dec 2018
14. United Nations E-Government Survey 2016: E-Government In Support Of Sustainable Development. United Nations (2016)
15. Goloventchik, G.G.: Rating analysis of the level of digital transformation of the economies of EAEU and EU the countries. Digit. Transform. **2**(3), 5–18 (2018)
16. Tap Into New Growth With Intelligent Connectivity: Mapping your transformation into a digital economy with GCI, Oxford Economics, Huawei (2018). http://www.huawei.com/minisite/gci/assets/files/gci_2018_whitepaper_en.pdf?v=20180605. Accessed 20 Dec 2018
17. The Global Innovation Index 2018: Energizing the World with Innovation. Dutta, S., Gurry, F., Lanvin, B. (eds.) Cornell University, Geneva, INSEAD, WIPO (2018)
18. IMD World Digital Competitiveness Ranking: 2018 IMD World Competitiveness Center (2018)
19. Digital Society Index: 2018. Full Report. Oxford Economics. https://www.oxfordeconomics.com/recent-releases/the-digital-society-index-2018. Accessed 20 Dec 2018
20. Digital life of Russian megapolises 2016. (in Russian). https://iems.skolkovo.ru/downloads/documents/SKOLKOVO_IEMS/Research_Reports/SKOLKOVO_IEMS_Research_2016-11-30_ru.pdf. Accessed 20 Dec 2018
21. Design: The Digital Russia Index. https://finance.skolkovo.ru/ru/sfice/research-reports/1779-2018-10-001-ru
22. Digital Literacy Index, 2017 (ROCIT). (in Russian). http://xn–80aaefw2ahcfbneslds6a8jyb.xn–p1ai/media/Digital_Literacy_Index_2017.pdf. Accessed 19 Dec 2018

23. Datyev, I.O., Fedorov, A.M.: Information monitoring of community's territoriality based on online social network. In: Silhavy, R., Silhavy, P., Prokopova, Z. (eds.) Computational and Statistical Methods in Intelligent Systems, CoMeSySo 2018. Advances in Intelligent Systems and Computing, vol. 859. Springer, Cham (2019). https://doi.org/10.1007/978-3-030-00211-4_10

24. Zamyatina, N.Yu., Yashunsky, A.D.: Virtual geography of virtual population. Monitoring of Public Opinion: Economic and Social Changes. **1**, 117–137 (2018). https://doi.org/10.14515/monitoring.2018.1.07

25. Smirnov, I., Sivak, E., Kozmina, Y.: In search of lost profiles: the reliability of VKontakte data and its importance in educational research. Educ. Stud. **4**, 106–122 (2016)

26. Federal Law "On Personal Data" No. 152-FZ. https://pd.rkn.gov.ru/authority/p146/p164. Accessed 21 Dec 2018

27. EU Regulation 2016/679 of April 27, 2016, GDPR. https://ec.europa.eu/info/law/law-topic/data-protection_en. Accessed 21 Dec 2018

28. VTsIOM: About Half of Russians Use Social Networks Almost Every Day. https://tass.ru/obschestvo/4950095. Accessed 15 Jan 2019

29. Can, U., Alatas, B.: Big Social Network Data and Sustainable Economic Development. Sustainability **9** (2017). https://doi.org/10.3390/su9112027

A Simulink Network Communication

Martin Sysel[(✉)]

Tomas Bata University in Zlín,
Nad Stráněmi 4511, 760 05 Zlín, Czech Republic
Sysel@utb.cz

Abstract. This paper describes TCP/IP communication blocks in the program MATLAB/Simulink. The developed blocks have been modified and they support implementing any type of communication protocol by using Dynamic-link library. The communication protocol can be easily changed. In addition, a sending buffer can be implemented in a case of high frequency. This server and client blocks allow Simulink models to communicate between Simulink and remote applications or devices over TCP/IP communications. S-functions provide interaction with Simulink and extend capabilities but have to be compiled as C MEX files using the mex utility. In the S-function algorithm is implemented a socket as a client or a server and enables to send or receive data. This paper describes these blocks and brief instructions for building these blocks. The code and behavior of the client is presented in the sample example.

Keywords: Simulink · Communication · Network · TCP/IP

1 Introduction

Simulink is a graphical programming environment for simulating systems. Its interface uses a graphical block diagrams. The developed Simulink blocks allow communication with remote applications and devices over TCP/IP. The first block enables sending data on-line from Simulink model to another application or devices using TCP/IP. It is also possible to send data to the developed Simulink TCP/IP Server block. The base element of the blocks is an S-function, which use C MEX file. The first, Simulink model and S-function stages are described together with callback methods. Used Windows Socket API implementation is briefly presented and finally the procedure how to compile communication block is described.

1.1 S-Function

S-functions (system-functions) provide a powerful mechanism for extending the capabilities of the Simulink environment. An S-function is a computer language description of a Simulink block written in MATLAB, C, C++, Ada, or Fortran. S-functions are dynamically linked subroutines that the MATLAB interpreter can automatically load and execute. S-functions use a special calling syntax called the S-function API that enables to interact with the Simulink engine. A handwritten C MEX S-function provides the most programming flexibility. By following a set of

© Springer Nature Switzerland AG 2019
R. Silhavy (Ed.): CSOC 2019, AISC 984, pp. 403–414, 2019.
https://doi.org/10.1007/978-3-030-19807-7_39

simple rules can be implemented an algorithm in an S-function and used as the Simulink block in the model [1, 2].

S-Function Simulation Stages. Execution of a Simulink model proceeds in stages (Fig. 1). First comes the initialization phase. In this phase, the Simulink engine incorporates library blocks into the model, propagates signal widths, data types, and sample times, evaluates block parameters, determines block execution order, and allocates memory. The engine then enters a simulation loop, where each pass through the loop is referred to as a simulation step. During each simulation step, the engine executes each block in the model in the order determined during initialization. For each block, the engine invokes functions that compute the block states, derivatives, and outputs for the current sample time [1,3]. The entire simulation loop then continues until the simulation is complete.

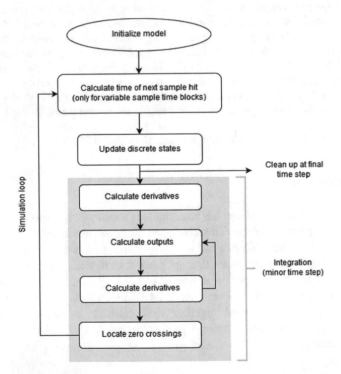

Fig. 1. Simulink simulation stages (Created using a source: [1])

A C MEX S-function consists of a set of callback methods (Fig. 2) that the Simulink engine invokes to perform various block related tasks during a simulation. Because the engine invokes the functions directly, C MEX S-functions must follow standard naming conventions specified by the S-function API [2]. A C MEX S-functions provide many sample time options, which allow for a high degree of flexibility in specifying when an S-function executes. If the behavior of S-function is a function of discrete time intervals, it can be defined a sample time to control when the Simulink engine calls the S-function mdlOutput and mdlUpdate [2].

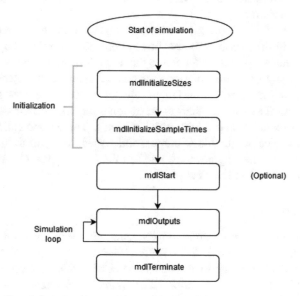

Fig. 2. Simulink S-function callback methods (Created using a source: [2]).

The Simulink engine interacts with a C MEX S-function by invoking callback methods that the S-function implements. The general format of a C MEX S-function is shown below:

```
#define S_FUNCTION_NAME sfunction_name
#define S_FUNCTION_LEVEL 2
#include "simstruc.h"
static void mdlInitializeSizes(SimStruct *S) { … }
static void mdlInitializeSampleTimes(SimStruct *S) { … }
static void mdlStart(SimStruct *S) { … }
static void mdlOutputs(SimStruct *S, int_T tid) { … }
static void mdlTerminate(SimStruct *S) { … }
#ifdef MATLAB_MEX_FILE /* Is compiled as a MEX-file? */
#include "simulink.c" /* MEX-file interface mechanism */
#else
#include "cg_sfun.h" /* Code generation registr. func. */
#endif
```

[The general format of a C MEX S-function from [2]]

The structure example starts with the define statements. After defining the name of the S-function and level format, the code includes simstruc.h, which is a header file that gives access to the SimStruct data structure and the MATLAB Application Program Interface (API) functions. The simstruc.h file also defines macros that enable your MEX file to set values in and get values (such as the input and output signal to the block) from the SimStruct [2].

The next part of the S-function contains implementations of required callback methods. The mdlInitializeSizes is the first routine the Simulink engine calls when interacting with the S-function. This define the number of input and output ports, sizes of the ports, and any other information (such as the number of states) needed by the S-function. The mdlInitializeSampleTimes callback method specifies the actual value of the sample time. The call mdlStart is an optional method, It is invoked once at the beginning of a simulation and could initialize states. The engine calls mdlOutputs at each time step to calculate the block outputs and writes the result to the output. At the end of a simulation, the engine calls mdlTerminate [2]. This C MEX structure is described in more detail in the Sect. 4.

1.2   Windows Sockets

Traditional network programming implemented in Windows environment uses Windows Sockets API (Winsock API - WSA). WSA is similar to UNIX/Linux Sockets programming with a few exceptions such as header files, that provided to suit Windows environment and enhances the functionalities. Windows Sockets 2 (Winsock) enables programmers to create advanced network capable applications to transmit application data across the wire, independent of the network protocol being used. With Winsock, programmers are provided access to advanced Microsoft Windows networking capabilities centered around TCP/IP [5]. There are two distinct types of socket network applications: Client and Server. Clients and servers have different behaviors; therefore, the process of creating them is partially different [5]. Some of the steps are the same for a client and a server and some of them are specific. The developed Simulink blocks are both kinds, client and server. Next follows the general steps for creating a streaming TCP/IP Client and Server [5].

Client
- Initialize Winsock.
- Create a socket.
- Connect to the server.
- Send and receive data.
- Disconnect.

Server
- Initialize Winsock.
- Create a socket.
- Bind the socket.
- Listen on the socket for a client.
- Accept a connection from a client.
- Receive and send data.
- Disconnect.

2 Block Description

2.1 A Client Block

The Client block calls callback method mdlStart for initialization Winsock. It creates socket, connects to the server and load library. In the callback method mdlOutputs repeatedly sends the block input values in the specified format to the connected server using the TCP/IP socket. The specified data format is defined in the previously loaded Dynamic-link library. Disconnection from the server will take place at the end of the simulation callback method mdlTerminate, where also calls the FreeLibrary function to unload the DLL. A class named socketClient is implemented in the source code. The class contains a constructor including Winsock Initialization and socket creation. Also, the static variable is incremented in the constructor, which is used as a counter of the number of blocks in the Simulink schematic. The using of Winsock is described in [5]. All processes that call Winsock functions must initialize the use of the Windows Sockets DLL before making other Winsock functions calls. This also makes certain that Winsock is supported on the system. To initialize Winsock is necessary create a WSADATA object called wsaData, Call WSAStartup and return its value as an integer and check for errors. The WSAStartup function is called to initiate use of WS2_32.dll. Then It calls the socket function and return its value to the ConnectSocket variable. A TCP stream socket was specified with a socket type of SOCK_STREAM and a protocol of IPPROTO_TCP [5].

```
WSADATA wsaData;
int iResult;
numBlock = countObj;
++countObj;
// Initialize Winsock
iResult = WSAStartup(MAKEWORD(2,2), &wsaData);
if (iResult != 0) {
    printf("WSAStartup failed: %d\n", iResult);
    return 1;
}
// Create a SOCKET
ConnectSocket = socket(AF_INET, SOCK_STREAM,
IPPROTO_TCP);
if (ConnectSocket == INVALID_SOCKET) {
    printf("Error at socket(): %ld\n", WSAGetLastEr-
ror());
    freeaddrinfo(result);
    WSACleanup();
    return 2;
}
```

[The initialization and creation a socket]

Another method of the socketClient class used in the callback method mdlStart is named sConnect and provides connection to the server using a socket. It calls the connect function, passing the created socket and the sockaddr structure as parameters. The getaddrinfo function is used to determine the values in the sockaddr structure.

```
iResult = getaddrinfo(myHost, port, &hints, &result);
    if ( iResult != 0 ) {
        ssPrintf("getaddrinfo failed: %d\n", iResult);
        WSACleanup();
    }
// Connect to the server.
iResult = connect(ConnectSocket, result->ai_addr,
(int)result->ai_addrlen);
        if (iResult == SOCKET_ERROR) {
            closesocket(ConnectSocket);
            ConnectSocket = INVALID_SOCKET;
        }
```

[The socket connect]

The data are sent in the cycle every callback of method mdlOutputs, using the socketClient method called sWrite, which sends the data in a format that is defined in the dynamic-link library.

2.2   A Server Block

A server block is very similar to the client block. In the callback method, mdlStart initializes Winsock, create a socket and bind a socket. Then block listen on the socket for a client communication. In the callback method mdlOutputs it repeatedly receives data that sends as a Simulink block output values for further processing in the Simulink schema. It is possible to specify a port and sample time period. The sample time period is the rate at which the block expects the data on the specified port during the simulation it uses blocking mode. The disconnection is performed at the end of the simulation (callback mdlTerminate).

3   Dynamic-Link Library

The following source code is needed to create a simple DLL named protocol.dll. It defines a simple data formatting function called getFormat. The protocol DLL does not define an entry-point function, because it is linked with the C run-time library and has no initialization or cleanup functions of its own to perform. If calling function is successful, the total number of characters written is returned excluding the null-character appended at the end of the string, otherwise a negative number is returned in case of failure. It is necessary note that the MATLAB shared library interface supports C library routines only [4]. The shared library interface does not support C++ classes or overloaded functions elements. Functions must be declared as extern "C". Other limitations to shared Library support is described in [4]. Presented format of sent data is very simple, it contains four colon-separated fields for number of the block (it supports more communication blocks in the Simulink schema), number of the signal (it supports multiplexed signals), current sample time and value of the signal. Each line is ended with semi-colon and escape sequence for new line. Data are stored into char array variable.

```
#include <windows.h>
#include <stdio.h>
#include "protocol.h"

#ifdef __cplusplus
extern "C" {
#endif

__declspec(dllexport) int getFormat(char myFormat[], int
numBlock, int numSignal, double sampleTime, double value)
{
return sprintf_s(myFormat,64,"%u:%u:%f:%f;\n", numBlock,
numSignal, sampleTime, value);
}

#ifdef __cplusplus
}
#endif
```

[The source code of protocol DLL]

4 A C MEX S-Function

This chapter contains simplified description of the source code of the developed Simulink client block. The base element of the block is S-function block, which use C MEX file. Finally, it is necessary compile source code.

The S-function code starts with the define and include statements. It is necessary to define statement which specifies the name of the S function. The code includes sim-struc.h, which is a header file that gives access to the SimStruct data structure and the MATLAB API functions. There are included other headers (e.g. winsock2.h for using sockets) together with global declaration. The typedef below defines a new function type FUNCT. Variable of this type points to function from DLL library that takes parameters as input and return type int. The parameters have been described in Sect. 3. The simplified implementations of S-Function callback methods are described in next subsections.

```
typedef __declspec(dllexport) int (*FUNCT)(char myFor-
mat[], int numBlock, int numSignal, double sampleTime,
double value);
```

[A new function type. Variable of this type points to function from DLL.]

4.1   mdlInitializeSizes

The Simulink engine calls mdlInitializeSizes to inquire about the number of input and output ports, sizes of the ports, and any other information (such as the number of states) needed by the S-function. The client block implementation of mdlInitializeSizes specifies the following size information:

```
ssSetNumSFcnParams(S, 3);
```

It defines three input parameters accessible in the masked block:

- Address – (String input parameter) Name address of the server (IP address) - An Internet Protocol (IP) address is a numerical identification that is assigned to devices participating in a computer network.
- Port – (Integer input parameter) – Defines listening server port. In computer networking, a port is an application-specific or process-specific software construct serving as a communications endpoint used by Transport Layer protocols of the Internet Protocol Suite such as Transmission Control Protocol (TCP) or User Datagram Protocol (UDP). A specific port is identified by its number associated with the IP and the protocol used for communication.
- Sample time period – Sampling period of the signal output.

```
if (!ssSetNumInputPorts(S, 1)) return;
ssSetInputPortWidth(S,0,DYNAMICALLY_SIZED);
```

It defines one dynamically sized input port, wihich is dynamically sized. That's why TCP output is in the special format which is easy modifiable.

```
ssSetOptions(S,
SS_OPTION_WORKS_WITH_CODE_REUSE |
SS_OPTION_EXCEPTION_FREE_CODE |
SS_OPTION_USE_TLC_WITH_ACCELERATOR |
SS_OPTION_CALL_TERMINATE_ON_EXIT);
```

Specifying these options together with exception-free code speeds up execution of S-function. Option SS_OPTION_CALL_TERMINATE_ON_EXIT guarantees the engine calls the mdlTerminate method before destroying a block because it is necesary close socket and frees the loaded dynamic-link library (DLL) module.

```
ssSetSampleTime(S,0, mxGetScalar(ssGetSFcnParam(S, 2)));
```

The Simulink engine calls mdlInitializeSampleTimes to set the sample times of the S-function. A client block executes in specified period (the third input parameter).

```
ssSetNumPWork(S,3);
```

This function specifies the size of a block's pointer work vector. Block's pointer vector store pointers to persistent objects during Simulink life cycle. It is possible retrieve the pointer in any subsequent method invocation to access the object.

4.2   mdlStart

Simulink invokes this an optional method at the beginning of a simulation. It should initialize and connect the windows socket. A pointer work vector is used for socketClient object to holds pointer to the persistent object between Simulink method invocations. The method sConnect from the class socketClient is called with the parameters of the block. Input parameters Address, Port and Sample time are used for TCP communication. The code uses the LoadLibrary function to get a handle to the Protocol library and returns HINSTANCE type value stored in hDLL variable. If LoadLibrary succeeds, the program uses the returned handle in the GetProcAddress function to get the address of the DLL's getFormat function [6]. Function pointer to the DLL and the address of the DLL's function are stored to block's pointer work vector. Multiple LoadLibrary() calls from a process to a DLL always returns the same pointer as the first call to LoadLibrary(). This means that multiple instances of S-Function block cannot load separate instances of the DLL which link to. Since they are all using the same instance it is not possible unload the DLL (using FreeLibrary()) until all instances of S-Function are done using it. Possible solution is declaration a global counter variable which will count the number of S-Function instances and decrement the counter in mdlTerminate(), when the counter is zero then call FreeLibrary() [7].

```
ssGetPWork(S)[0] = (void *) new socketClient;
socketClient *newsock = (socketClient *)ssGetPWork(S)[0];
mxGetString(ssGetSFcnParam(S, 0),buf,buflen);
newsock->sConnect(buf, (int) mxGetSca-
lar(ssGetSFcnParam(S, 1)));
hDLL = LoadLibrary("protocol.dll");
if (hDLL != NULL){
    pFunkce = (FUNCT) GetProcAddress(hDLL,"getFormat");
}
ssGetPWork(S)[1] = (void*) hDLL;
ssGetPWork(S)[2] = (void*) pFunkce;
```

[The important fragment of source code in mdlStart]

4.3 mdlOutputs

The engine calls mdlOutputs at each time step to calculate the block outputs. The client implementation of mdlOutputs takes the input signal, uses data format specified in the DLL and writes the data to the created output socket. This means that client block does not have the output port because the output is sent to the socket.

```
int_T i; // input signal counter
InputRealPtrsType uPtrs = ssGetInputPortRealSig-
nalPtrs(S,0);
int_T width = ssGetInputPortWidth(S,0);
double  sampleTime;
char pBuffer[64];
std::stringstream text;

// retrieve C++ object
socketClient *newsock = (socketClient *)ssGetPWork(S)[0];
FUNCT pFunkce = (FUNCT) ssGetPWork(S)[2];

if(ssIsSampleHit(S,0,tid)) {
    sampleTime = ssGetTaskTime(S,0);
}
for (i=0; i<width; i++) {
    pFunkce(pBuffer,newsock->getNumBlock(), i, sam-
pleTime, (*uPtrs[i]));
    text << pBuffer; //append data of multiplex signals
}
// sending buffer can be implemented here
int size = newsock->sWrite(text.str());   // socket send
```

[The important fragment of source code in mdlOutputs]

4.4   mdlTerminate

The engine calls mdlTerminate to provide the S-function with an opportunity to perform tasks at the end of the simulation. This is a mandatory S-function routine. The client S-function terminate created socket and calls the FreeLibrary function to unload the DLL.

```
std::stringstream text;
HINSTANCE hDLL = (HINSTANCE) ssGetPWork(S)[1];
socketClient *newsock = (socketClient *)ssGetPWork(S)[0];
text << "Simulation Stop\n" ;
int size = newsock->sWrite(text.str());
delete newsock;            // object in the termination
FreeLibrary(hDLL);
```

[The source code in mdlTerminate]

4.5   Compiling

Compiling the C MEX-Files is similar to compiling with gcc or any other command line compiler. Because the program uses run-time dynamic linking for protocol data formatting, it is not necessary to link the module with an import library for the DLL. Applications that use Winsock must be linked with the Ws2_32.lib library file. The implementation is in a file called ws2_32.dll and there are 32-bit and 64-bit versions of the DLL in 64-bit Windows. This example supposes 64-bit Windows architecture (option –DWIN64). Following MATLAB command links the object code together with the library WS2_32.lib.

```
>> mex -O client.cpp WS2_32.lib  -DWIN64
```

The Simulink S-function with C MEX code is masked because encapsulation the block diagram allows to have its own parameter dialog box with its own block description. The Fig. 3 shows the simple example where two multiplexed signals enter to TCP/IP Client block. The Block Parameters dialog can be used for setting up communication parameters.

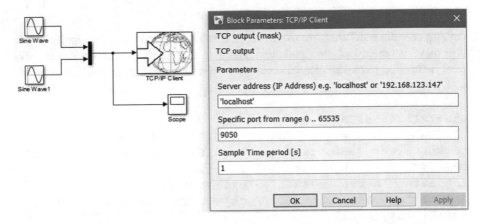

Fig. 3. A simple example of the TCP/IP client block with block parameters dialog.

5 Conclusion

MATLAB and Simulink support TCP/IP communication using commercial Instrument Control Toolbox which is more complex than described proof of concept. This developed TCP/IP client block is open-source and sends data to the network socket, it uses TCP/IP protocol specified in dynamic-link library. The data are sent at fixed intervals during a simulation and it is possible easily change communication protocol by replacing DLL. The TCP/IP client block has one input port. The input port has dynamic size (inherited from the incoming data). This block has no output ports because the output is sent to the socket. The second developed block named Server TCP/IP block enables Simulink models to accept network communication from remote applications and devices over TCP/IP network in blocking mode. The full source code will be published on the Mathworks website in the Community File Exchange section.

References

1. MathWorks: Simulink® User's Guide. The MathWorks, Inc., Natick, USA (2016)
2. MathWorks: Simulink: Writing S-Functions. The MathWorks, Inc., Natick, USA (2016)
3. MathWorks: S-Function Concepts. https://www.mathworks.com/help/simulink/sfg/s-function-concepts.html. Accessed 15 Jan 2019
4. MathWorks: Limitations to Shared Library Support. https://www.mathworks.com/help/matlab/matlab_external/limitations-to-shared-library-support.html. Accessed 15 Jan 2019
5. Microsoft: Windows Sockets 2. https://docs.microsoft.com/en-us/windows/desktop/winsock/windows-sockets-start-page-2. Accessed 15 Jan 2019
6. Microsoft: Dynamic-Link Libraries. https://docs.microsoft.com/en-us/windows/desktop/dlls/dynamic-link-libraries. Accessed 15 Jan 2019
7. Anonymous: Programming and Web Development Forums - linking .dll in C S-function. http://www.44342.com/matlab-f582-t132884-p1.htm. Accessed 15 Jan 2019

Transactions as a Service

Aerton Wariss Maia and Pedro Porfrio Muniz Farias[(⊠)]

University of Fortaleza, Av Whashington Soares, Fortaleza, CE 1321, Brazil
aerton.wariss@wariss.com.br, porfirio@unifor.br

Abstract. Multi-Tenant is a software-as-a-service (SaaS) model that allows the sharing of WEB systems by several customers (tenants) as if they were unique and exclusive to each one [4]. This paper proposes to see the Multi-Tenant systems as a set of transactions and periodically charge their users according to the number of times that each transaction is used. The cost of each transactions was set in proportion to its complexity measured by the Function Point Analysis. This pricing scheme is closer to the users' business needs and thus facilitates the evaluation of the relation between cost and benefit of using the software. A case study was constructed and applied in a real ERP with a hundred of clients to validate the proposal.

Keywords: Multi-tenant · Cloud computing · Software as a service

1 Introduction

Among the software-as-a-service (SaaS) models for the WEB environment, Multi-Tenant has been standing out in the area of WEB systems as a good option to reduce development costs [1]. Multi-Tenant is a software-as-a-service (SaaS) model that allows the sharing of WEB systems by several customers (tenants) as if they were unique and exclusive to each one [4]. The same system can be used by the most varied companies or institutions of the most varied sizes. From micro to small, medium and large companies.

Despite the rapid evolution of technologies, many service providers think that classical charging systems are sufficient for emerging networks [2]. In emerging information markets, companies can profit from low fixed penetration prices, but as these markets mature, the ideal price mix should expand to include a wider range of usage-based pricing options [3].

Although a Multi-Tenant system can serve several companies in the same branch of activity, each company may have its specific needs covering more or less resources than the system as a whole has [4].

This paper has two purposes: (1) First, it presents transactions as a Service. This approach of Saas considers each software as a set of transactions seen from the user's point of view. The concept of transaction used is the proposed by FPA (Function Point Analysis). For example, in a ERP (Enterprise Resource Planning) system, a transaction could be to register the sale of a product, another transaction could be to print a report.

© Springer Nature Switzerland AG 2019
R. Silhavy (Ed.): CSOC 2019, AISC 984, pp. 415–423, 2019.
https://doi.org/10.1007/978-3-030-19807-7_40

As the transaction is the unit of the service, this vision has effects on the pricing of the service. So, if the user prints 20 times a report then, proportionality, he will be charged 20 times the cost of printing the report. Transactions are adherent to the business vision. Therefore, charging the user by invoking each transaction allows a better cost-benefit analysis.

Transactions as a service is different from FaaS (Function as a Service).

Faas (Function as a Service) is an emergent alternative in cloud computing. FaaS was made available by AWS Lambda, Google Cloud Functions, Microsoft Azure Functions and many others. The Functions are internal components of the software normally not perceived by the final users. The division of software into functions aims to improve its scalability. Cloud Functions are priced according to how long your function runs, how many times it is invoked and how many resources one provisions for the function. Hence, the amount of memory, cpu cycles, and another hardware issues influence the price of the functions.

(2) The second purpose of this paper is to propose the pricing of transactions in terms of their complexity based on the Function Point Analysis (FPA). The FPA is typically used to estimate software size based on the amount of data manipulated by each transaction. Of course, the same transaction can be used with different amount of data, but more complex transactions normally handle more data that transactions of low complexity. This proportionality is used here to avoid the overhead of measuring the amount of memory, cpu cycles, and another hardware details.

Micro and small businesses often lack the resources that information technology can offer to benefit their enterprises due to the disproportionate cost imposed to them. As a result of the implementation of this metric, it is expected to provide a more transparent billing for customers who, although working in the same branch of activity, have different volumes of data, transactions and resources.

The rest of this paper is divided as follow: Sect. 2 reports the current billing methods for Multi-Tenant systems. Sect. 3 describes the application of FPA to measure the complexity of transaction. Section 4 presents the implemented framework that calculate the costs of using the system for each user. Section 5 brings a Case Study in a real Multi-Tenant system and Sect. 6 presents the conclusions.

2   Current Billing Methods for Multi-Tenant Systems

A key element in making Software-as-a-Service (SaaS) successful is the so-called multi-tenant, that refers to an architecture model in which a software instance serves a set of multiple clients from different organizations (tenants). Therefore, it reduces the number of application instances and thus operating costs in the WEB environment. The problem that SaaS providers face in the market today is how to define a charging model that enables profitability in such an environment. Being profitable with SaaS, the art is to design billing methods that cover the costs of resources allocated to the underlying PaaS/IaaS provider. Challenges with measuring tenant consumption is a prerequisite to defining a cost-effective charging model. [4].

In terms of calculating the amounts to be charged to Multi-Tenant customers for the software as a service, where the same system can be shared by numerous companies/customers that have totally different characteristics, the individualization of this calculation tends to become much more complex.

Multi-tenant applications are created and sold as highly configurable WEB services for a variety of clients where each has its own specific preferences and required configurations. To meet the unique requirements of different tenants, the application must be highly configurable and customizable. The current state of practice in SaaS development is that configuration [5,6] is preferred over customization that is considered too complex [7]. Configuration generally supports variance by configuring pre-defined parameters for the data model, user interface, and business rules of the application. Customization, on the other hand, involves software variations at the core of the SaaS application in order to meet specific tenant requirements that can not be resolved through configuration. Compared with configuration, customization is currently a much more expensive approach for SaaS suppliers as it introduces an additional layer of application engineering complexity and additional maintenance costs [8].

Some attempts were made to carry out this normally monthly calculation using mostly the following three basic models: number of users, a single price for all or a single price for each software version with more or less resources regardless of their use (bronze, silver, gold) [9–11]. Most of them are postpaid models. Thus the tenant receives an account and pays periodically. To calculate the costs of consumption, the use of each tenant is monitored and computed. The safest method from the perspective of the SaaS provider is to charge their renters in the same way that the PaaS/IaaS providers charge them for the use of their resources, that is, directly passing on to the tenants the cost of the PaaS/Iaas providers plus an additional fee. This model is very technical and not very transparent to tenants. The SaaS provider needs to estimate or even calculate the costs of the resources used by each tenant (costs such as consumed CPU or storage). This involves monitoring each tenant and logging how they use the application. They basically look at the use of application resources by each tenant and increase the amount charged based on usage metrics. Alternatively, collection models are based on factors that are best understood by tenants, such as the time of use [4].

Another type of charge used is a flat fee per month. However, it is difficult to predict the costs that a tenant's use will produce during this period. Excessive use by a single tenant could lower the profit margin of a SaaS provider or even make it negative.

There is also the prepaid model. Customers charge a value on their accounts before any consumption. However, some method of controlling resource consumption must be used by the tenant to see if their use limit has been reached.

Therefore, Multi-Tenant in the WEB environment is a booming reality in the ERP market, there is currently a growing demand for how to perform this calculation in a more fair and individualized way for each company/customer, respecting the size and particularities of each well such as the complexity of the transactions involved in their daily operations.

3   Function Point Analysis Applied to Transactions

There are now several consolidated metrics over the last decades used to measure and price software development costs. Software metrics are quantitative patterns of measures of various aspects of a software project or product [12].

Among the most widespread metrics is Function Point Analysis - FPA, created by Allan Albrecth of IBM in 1979, in order to measure software size from its functional specification. It was considered an advance over the counting method by Source Code Lines - SLOC [12].

Function Point Analysis (FPA) is a problem-oriented method that defines elements that can be accounted for by the system requirements on the transactions contained in them. It captures the size of a software through its functional characteristics, regardless the technology employed, which gave rise to a new category of metrics, called Functional Size Metrics - FSM, or simply Functional Metrics [12]. It is a measure of software scaling through the functionality implemented (or to be implemented) in a system from the point of view of the user [13].

Basically FPA measures the functional requirements in terms of:

- **Data Function** - There are two types of data functions: (ILF - Internal Logical File and EIF - External Interface File)
- **Transactions Functions** - There are three types of transaction functions: EI - External Input, EQ - External Query, EO - External output

As only the complexity of transactions will be compared, only will be considered the transaction functions. So, the data functions do not be analysed.

As usual, firstly, the transaction is classified in one of the three possible types (EI, EO or EQ) and, then, calculated its complexity (low, medium or high) in function of the RLF (Number of Referenced Logical Files) and Number of Data Types - DTs (Number of fields involved per transaction execution call).

3.1   Metric Application

For each transaction that exists in the Multi-Tenant system, a record is inserted informing the Type of Transaction Function (EI, EO or EQ), the number of Logical Files (RLF) and the amount of Data Types (DT) of that transaction.

When a tenant invokes a transaction, the respective amount of function points is calculated and also its cost is added to the payment invoice.

4   Calculation of the Cost of Using Each Transaction

A function was developed to record the use of each transaction and accumulate the cost assigned to its use. This function will be triggered every time the client performs the call to execute some transaction of the System.

The Fig. 1 presents the data diagram used.

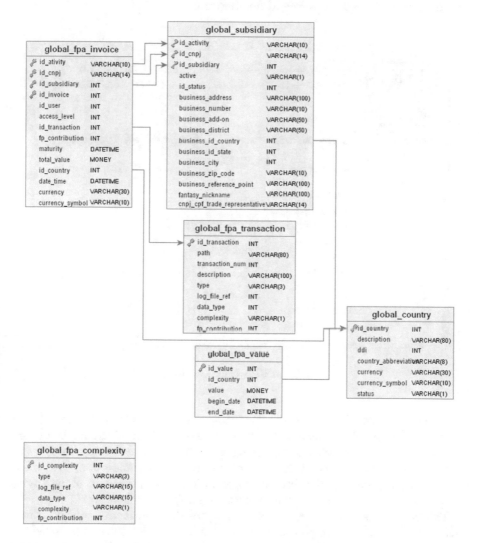

Fig. 1. Data model used to calculate and accumulate the cost of each transaction

The first parameter used by the function is the cost assigned to each function point. A screen allows the system administrator to record this values on the table called global_fpa_value.

To each transaction the associated function points must be calculated. This is calculated by setting the type of the functionality (EQ, EI or EO), the RLF and DT (Data type).

The system has in the administrative area a screen referring to the Settings Panel, as shown in Fig. 2, which enables the registration of information regarding each possible transaction in the system for calculating the complexity, type, and amount of contribution in terms of function point.

Fig. 2. FPA configuration

So, at the time of setting up a new transaction, through the Settings Panel, one will be informed the path to the folder where the PHP script file is located; the order number of the transaction being executed (one feature may have several different transactions); the description of the respective transaction; the type of transaction, the number of referenced logical files, and the amount of data types in it. All these fields compose the dataset global_fpa_complexity.

The system also calculates the complexity of this transaction and its contribution number in function points. With all this information the system generates a new transaction ID for this new transaction being set up and stores everything in a new tuple in the table global_fpa_complexity.

From this moment on, every time this transaction is executed on the system, the table global_fpa_transaction is queried and its contribution in function points will be informed and stored by means of a new insertion in the table global_fpa_invoice.

The currency value used in the case study was R\$ 0.017 for each 1 (one) function point. This value was arbitrated after some months of application of the developed method, in some clients in order to find a monthly value that corresponded approximately to the fixed value that they had been paying before the development of the function.

5   Case Study

5.1   ERP and Company Characterization

This proposal was tested in an real Multi-Tenant SaaS ERP system that segments the client companies by subject and branches of activity and facilitates their management, streamlines the activation of new clients and allows a more dynamic and constant evolution for all in terms of existing and new transaction still to be developed.

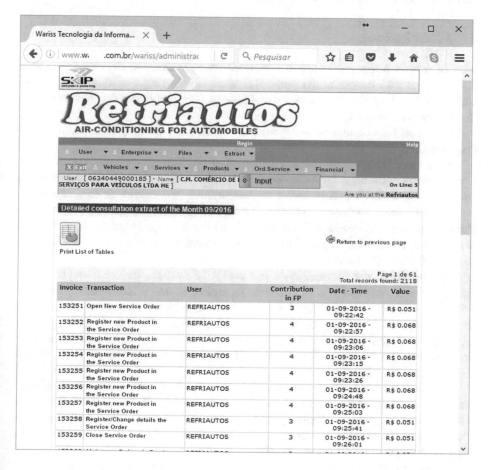

Fig. 3. Example of detailed account statement for client "Refriauto"

5.2   Use of the Metric Proposed in the Case Study Application

Based on the metric described in this article, a PHP function was developed and associated with each transaction performed by each functionality of the systems. This Function permits to account individually the consumption of each customer and calculate the subscription charge in a certain period of time.

For each transaction, this PHP function should be included in the PHP script right after the completion of any transaction (inclusion, change, exclusion or deletion) to account its execution.

The Fig. 3 presents an Example of detailed account statement, for the client "Refriautos", for a month of transactions. For each transaction carried out in the system is presented the description of the transaction, the user who performed it, the amount of contribution of that transaction in function points, the exact date and time of the transaction and the calculated cost value of this transaction according to its contribution amount of function points and, finally, the cost value of 1 (one) function point registered in the function settings.

6   Conclusions

The presented approach suposes the multi-tenat system as divided in a set of transactions and proposes charging the users in the proportion of use of each functionality. The cost associated to a functionality was based on their complexity estimated using FPA.

More complex transactions, with more function points, manipulate more complex data and, therefore, demand more computational resources.

In this way, the metric used provides an approximate estimate, correlated to the resources demanded by each user in a multi-tenant system. On the one hand, this FPA based estimate is easier to accumulate and calculate than measures such as cpu time, memory, network traffic and Gb stored. On the other hand, the transactions analyzed in FPA are defined from the point of view of the user and therefore allow him to analyze better the relation between the cost and the benefit of using the system.

This proposal was implemented in a real ERP system. Before using this metric the clients received a fixed charge according to the client's size There has been a significant variation in the amounts charged since then. Some continued to pay roughly the same amount they paid earlier. Others had their monthly payments reduced significantly and in contrast others started to pay more effectively due to the high number of invoked transactions that the previous metric did not allow to identify.

For those who have paid less for the same services, cost reduction is always welcome in terms of improving competitiveness and digital inclusion. For those who started to pay more, it was easy to demonstrate, through the detailed extract that the metric generates, the real need for adjustment of values through the high consumption of resources by them.

A more rational use of the system was perceived because the amounts to be charged to each client now is proportional to the use of the system.

In case there is a need to reformulate some functionality already existing in the system in order to increase or decrease its complexity and, consequently, the value to be charged for the use of it; the recalculation will automatically be redone only by changing the settings inherent in that feature.

The use of transaction as a service provides greater digital inclusion to micro and small businesses, making information systems and management assistance more accessible. This metric can be extended to any segment or activity that involves the use of an information system on WEB.

References

1. Andrikopoulos, V., Binz, T., Leymann, F., Strauch, S.: How to adapt applications for the cloud environment challenges and solutions in migrating applications to the cloud how to adapt applications for the cloud environment (2013)
2. Ruiz-agundez, I., Ruiz-agundez, I., Penya, Y.K., Bringas, P.G.: A flexible accounting model for cloud computing a flexible accounting model for cloud computing, no. May 2011 (2014)
3. Sundararajanhttp, A.: Nonlinear pricing of information goods nonlinear pricing of information goods, no. July 2015 (2004)
4. Schwanengel, A., Hohenstein, U.: Challenges with tenant-specific cost determination In: Multi-Tenant Applications, no. c, pp. 36–42 (2013)
5. Chong, F., Carraro, G.: Architecture strategies for catching the long tail. MSDN, vol. 479069, pp. 1–20 (2006)
6. Guo, C.J., Sun, W., Huang, Y., Wang, Z.H., Gao, B.: A framework for native multi-tenancy application development and management (2007)
7. Sun, W., Zhang, X., Guo, C.J., Sun, P., Su, H.: Software as a service: configuration and customization perspectives (2008)
8. Walraven, S., Truyen, F., Joosen, W.: A middleware layer for flexible and cost-efficient *multi-tenant* applications. LNCS(LNAI), vol. 7049, no. i, pp. 370–389 (2011)
9. Hohenstein, U., Appel, S.: The impact of public cloud price schemes on multi-tenancy, no. c, pp. 22–29 (2016)
10. Ojala, A.: Software-as-a-service revenue models. IT Prof. **15**(3), 54–59 (2013)
11. Chappell, D.: How SaaS changes an ISVs Business. Chappell Assoc., pp. 1–20 (2012)
12. Abrahao, S., Pastor, O.: Measuring the functional size of web applications. Int. J. Web Eng. Technol. **1**(1), 5 (2003)
13. Cleary, D.: Web-based development and functional size measurement. In: Proceedings of the IFPUG Annual Conference, pp. 1–55 (2000)

Mathematical Dungeon: Development of the Serious Game Focused on Geometry

Marián Hosťovecký[1(⊠)], Ferdinand Fojtlín[1], and Erik Janšto[2]

[1] Department of Applied Informatics and Mathematics,
University of SS. Cyril and Methodius, Trnava, Slovak Republic
marian.hostovecky@ucm.sk,
ferdinand.fojtlin@google.com
[2] Department of Informatics, Slovak University of Agriculture in Nitra,
Nitra, Slovak Republic

Abstract. With the current spread of information and communication technologies (ICT), games can be even more implemented into school curriculums. Therefore, serious games (SG) may be more attractive not just for pupils, but also for teachers and lectors. The main purpose of this research is to introduce the first stage of our research: developing a 3D SG called Mathematical Dungeon focused on geometry for second-level primary school pupils. The SG consisted of several activities to learn basic geometric shapes (cubes and rectangular prism - cuboid). The Unity3Dgame engine was used as a development tool. We discuss the process of design and development such as modeling, scripting, creating scenes, rooms etc.. Pupils can interact with those scenes using personal computer or laptops. Game-based learning could be promoted as a way of offering something new and interesting and could enhance science, technology, engineering and mathematics education [1].

Keywords: Serious game · Geometry · Design · Development

1 Introduction

Education and its quality belong to frequently discussed topics in the context of the transformation of educational system at all levels – primary, secondary and university level [2]. Present educational process requires active participation of students to acquire new knowledge. Teachers stress and apply individual study methods, for example, the use of electronic study materials is on the rise. A modern teacher of mathematics has new roles in the education – to master available didactic means and via them to establish new specifications for creative and individual study [3]. The dynamics of changes in the field of information technologies opens the doors to use of new methods in education. Communication bandwidth (networks), computers, laptops, tablets and mobiles (hardware) and new generation of operating systems, program languages and game engines (software) offer possibilities of developing more attractive, funnier and more educating e-material or educational software. Research works from all over the world confirm that application of the computer in education has had positive results from primary schools to universities. Devices and networks are still improving, getting

© Springer Nature Switzerland AG 2019
R. Silhavy (Ed.): CSOC 2019, AISC 984, pp. 424–432, 2019.
https://doi.org/10.1007/978-3-030-19807-7_41

faster and prices are falling, which means that ICT devices are becoming more affordable. These attributes started to support more digital form of education. Digital games in education, also known as serious games (SG), are one of those areas that are still increasing.

The use of educational videogames can be effective only if elements such as goals, competition, challenges, and fantasy influence motivation and facilitate learning. Motivation refers to the initiation, intensity and persistence of behaviour. Nevertheless, students are not always highly motivated. Previous research claims that a game's story can motivate students to use an educational game [4, 5].

There are many definitions of SG based on the opinions of their nature, their benefits, their potential to develop knowledge, or personal attitude towards them, etc. Miloš Zapletal, one of the most important Czech writers dealing with the phenomenon of games, described the game as "an active dynamic process employing lesser or greater mental and physical abilities that are simultaneously trained and developed [6]." Definition of didactic games by Houška: "Activity in which the child spontaneously applies cognitive activities carried out under the primary influence of the relevant rules, which make the learning and teaching take place peacefully and in the second plan [6]." Serious games are not just developed and employed exclusively for entertainment purposes, but have successfully been incorporated as learning and training tools for a broad range of different areas [7, 8]. Following the rise of digitalization, games can be composed with the help of the computer, or can even be adopted for and with the computer, in order to "learn by playing" [9]. Serious game (SG) can bring new perspectives of education. It can be described as a tool for improving specific skills in a particular subject or a field [10] and motivation in education [11, 12]. It also constitutes a great tool for distance education [13], for improving communication skills [14], managerial education [14, 17], math competences [15]. The use of serious game in education also has to meet the requirements and criteria of the concept of computer network and its management in a given locality, on the Internet as well as safety and reliability of computer networks [16–18].

2 Design and Development of the Serious Game

Our main interest was to create a serious game Mathematical Dungeon focused on supporting math skills for pupils of 7th grade of primary school. The game focuses on basic geometry shapes.

2.1 Game Description

In the game, we tried to focus on understanding of two elementary shapes (Fig. 1a, b). It was necessary for us to develop a game that can improve this piece of knowledge by solving simple geometry tasks. Thus, we chose:

- *The volume and the surface area of a rectangular prism – cuboid.*

Cuboid is a three-dimensional shape whose walls are formed by six rectangular quadrilateral rectangles (mostly, but there are also special cases). Its base is a rectangle.

– *The volume and the surface area of a cube.*

A cube or regular six-quarter is a symmetrical three-dimensional shape. Its walls consist of six equal squares. Cube belongs to so-called platonic solids, because all sides and edges are identical. The only difference between cubes and cuboids is the shape of the six faces. Each face of a cube is a square and all of these squares are of equal size.

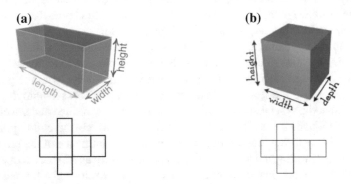

Fig. 1. (a) Rectangular prism (cuboid) Source:www.mathsisfun.com/geometry (b) Cube Source: http://maths.nayland.school.nz

2.2   Game Design

When a game designer proposes a game, they analyze all aspects and elements of the game. The concept of the whole design must reflect the final form of the game. The idea at the beginning of the design is therefore the main driving engine. When we decided to design this SG, the idea was to improve understanding of geometry, i.e. spatial shapes (cube and cuboid). In addition, it was important to design the SG in a way that will not just support a specific piece of knowledge (in our case, spatial imagination), but also player's personal skills, social skills, and the notion of *playing* for *fun* or *playing* to win. It was necessary to adjust not only to the game as such, but also to its design, playfulness, as well as graphic elements of the game and the player, etc. When designing the serious game, we needed to identify the needs (*what design*), the target group (*for whom*), to determine *goals of the SG,* to choose development tools, to obtain the feedback and to provide the support.

The basic element of the educational game was to design a simple story based on solving tasks (focusing on the basics of geometry) to meet the goal of the game. At this stage, we have set the goal to choose the genre for the game and, at the same time, the way in which our story would be interpreted to the player (pupil). Among the fundamental objectives was also the robustness of the game. Therefore, it was important to choose the right genre of the game. Based on this, we chose an adventure. The genre offers the player movement and interaction. The player moves in an environment and communicates with the surroundings, collecting and touching objects. The main motive of the game is to solve the primary task that originates from the definition of the genre.

2.3   Development Tools

After analyzing software options, we found a suitable development tool - the cross-platform *game engine* and a real-time *engine* called Unity. For graphic design, we used other graphic programs like GIMP, Blender Studio (space/objects modeling, creating animations). Each development tool will be described below.

Unity Game Engine
Unity comes with tools to support the workflow for 2D and *3D*. It is a suitable, powerful and user-friendly game development tool. It is simple in operation and compatible with a wide variety of media format types such as graphics, audio, video and text. This game engine is a freeware. It offers several possibilities to create objects that are equivalent in practical use and their utilization depends on the user's habits. One way is through the "GameObject" button on the toolbar that shows the object types that can be created. The basic types of objects are 3D objects, 2D objects, Audio, User interface and Light. Another way is to use the "Create" button at the object Hierarchy. Created objects are visible on the Scene and it is possible to work with them instantly. The second option was taking the opportunity of free models that are available at the Asset store. These were adapted according to the scene. Besides models, assets store offers also other content such as animations, sounds, scripts, etc. Some of these attributes were also used and adjusted to meet our needs. Other reasons for our choice was the fact that the software has been around for several years and it is connected to a plenty of means that serve to educate users about working in this environment, whether they are training videos or entire online documentation of scripts and programming methods useful in the creation of any content.

GIMP
GIMP is an open-source raster image editor. It offers plenty advanced features including filters, Beziers curves, supported pre-installed plug-ins, layers and scripting. It is a multiplatform graphic software available for many operating systems.

Blender Studio
Blender Studio is a professional application for 3D graphics used for modeling, animation and visualization of created models. By this application, photo-realistic images and animations can be created. It is widely used among architects for building models and visualization. The reason for using multiple programming software was the fact that the player rather enjoys good graphic design than other effects (music and sound). We wanted to achieve the best impression from the game and, therefore, we considered this when designing the game.

2.4   Game Development

When creating the game design, we had to think especially about the implementation and development. We had to adapt the visual of scenes, objects and structure to adventure design.

Scenes and Rooms

Once we began developing and programming the game, it was very important to decide about the number of scenes we would use and the way we will be switching between them. After consulting, we decided to use just one scene. However, this scene contains five different rooms. Every room contains several different objects and components (to increase the attractiveness). It was necessary to model the objects (some of them were used from the Asset store). Each room was modeled individually for the simple movement of the character (Fig. 2a, b):

(a)
Cube:
Volume(V): a^3
Surface area (SA): $6a^2$
Space diagonal (D): $\sqrt{3}a$

(b)
Cuboid:
Volume(V): length * width * height
Surface area (SA): $2(lw) + 2(hl) + 2(hw)$
Space diagonal (D): $\sqrt{(l^2 + w^2 + h^2)}$

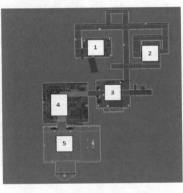
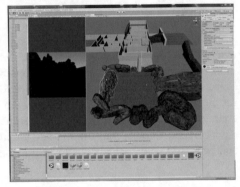

Fig. 2. (a) Scene design view (b) 3D scene design view Source: own production

- *The first room* contains an NPC (Non-Player-Controller) specifying the task. The environment is matched with appropriate objects.
- *The second room* contains an NPC that uses an interview to interact. The room is complemented by objects.
- *The third room* is the starting/opening room of the scene. It contains a player (3D Controller), which means we see the player's back. The room has many other objects to improve the overall impression.
- *The fourth room* contains enemies. Players need to fight off the enemies' attacks. It is an adjoining room, i.e. the rooms are next to each other and there is no door inside connecting them. There is a possibility to move to the fifth room.
- *The fifth room* – there are stronger enemies and after the player defeats them, there are some objects needed to complete some geometry tasks, e.g. volumes, surface area:

Creating objects is among the most basic activities in any development environment and its definition is largely unchanged, although it might look differently in each area (Fig. 3). In the game engine, the object is any object on the scene that can be manipulated or modified in its appearance or behaviour towards surrounding or other objects.

(a)   **(b)**

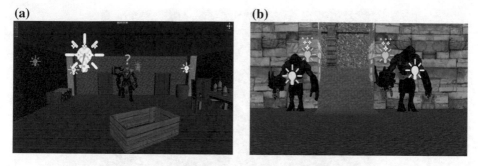

Fig. 3. (a) Scene design view (b) Enemies – motivation in game

Scripting
Unity generally supports several programming languages used in programming scripts. These scripts contain encoded logic of objects' behaviour present in the game. The programming languages supported by Unity include:

- C#;
- Java Script;
- Boo;

For C# and Java Script, there is online documentation of all programming methods where methods of using, properties, and parameters used when working with them can be found. Each programming script can be saved to a separate file and managed in the Project window like any other file. Unity also supports Microsoft Visual Studio for programming.

Gaming Interface
Graphic User Interface (GUI) plays an important role in presenting the various game elements. GUI design elements were created using external software for editing 2D graphics. We used mainly freely available programs, such as Gimp, in which we drew pictures to GUI representing key objects in the game. When creating the menu, we used a new system of Unity, which allows intuitive editing of individual windows and modifying their appearance and functionality by adding components. This way, it is possible to adjust a menu item, which normally acts as a text, by adding a button. The menu (flash screen) of the game was created using similar graphics as the environment of the game. It is a basic prototype of the intro screen of how the game will look like (Fig. 4). The screen contains the following buttons:

- *Play Game* – launches a new game.
- *Load Game* – loads saved games.
- *Save Game* – saves the game progress of a player.
- *Help* – a short guide to game controls and the goal of the game.
- *Options* – settings (including graphic and audio settings).
- *Credits* – creators, acknowledgments.
- *Quit Game* – leaves the game.

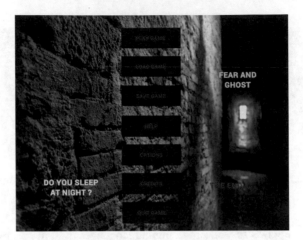

Fig. 4. Flash menu of the Mathematical Dungeon

2.5   Game Testing

Testing was done using the white box method based on the knowledge of the internal structure and source code of the tested system - the game. While testing this first phase, we encountered bugs that caused malfunction of one of the features. Errors were then removed. We are aware that we will encounter other iterations of testing, but we will reveal further shortcomings in the code.

3   Conclusion and Future Work

In the theoretical part, we dealt with the characteristics of serious games, advantages, and benefits of using serious games. We then described and characterized the process of designing the game. It was necessary to comply with the rule making for the game so that the fun factor does not outweigh the educational factor of a serious game.

The game was designed and developed in association with elementary school math teachers. At present, the game is being implemented at selected primary schools and we need to find out about the effectiveness (i.e. if the game increases cognitive levels of pupils, motivation and understanding in the field of geometry in elementary schools). The results obtained from schools will be evaluated, published, and compared with other similar research works. Pupils will be more involved in the process of education. The game is designed so that it can be further developed by adding objects and settings. The game could offer a number of illustrative examples focused on geometry (other geometric shapes such as pyramid, cylinder, sphere, cone, pentagonal prism will be added to the game). Further research will be conducted to find out what kind of role teachers will have and how games can help pupils improve their knowledge in spatial imagination. After the development of the serious game is completed, the game will be available on a web portal.

Using a combination of traditional and ICT (serious game, virtual reality, mixed reality, etc.), we can stimulate children's interest in learning mathematics, create appropriate (and verified) conditions for the individualisation of education, verify their effectiveness in practice, and consequently improve conditions for improving the quality of science education.

Acknowledgements. This paper was supported by KEGA project No.011UCM-4/2018.

References

1. Misfeldt, M., Gjedde, L.: What is a game for geometry teaching: creative, embodied and immersive aspects (2015)
2. Országhová, D., Gregáňová, R.: The analysis of exams outcomes of the mathematical compulsory and optional subject. In: Forum Statisticum Slovacum, vol. 7, pp. 133–137 (2011). ISSN 1336-7420
3. Országhová, D., Gregáňová, R., Majorová, M.: Professional training of economists and managers in the context of the transformation of university education. In: Scientific papers (CD) "The Path of Internationalization and Integration in the Europe of Regions", Curtea de Arges, Romania, pp. 259–264 (2007). ISBN 978-80-8069-857-7
4. Bopp, M.: Storytelling as a motivational tool in digital learning games. In: Didactics of Microlearning. Concepts, Discourses and Examples, pp. 250–266 (2007)
5. Chorianopoulos, K., Giannakos, M., Chrisochoides, N.: Design principles for serious games in mathematics. In: Proceedings of the 18th Panhellenic Conference on Informatics (2014). https://doi.org/10.1145/2645791.2645843
6. Zapletal, M.: Hry v klubovně. 1. vyd. Olympia, Praha, p. 567s (1986). ISBN 27-053-86
7. Breuer, J., Bente, G.: Why so serious? On the relation of serious games and learning. J. Comput. Game Cult. **4**(1), 7–24 (2010). Eludamos
8. Ritterfeld, U., Cody, M.J., Vorderer, P.: Serious Games: Mechanisms and effects. Routledge, New York (2009)
9. Ritterfeld, U.: Beim Spielen lernen? Ein differenzierter Blick auf die Möglichkeiten und Grenzen von Serious Games. In: Computer + Unterricht, vol. 84, pp. 54–57 (2011)
10. Mišútová, M., Mišút, M.: Impact of ICT on the quality of mathematical education In: IMSCI 2012 - 6th International Multi-Conference on Society, Cybernetics and Informatics, Proceedings, pp. 82–86 (2012)
11. Štubňa, J.: Selected determinants influencing on student motivation in creating a relationship towards of science. In: Acta Humanica, vol. 13, no. 1, pp. 68–77 (2016)
12. Štubňa, J.: The importance of educational game as a part of teacher students preparation in the context of developing intersubject relationships in science subject. In: Acta Humanica, vol. 13, no. 2, pp. 39–45 (2016)
13. Toman, J., Michalík, P.: Possibilities of implementing practical teaching in distance education. Int. J. Modern Educ. Forum **2**(4), 77–83 (2013)
14. Mišút, M., Pribilová, K.: Communication impact on project oriented teaching in technology supported education. Lecture Notes in Electrical Engineering, LNEE, vol. 152, pp. 559–567 (2013)
15. Gregáňová, R., Országhová, D.: K novým kompetenciám učiteľa matematiky v kontexte elektronického vzdelávania. In: Zborník vedeckých príspevkov z medzinárodnej vedeckej konferencie "The 6rd international conference APLIMAT", pp. 337–343. STU, Bratislava (2007) (2007). ISBN 978-80-969562-8-9

16. Šimon, M., Huraj, L., Siládi, V.: Analysis of performance bottleneck of P2P grid applications. J. Appl. Math. Stat. Inf. **9**(2) (2013). ISSN 1336-9180. (IET Inspec)
17. Polakovič, P., Hennyeyová, K., Šilerová, E., Hallová, M., Vaněk, J.: Managerial ICT education, innovation objectives and barriers to education of agricultural enterprises managers. In: Agrarian perspectives XXVII, pp. 243–251. Czech University of Life Sciences, Praha (2018). ISBN 978-80-213-2890-7
18. Pribilová, K., Gazdíková, V., Horváth, R.: Use of virtual excursions at secondary schools in Slovakia. In: 12th IEEE International Conference on Emerging eLearning Technologies and Applications (ICETA), Stary Smokovec, 2014, pp. 163–167 (2014). https://doi.org/10.1109/iceta.2014.7107578
19. Tóthová, D., Fabuš, J.: Portal of Slovak universities. In: ICABR 2015, pp. 1062–1068. Mendel University, Brno (2015). ISBN 978-80-7509-379-0. http://www.icabr.com/fullpapers/icabr2015.pdf

Intelligence-Software Cost Estimation Model for Optimizing Project Management

Praveen Naik[1](✉) and Shantharam Nayak[2]

[1] Research Scholar VTU, Department of Computer Science and Engineering, Faculty of Engineering, CHRIST (Deemed to be University), Bangalore, India
praveen.research.se@gmail.com
[2] Department of Information Science and Engineering, RV College of Engineering, Bangalore, India

Abstract. With the evolution of pervasive and ubiquitous application, the rise of web-based application as well as its components is quite rising as such applications are used both for development and analysis of the web component by developers. The estimation of software cost is controlled by multiple factors right from human-driven to process driven. Most importantly, some of the factors are never even can be guessed. At present, there are no records of literature to offer a robust cost estimation model to address this problem. Therefore, the proposed system introduces an intellectual model of software cost model that is mainly targets to perform optimization of entire cost estimation modeling by incorporating predictive approach. Powered by deep learning approach, the outcome of the proposed model is found to be cost effective in comparison to existing cost estimation modeling.

Keywords: Cost estimation · Deep learning · Intelligence · Predictive · Software cost · Uncertainty · Machine learning

1 Introduction

A development of software has various cycles depending upon the type of adoption by every IT organization [1]. At present, there are various cost modeling and frameworks used in software project management [2–5]. Usage of software cost models are witnessed prior stage as well as post stage of the software project development [6]. The pre-stage modeling focuses on performing a pilot assessment of probable cost involved in the software projects whereas post-stage modeling of cost emphasizes on ensuring the success rate of the cost model followed. However, existing system has seen mainly importance towards pre-stage modeling as maximum amount of risk factors can be visualized as well as addressed [7, 8]. In current era of cloud computing, it is known that majority of the software applications are in the form of web-components, which is not only bigger in dimension but also involves lots of complicated design issues [9]. It is because majority of the cloud-based products uses web-components that are accessed by both users as well as developers. At present, there is no such standard cost modeling for controlling software cost involvement in development of such web components and hence requires a serious investigation. Hence, the proposed manuscript introduces a

© Springer Nature Switzerland AG 2019
R. Silhavy (Ed.): CSOC 2019, AISC 984, pp. 433–443, 2019.
https://doi.org/10.1007/978-3-030-19807-7_42

comprehensive model towards cost estimation and incorporates intellectual feature for offering maximum applicability. Section 2 discusses about the existing research work followed by problem identification in Sect. 3. Section 4 discusses about proposed methodology followed by elaborated discussion of algorithm implementation in Sect. 5. Comparative analysis of accomplished result is discussed under Sect. 6 followed by conclusion in Sect. 7.

2 Related Work

At present, there are various existing approaches towards facilitating the estimation of the cost involved in software project management [10]. The work carried out by Ghasemabadi et al. [11] has presented a cost minimization model using *linear programming approach* as a part of implementation. Bagheri and Sendi [12] have presented a risk-involved cost estimation model for facilitating cost incurred in effort. Boehm [13] has presented a theoretical discussion associated to estimate software cost using diversity concept. Discussion from the archives of literature associated with all intelligent approaches has been discussed by Elfaki et al. [14] considering the case study of construction engineering. Consideration of such commercial case study was also carried out by Hao and Guo [15]. Although, the idea is about financial cost estimation but it is helpful for software project cost estimation too. Essential information associated with cost management with project lifecycle was discussed by Jiang et al. [16]. Usage of neural network towards cost estimation was carried out by Juszczyk et al. [17] considering sports sector. Adoption of fuzzy logic was also reported towards contributing the software cost considering case study of defects removal (Khalid and Yeoh [18]). The work done by Kumar and Sharma [19] has also introduced a cost modeling considering defect analysis of software focusing on software reliability issue. Inventory-based cost analysis was carried out by Li et al. [20] where cost estimation towards effort management in software cost has been presented. The work of Najadat et al. [21] have used data mining approaches towards prediction of software cost. Cost is also studied with respect to bug identification and removal as seen in the work carried out by Razzaq et al. [22]. Theoretical discussion of the cost models was seen in work of Rosa et al. [23] where statistical results shows good accuracy in prediction for agile based management of project life cycle. The study of cost was also carried out by Shrestha et al. [24] using inferential statistics where the outcome highlights duration as significant indicator of cost. A unique cost estimation model was presented by Smite et al. [25] where emphasis was given to latent cost involved in offshoring the projects. There are various parameters involved in controlling the cost and such emphasis could be finding in literature written by Suliman and Kadoda [26] using mixed-method study. Usage of bio-inspired algorithm and classifier towards cost modeling was reported in work of Yi et al. [27]. Zhang and Liu [28] have presented an analytical modeling towards cost computation where the modeling is more focused on analysis rather than controlling. Zhou et al. [29] have used evolutionary approach as well as fuzzy logic towards cost modeling approach using quantitative approach. The next section discusses about the research problems being identified after reviewing the existing system.

3   Problem Description

The problems associated with the existing research work are as follows viz. (i) Majority of the existing approaches are less focused considering the complexities associated with IT industry and software project management, (ii) none of the existing study has ever focused on optimization of software cost models for improving productivity, (iii) consideration of web-components and their associated complexities are never reported in any existing approaches, and (iv) there was no intellectual incorporation being ever carried out towards optimizing the performance of software cost modeling in any existing literatures. Therefore, the problem statement is *"Constructing a comprehensive model with capability to incorporate intellectual features is quite challenging for achieving optimized cost modeling concentrating on complexities of web-components."*

4   Proposed Methodology

This part of the study will perform further optimization of the outcome obtained in prior research state of mathematical modeling [30]. Assuming the completion of research work till core research objective, it can be said that (i) there is an analytical framework that can offer cost effective software cost estimation by inclusion of uncertainty and (ii) a comprehensive mathematical model that is capable of investigating the effect of different factors of web components using CCF. However, both the process lacks integration without which it is quite challenging to assess the practicality of the outcome. Hence, for this purpose, the study will re-formulate multiple non-linear constraints as well as essential performance metric of dependability and steadiness. Figure 1 highlights the tentative schematic diagram of implementation.

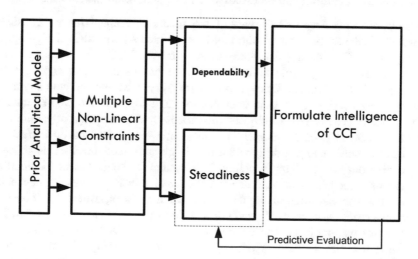

Fig. 1. Adopted scheme of proposed system

5   Algorithm Implementation

The prime motive of the proposed algorithm is basically to incorporate intelligence towards performing optimization of the usage of Cost Curtailing Factor (CCF). A statistical-based approach using analytical methodology is basically implemented in proposed algorithm for ensuring that it performs dual task viz. (i) computation of dependability cum steadiness of the proposed model towards optimization of software cost estimates and (ii) a capability of building intelligence towards identifying risk factors towards cost effectiveness of software project management concerning with web components. The steps of the algorithm are as follows:

> **Algorithm for optimizing software cost estimates**
> **Input**: $p_{.N}$ (N number of real problems p), q_{met} (quality metric)
> **Output**: D (dependability), S (steadiness)
> **Start**
> 1. *init* $p_r = \{p_{.N}\}^c$
> 2. extract $q_{met} \rightarrow (m_k)$
> 3. $c_{anal} = f(p_r, q_{met})$
> 4. CCF $\rightarrow (c_{anal})^{th}$
> 5. $\alpha = g(CCF) \rightarrow (ci_j)$
> 6. *apply* $h(x) \rightarrow [(ci_j)|Err]$
> 7. *compute* D and S
> **End**

The discussion of the proposed algorithm is as follows: The algorithm takes the input of $p_{.N}$ (N number of real problems p) and q_{met} (quality metric) which after processing yields an outcome of D (dependability) and S (steadiness). The core operation of the algorithm can be discussed in two significant processes as follows:

- *Pre-Optimization Stage:* This part of the modeling emphasizes on considering mathematical variables for all the real-time problems evolved in designing web-components. Considering that there are N number of real-time problems that are bounded by $c = (c_N)$ constraint (Line-1), the proposed system can be said to consider all non-linear multiple (N) constraint that can be mapped with any number of web-based software projects with multiple and sophisticated web-components. This stage consist of two significant sub-process that is executed in parallel:
  - *Extraction of Quality Metric:* This stage of work is focused on extraction of quality metric using a design tool that is globally recognized. This process is carried out for involving benchmarking figures for better form of validation of proposed model. As the use case is set of massive web components, hence, there are good possibilities of higher number of metrics m_k where k is considered number of web-metrics (Line-2). The value of k may vary based on different types of web-projects. These web-metrics are associated with coupling, inheritance, tree size, methods used, and many more. Generally all these metrics are related to object-oriented logic that is frequently used in developing web components.

- *Computing CCF:* This part of the study is focused on applying an integrated function $f(x)$ which takes into consideration both real-time problems p_r and quality metric q_{met} (Line-3). The function generates a matrix canal which uses a threshold in order to compute CCF (Line-4). The empirical representation of the CCF is expressed as,

$$CCF = \frac{\Delta \delta}{\varphi}$$

  In the above expression, numerator $\Delta \delta$ is basically a difference between two mathematical components where the first component is a scalar multiplication of (i) quality metric q_{met}, (ii) unit duration factor, and (iii) total duration of effort without optimization consideration. The second component is a scalar multiplication of (i) quality metric q_{met}, (ii) unit duration factor, and (iii) total duration of effort with optimization consideration. The denominator ϕ is basically a difference between total day count for developing assigned web components and duration needed with cost curtaining factor. This operation is further followed by applying a function $g(x)$ which results in outcome of CCF as cost indicator ci_j where j is the number of cost indicators (Line-5). These cost indicators are also non-linear in forms and offers discrete information about all possible real-time cost factors that can be modeled as per the need of any web-based project management.

- *Post-optimization Stage*: This part of the analysis is responsible for performing predictive operation in order to build intelligence towards ensuring better optimization of the software cost estimation. The proposed system make use of deep learning mechanism, which offers significant predictive performance with a flexibility of applying any form of The proposed system applies a machine learning function $h(x)$ in order to check the consistency of its accuracy. The obtained cost indicators ci_j are concatenated with error *Err* in order to include various possible of other real-time problems that could have chances to evolve in future (Line-6). The reason behind error consideration is carried out in order to offer more practicability to the proposed model. The proposed system computes dependency factor and steadiness factor for assessing the effectiveness of the proposed intellectual system (Line-7). The dependency factor is computed by dividing $h(x)$ subtracted by lower limit of variances by maximized limit of variances over the input ci_j and *Err*. The dependency factor is also multiplied with specific weight. The steadiness factor is responsible for computing the scale of variances of the dependency factor. Increased variance of dependency factor could mean defective modeling while streamlined and reduced variance of dependency factor represents robust modeling.

The pre-optimization stage of proposed system is focused on CCF while the post-optimization stage emphasizes on incorporating intellectual using deep learning mechanism towards optimizing the estimates of the software cost associated with web component design management. The next section discusses outcome accomplished.

6 Results Discussion

The assessment of the proposed study is carried out using a unique sequence of implementation where focus is laid over evaluating the cost factor to claim the optimization performance owing to proposed intellectual concept incorporation. An open source sample code for web components are collected which comprises of 140 classes, 2000 methods, and 35,000 lines of code. Open source tool e.g. MyEclipse is used for assessing 5 different metrics which are associated with methods, class response, inheritance length, coupling, sub-classes. Higher number of this will result in more cost involvement. The study outcome is also compared with the nearly similar study carried out by Herbold [31] and Subramanyam [32].

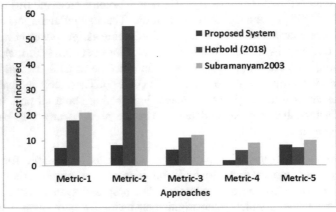

(a) Cost Estimation before Optimization

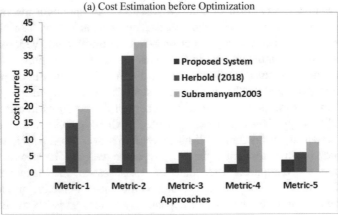

(b) Cost Estimation After Optimization

Fig. 2. Comparative analysis of cost estimation

The outcome exhibited in Fig. 2 shows two inferences viz. (i) proposed system offers lower cost involvement in contrast to existing system of cost control towards ensuring software quality and (ii) optimization using deep learning approach has positive effect on cost minimization. The outcome is a direct indication that proposed system could be used for curtaining significant cost if the information of real-time problems is clearly defined which is not in the case of existing approach. The performance of cost is quite inferior for class response (Metric-2) as existing system considers higher number of methods increasing the cost, while proposed system uses its training algorithm (Feed forward Backpropagation) which eliminates the replicates methods and retain only superior methods to perform better prediction of cost. Hence, cost incurred for proposed system is always low. This outcome is also supported by accuracy analysis.

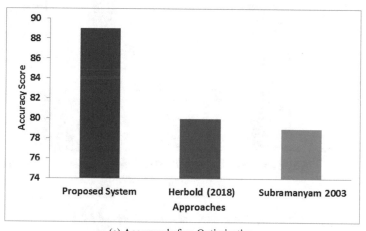

(a) Accuracy before Optimization

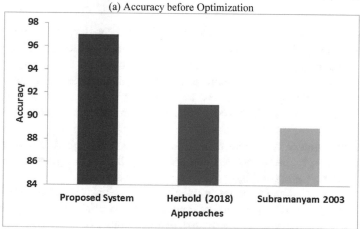

(b) Accuracy after Optimization

Fig. 3. Comparative analysis of accuracy

Similar trend of performance can be seen in Fig. 3 where proposed scheme excels better outcome in contrast to existing approach. The root causes of accuracy for proposed system are (i) usage of advanced machine learning, (ii) clear definition of real-problems and its associated problems, (iii) validation by benchmarked tool for quality indexers. In order to perform similar test environment, the work of existing approach is also trained with similar training algorithm used in proposed study. Such adoption and strategy is not present in existing system, which results in degradation of accuracy. It should be noted that after consideration of uncertainty factor, the proposed system offers better accuracy performance which is quite practical in nature.

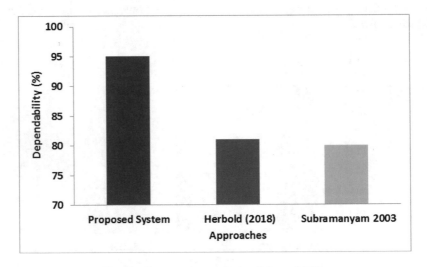

Fig. 4. Comparative analysis of dependability

The performance parameter of dependability relates to the extent of quality assurance of proposed intelligence-based estimation of software cost. Figure 4 shows that proposed system offers more dependability as compared to existing approaches. The prime reason behind this is inclusion of uncertainty parameter over a statistical range allows the training operation to offer increasing accuracy score in very practical way. This operation is not involved in existing approaches and hence dependability reduces.

The computation of steadiness factor is calculated observing 10 different iterations of proposed model to check variance in dependency factor. The graphical outcome in Fig. 5 exhibits that proposed system offers superior steadiness performance whereas there is no significant improvement in existing approaches. The reason behind higher steadiness is usage of both feed-forward and back-propagation in training which offers highly updated predictive outcome irrespective of number of epoch. Hence, proposed solution offers uniformity in its outcome towards software cost estimation.

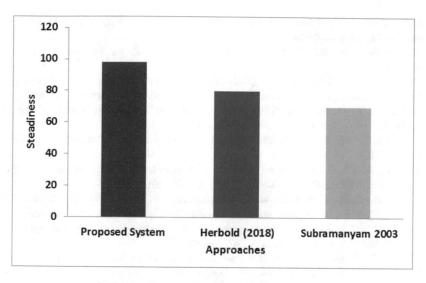

Fig. 5. Comparative analysis of steadiness

7 Conclusion

Software cost estimation is one of the essential and highly sensitive part of the operation in project management; however, it includes various challenges especially if the complex web-based component development is considered. The proposed system hypothesizes that inclusion of intellectual property in the software cost modeling will significantly assists in estimation of the software cost along with significant scale of optimization. The contribution of the proposed model are as follows: (i) the proposed model offers higher accuracy and lower cost involvement even if the project size is bigger, (ii) it doesn't require any specialized skill to manage/handle and hence it if free from any re-engineering process, (iii) the tool used for designing the model is publically available tool and is highly affordable (MATLAB, JAVA, OriginPro, SPSS, Minitab, etc.). Apart from above mentioned contribution, the study outcome is found to show that proposed system offers better optimization performance in comparison to existing system.

References

1. Mahmood, Z.: Software project management for distributed computing: life-cycle methods for developing scalable and reliable tools. Springer (2017)
2. Anand, A., Ram, M.: System Reliability Management: Solutions and Technologies. CRC, Boca Raton (2018)
3. Trendowicz, A., Jeffery, R.: Software Project Effort Estimation: Foundations and Best Practice Guidelines for Success. Springer (2014)
4. Lu, W., Lai, C.C., Tse, T.: BIM and Big Data for Construction Cost Management. Routledge, Abingdon (2018)

5. Blokdyk, G.: Cost Management Cost Optimization the Ultimate Step-By-Step Guide. Emereo Pty Limited (2018)
6. Heusser, M., Kulkarni, G.: How to Reduce the Cost of Software Testing. CRC Press, Boca Raton (2018)
7. Trendowicz, A.: Software Cost Estimation, Benchmarking, and Risk Assessment: The Software Decision-Makers' Guide to Predictable Software Development. Springer (2013)
8. Kumar, R., Tayal, A., Kapil, S.: Analyzing the Role of Risk Mitigation and Monitoring in Software Development. IGI Global, Hershey (2018)
9. Bhargava, S., Jain, P.B.: Software Engineering: Conceptualize. Educreation Publishing, New Delhi (2018)
10. Naik, P., Nayak, S.: Insights on research techniques towards cost estimation in software design. Int. J. Elect. Comput. Eng. (IJECE) **7**, 2088–8708 (2017)
11. Ghasemabadi, M.A., Ashtiani, M.G., Mohammadipour, F.: PMBOK five process plan for ISMS project implementation considering cost optimization for a time constraint: a case study. In: 2011 2nd IEEE International Conference on Emergency Management and Management Sciences, Beijing, pp. 788–791 (2011)
12. Bagheri, S., Shameli-Sendi, A.: Software project estimation using improved use case point. In: 2018 IEEE 16th International Conference on Software Engineering Research, Management and Applications (SERA), Kunming, pp. 143–150 (2018)
13. Boehm, B.W.: Software cost estimation meets software diversity. In: 2017 IEEE/ACM 39th International Conference on Software Engineering Companion (ICSE-C), Buenos Aires, pp. 495–496 (2017)
14. Elfaki, A.O., Alatawi, S., Abushandi, E.: Using intelligent techniques in construction project cost estimation: 10-year survey. Adv. Civ. Eng. **2014**, 11 (2014)
15. Hao, S., Guo, P.: The researched on water project cost accounting based on activity-based costing. In: 2012 International Conference on Information Management, Innovation Management and Industrial Engineering, Sanya, pp. 146–149 (2012)
16. Jiang, C., et al.: Some thoughts about the whole life-cycle cost management of civil engineering projects. In: 2009 16th International Conference on Industrial Engineering and Engineering Management, Beijing, pp. 496–499 (2009)
17. Juszczyk, M., Leśniak, A., Zima, K.: ANN based approach for estimation of construction costs of sports fields. Complexity **2018**, 11 (2018)
18. Khalid, T.A., Yeoh, E.: Early cost estimation of software reworks using fuzzy requirement-based model. In: 2017 International Conference on Communication, Control, Computing and Electronics Engineering (ICCCEE), Khartoum, pp. 1–5 (2017)
19. Kumar, G., Sharma, R.: Analysis of software reliability growth model under two types of fault and warranty cost. In: 2017 2nd International Conference on System Reliability and Safety (ICSRS), Milan, pp. 465–468 (2017)
20. Li, Q., Yang, R., Li, J., Wang, H., Wen, Z.: Strength and cost analysis of new steel sets as roadway support project in coal mines. Adv. Mater. Sci. Eng. **2018**, 9 (2018)
21. Najadat, H., Alsmadi, I., Shboul, Y.: Predicting software projects cost estimation based on mining historical data. ISRN Softw. Eng. **2012**, 8 (2012)
22. Razzaq, S., Li, Y., Lin, C., Xie, M.: A study of the extraction of bug judgment and correction times from open source software bug logs. In: 2018 IEEE International Conference on Software Quality, Reliability and Security Companion (QRS-C), Lisbon, pp. 229–234 (2018)
23. Rosa, W., Madachy, R., Clark, B., Boehm, B.: Early phase cost models for agile software processes in the US DoD. In: 2017 ACM/IEEE International Symposium on Empirical Software Engineering and Measurement (ESEM), Toronto, ON, pp. 30–37 (2017)

24. Shrestha, P.P., Leslie, A.B., David, R.S.: Magnitude of construction cost and schedule overruns in public work projects. J. Constr. Eng. **2013**, 9 (2013)
25. Šmite, D., Britto, R., van Solingen, R.: Calculating the extra costs and the bottom-line hourly cost of offshoring. In: 2017 IEEE 12th International Conference on Global Software Engineering (ICGSE), Buenos Aires, pp. 96–105 (2017)
26. Suliman, S.M.A., Kadoda, G.: Factors that influence software project cost and schedule estimation. In: 2017 Sudan Conference on Computer Science and Information Technology (SCCSIT), Elnihood, pp. 1–9 (2017)
27. Yi, T., Zheng, H., Tian, Y., Liu, J.: Intelligent prediction of transmission line project cost based on least squares support vector machine optimized by particle swarm optimization. Math. Probl. Eng. **2018**, 11 (2018)
28. Zhang, H.Y., Liu, Y.: A study of cost control system in the construction project of removing danger and reinforce engineering in KeKeYa reservoir in Shanshan County of Sinkiang. In: 2011 International Conference on Mechatronic Science, Electric Engineering and Computer (MEC), Jilin, pp. 2181–2185 (2011)
29. Zhou, W., Li, Y., Liu, H., Song, X.: The Cost management on the quantification of responsibility. Math. Probl. Eng. **2018**, 12 (2018)
30. Naik, P., Nayak, S.: A novel approach to compute software cost estimates using adaptive machine learning approach. J. Adv. Res. Dyn. Control Syst. (2017)
31. Herbold, S.: Benchmarking cross-project defect prediction approaches with costs metrics. arXiv:1801.04107v1 [cs.SE], 12 January 2018
32. Subramanyam, R., Krishnan, M.S.: Empirical analysis of CK metrics for object-oriented design complexity: implications for software defects. IEEE Trans. Softw. Eng. **29**(4), 297–310 (2003)

Outliners Detection Method for Software Effort Estimation Models

Petr Silhavy⬛, Radek Silhavy$^{(\boxtimes)}$⬛, and Zdenka Prokopova⬛

Faculty of Applied Informatics, Tomas Bata University in Zlin,
nam. T.G.M 5555, 76001 Zlin, Czech Republic
{psilhavy, rsilhavy, prokopova}@utb.cz

Abstract. Outliner detection methods are studied as an approach for simulated in-house dataset creation. In-house datasets are understood as an approach for increasing the estimation accuracy of the functional points-based estimation models. The method which was selected as the best option for outliners' detection is the median absolute deviation. The product delivery rate was used as a parameter for the median absolution deviation method. The estimation accuracy was compared for a public dataset and simulated in-house datasets, using stepwise regression models. Results show that in-house datasets increase estimation accuracy.

Keywords: Outliner detection · Function point analysis ·
Product delivery rate · Stepwise regression ·
Software development effort estimation

1   Introduction

The software development industry is facing an issue of estimation development effort, which is needed for cost elicitation and price negotiation. Several methods can be adopted for software development effort estimation.

The estimation methods which are adopted includes Use Case Points [1, 2] or Functional Points Analysis (FPA) [3, 4]. The several modifications for effort estimation model were studied. The common improvements are based on regression analysis [2, 5]. Some authors study a soft computing technique for estimation models adaptation [6–8].

The several authors study a productivity delivery rate (PDR) which is used for the final transformation of the estimated size of a software project. Azzeh and Nassif [9] discuss a PDR in the situation, that there is historical dataset available, or there is no public dataset. Urbanek et al. [10] recommend setting a PDR in interval from 10 to 30 person-hours. PDR describe a ration between effort and working time. It is measured in person-hours.

In this study the methods for outliner detection are tested. Authors design an approach, which allow creating a simulated in-house dataset. The effect of in-house datasets is not widely studied. Therefore, evaluation of the impact of in-house datasets for software effort estimation is performed in this paper. Because of the in-house datasets are unavailable, the simulation methods based on weighting the observation (Weights) or based on interquartile range (IQR) and finally median absolute deviation

© Springer Nature Switzerland AG 2019
R. Silhavy (Ed.): CSOC 2019, AISC 984, pp. 444–455, 2019.
https://doi.org/10.1007/978-3-030-19807-7_43

(MAD) method were investigated. All methods used in the adapted form and described later with a paper.

The application of the weighting approach for outliner detection can be seen in [11], where a similar weight application of weighting is used. A similar approach is used also in [12]. Both papers do not use a weighting based on the regression error, as is done in this paper. The weighting regression can be a priori adapted, when used in the two-stage design. The outliner effect is studied in [13].

The MAD methods are discussed in [14]. Authors present a method usage and setting. Further investigation can be found at [15], where MAD effects in time-series data are studied. IQR is described including a configuration in [16]. Authors introduce a general approach of handling outliners in datasets.

The rest of this study is structured as follows. Section 2 defines the datasets and how the dataset was processed. Section 3 describes a research questions. Section 4 is dedicated to the experiment design and discussing a problem statement.

The results are presented and discussed in Sect. 5. Finally, the Conclusion is set out in Sect. 6.

2 Dataset Description and Pre-processing Approach

In this study, an International Software Benchmarking Standards Group (ISBSG) dataset was adopted [17]. ISBSG dataset was pre-processed to filter only observations are estimated using an IFPUG methodology. In the ISBSG dataset the quality is rated as A-D labels. In this study only observations labelled as A or B are involved. Furthermore, only the observation, which has no empty values in the following variables were used (see Table 1).

Table 1. Variables involved in the study

Variable	Abbreviation	Description
External inputs	EI	Independent variable
External outputs	EO	Independent variable
External inquiry	EQ	Independent variable
Internal logical files	ILF	Independent variable
External interface file	EIF	Independent variable
Normalized work effort (person-hours)	Effort	Dependent variable
Product delivery rate	PDR	Data pre-processing variable

The obtained dataset (DS, n = 809) was split into training and testing parts using the hold-out approach in 2:1 ratio. The dataset descriptive statistics of training and testing can be seen in Table 2.

Table 2. DS dataset descriptive statistics

Dataset	n	Median effort	SD effort	Min effort	Max effort	Range effort
Training part	567	2,475	8,021	51	63,840	63,723
Testing part	242	2,435	8,986	31	71,729	71,698

As can be seen (Table 2), all descriptive statistics are nearly equal for training and testing part of the dataset. The hold-outed testing part was used for evaluating all models, including those, which are run on the simulated in-house dataset – as is explained later in the experiment design chapter.

3   Research Questions

In this paper research questions are focused on evaluation adapted outliner detection approaches, which allow simulating the in-house dataset if such is lacking. Therefore, research questions where set as follows:

RQ1: Are tested methods for outliners' detection helpful for creating a simulated in-house dataset?

RQ2: Is there any difference in accuracy when each method is compared to each other?

The response to those research questions is based on comparing the accuracy of FPA models to stepwise models. This comparison is carried-out on 4 datasets, which represent public data (DS) and simulated in-house datasets (Weights, IQR, MAD).

4   Experiment Design and Problem Statement

The FPA analysis research is focusing on improving the estimation accuracy. Dataset, which available for experiments contains no information about data vendors. Therefore, is not possible to perform an experiment on in-house data. When a PDR achieves a huge range. This range illustrates huge variably of projects, which was not expected within the software developers.

The datasets as is presented in the previous chapter are pre-processed and cleaned using standard data analysis approach. This leads to preparing a baseline dataset for further experiments, but the presented dataset cannot be presented as a simulated in-house dataset. It may be expected that in-house data will be more consistent. In this paper the following methods of increasing consistency are evaluated:

- Outliner detection based on observation's relative percentage error (Weights),
- Outliner detection based on the adapted interquartile range (IQR),
- Outliner detection based on the adapted median absolute deviation (MAD).

The evaluation process is based on comparing the accuracy of software effort development effort estimation models trained on baseline dataset (DS) and on simulated in-house datasets (Weighs, IQR, MAD). For such evaluation a standard set of evaluation criteria were used.

4.1 Outliner Detection Based on Observation on Relative Percentage Error

This method is using a Relative Percentage Error as the weighting factor. This weighting approach allows investigation influence of an individual observation to the model estimation quality. The estimation error $\varepsilon_i(1)$ is the absolute value of the difference of known effort value (y_i) and estimated effort value $(\widehat{y}_l)$ divided by known effort value (y_i).

$$\varepsilon_i = \frac{|y_i - \widehat{y}_l|}{y_i} \tag{1}$$

The weights are applied in interval $\langle 0; 1 \rangle$. When the observation indicates error higher then 50%, then a weight equals 0 (not included). If the observation indicates error less then 50%, than a weight equal to the difference between ε_i and 1. (2).

$$w_i = 1 - \varepsilon_i, \varepsilon_i \leq 1$$
$$w_i = 0, \varepsilon_i > 0.5 \tag{2}$$

The dataset cleaning workflow was based on estimation errors. That error was obtained, as described (1), from a weighted linear regression model, where initial weights are set to 1. An observation which was associated to $w_i = 0$, was removed from the training dataset. The graphical representation of all observation and identification of outliners can be seen in Fig. 1.

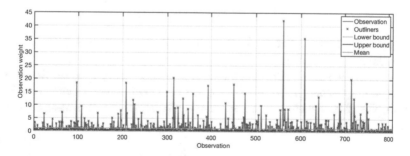

Fig. 1. Outliner detection based on relative percentage error

4.2 Outliner Detection Based on the Adapted Interquartile Range

The interquartile range method, a known method for removing outliner from numerical data. In this paper the adapted IQR method is used, to keep a PDR range close. In the cleaned dataset (IQR dataset) only the observations where PDR is less then PDR_{max} (3) and more then PDR_{min} (3) were kept.

$$PDR_{min} = Q_1 - 0.08 \times IQR$$
$$PDR_{max} = Q_3 + 0.08 \times IQR \tag{3}$$

$$IQR = Q_3 - Q_1 \tag{4}$$

The cleaning workflow is based on a calculation of the interquartile range between Q_1 (the first quartile) and Q_3 (the third quartile). Than a constant (0.08) is used to expand an IQR. The outliners, which were removed are represented in Fig. 2.

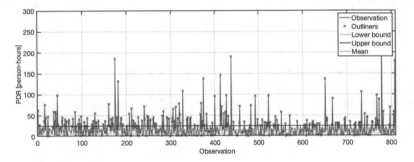

Fig. 2. Outliner detection based on interquartile range

4.3   Outliner Detection Based on the Adapted Median Absolute Deviation

Mean Absolute Deviation (5) method is based on evaluating a distance from the mean value of PDR.

$$\text{MAD} = median\{|x_i - \tilde{x}|\} \tag{5}$$

Where x_i represents a PDR value of an observation and $\tilde{x}$ is a median value of PDR, based on the whole set of observations.

The cleaning workflow allows to include only observation, which PDR is from the interval $\langle PDR_{min}; PDR_{max} \rangle$. In this case a PDR_{min}, PDR_{max} are calculated using formulas (6). The PDR values of all observations and outliners identification are shown in Fig. 3.

$$PDR_{min} = \tilde{x} - 0.6 \times MAD$$
$$PDR_{max} = \tilde{x} + 0.6 \times MAD \tag{6}$$

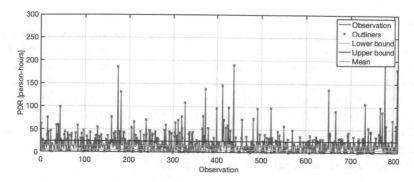

Fig. 3. Outliner detection based on mean absolute deviation

4.4   Models' Design and Accuracy Evaluation Criteria

In this paper an estimations accuracy is measured for the estimation model, which is based on a stepwise regression approach (SR). The method principal design is used as in [1, 18, 19]. SR of multiple linear regression is based on forward - and backward selection that involves an automatic process for the selection of independent variables; and can be briefly described as follows [1]:

- Set a starting model that contains predefined terms (backward); or set a null-model (forward)
- Set limits for the final model – determine the requested model complexity and which terms have to be included – linear, quadratic, interaction etc.
- Set an evaluation threshold – the sum of residual errors is used to determine whether to remove or add a predictor
- Adding or removing terms; re-testing the model
- Stepwise regression halts when no further improvement in estimation occurs.

Forward selection starts as a null-model and then iterates to add each variable which meets a given condition. When a non-significant variable is found, it is removed from the model. Backward selection works in a similar way but removes variables when they are found to be non-significant. Therefore, stepwise regression requires two significance levels: the first - for adding variables; and the second - for removing variables.

SR is a method of building many models from a combination of predictors. Therefore, Multiple Linear Regression (MLR) assumptions have to be fulfilled.

The models are tested were trained on training part of datasets DS, Weights, IQR and MAD. The model evaluation is based on estimation produced by all models on an identical set of data (testing fold from DS dataset). The following models were evaluated:

- FPA model – based on FP in IFPUG version, which using a median PDR valued for obtaining an estimation.
- SR_{DS} model – based on the SR approach and trained DS dataset.
- $SR_{Weights}$ model – based on the SR approach and trained on Weights dataset.
- SR_{IQR} model – based on the SR approach and trained on IQR dataset.
- SR_{MAD} model – based on the SR approach and trained on the MAD dataset.

Evaluation Criteria

All the tested models were evaluated using a Mean Absolute Percentage Error (MAPE), Mean Squared Error (MSE), Mean Magnitude of Relative Error (MMRE) and PRED(0.25), which represents an estimation with an error lower than 25%. MAPE was selected, because of in [20] authors proofs that MAPE (7) has practical and theoretical relevance for evaluation of regression models and its intuitive interpretation in terms of relative error. MMRE (8) is used because it allows comparison models, which are tested on non-equal datasets. MSE (9) is used because allows again comparing datasets not-equal in a number of observations. Finally (10) the PRED(0.25) evaluates how many observation is estimated with error lower then 25%.

$$\text{MAPE} = \frac{1}{n}\sum_{i=1}^{n}\frac{|y_i - \widehat{y_l}|}{y_i} \times 100 \tag{7}$$

$$MMRE = \frac{1}{n}\sum_{i=1}^{n}\frac{|y_i - \widehat{y_l}|}{y_i} \tag{8}$$

$$MSE = \frac{1}{n}\sum_{i=1}^{n}\varepsilon_i^2 \tag{9}$$

$$PRED(0.25) = \frac{1}{n}\sum_{i=1}^{n}\begin{cases} 1\,where\,\frac{|y_i-\widehat{y_l}|}{y_i} \leq 0.25 \\ 0\,where\,\frac{|y_i-\widehat{y_l}|}{y_i} > 0.25 \end{cases} \tag{10}$$

Where n is the number of observations, y_i is the known real value, $\widehat{y_l}$ is the estimated value and ε represents an estimation error.

5   Results and Discussion

In this section a results of models' evaluation are presented. The models were performed on DS dataset (SR_{DS} model) and then on simulated in-house datasets – $SR_{Weights}$ model, SR_{IQR} model and finally SR_{MAD} model.

All those SR based models were compared to each other and then to FPA method in which a median value of PDR was used for effort calculation.

In Table 3 there can be seen the comparison of DS and simulated in-house datasets. Mean, Standard Deviation (SD), minimum (min) and maximum (max) characteristics for PDF, Functional points (FP) and Effort variables are presented. All three methods, which were adopted for the creation of the simulated in-house dataset decrease variability and range of the PDR and other variables. More impact can be seen when IQR or MAD datasets are discussed.

Table 3. Comparison of DS dataset and simulated in-house datasets

Dataset	n	Mean			SD		
		PDR	FP	Effort	PDR	FP	Effort
DS	809	19.79	358.16	5291.8	22.94	651.43	8,317.07
Weights	328	24.62	422.3	6367.53	21.59	892.07	6,800.07
IQR	463	13.92	346.01	4488.27	5.18	476.22	5,875.35
MAD	362	12.93	362.51	4418.7	3.83	509.83	5,864.72

Dataset	n	min			max		
		PDR	FP	Effort	PDR	FP	Effort
DS	809	0.4	6	31	259.7	13,580	71,729
Weights	328	4	7	1,382	259.7	13,580	54,620
IQR	463	6.1	10	97	24.6	3,886	42,080
MAD	362	7.1	13	97	20.1	3,886	42,080

The estimation accuracy was studied on three models and compared to a model based on DS dataset and to FPA. FPA was used in pair with a median value of PDR. The median PDR is derivate from DS dataset.

The accuracy results were evaluated using MAPE, MMRE, MSE and PRED(0.25). In Table 4 there can be seen results achieved by the FPA method in IPFUG standard, when the effort estimation is done by median value of PDR (from finished projects).

Table 4. FPA + median PDR accuracy for all tested datasets (baseline models)

	MAPE	MMRE	MSE	PRED(0.25)
DS	194.829	1.948	1.297e+08	0.256
Weights	53.709	0.537	1.774e+08	0.411
IQR	95.508	0.955	2.153e+07	0.304
MAD	92.889	0.928	3.280e+07	0.388

Table 5. Stepwise regression-based model accuracy for all tested datasets

	MAPE	MMRE	MSE	PRED(0.25)
DS	176.110	1.761	4.721e+09	0.256
Weights	32.353	0.323	1.675e+08	0.460
IQR	45.146	0.451	4.421e+06	0.355
MAD	30.512	0.305	1.019e+07	0.462

In Table 5 an accuracy evaluation of the stepwise models is presented. It can be concluded that stepwise models achieved a significantly higher accuracy when compared to FPA method.

When FPA or stepwise models accuracy are compared on DS dataset, it can be concluded, that both behaved similarly. In the case of simulated in-house datasets, there can be seen an increase in estimation accuracy for stepwise models.

To answer a research questions following bar graphs comparing evaluated models were carried-out.

When RQ1 (Are tested method for outliners detection helpful for creating a simulated in-house dataset?) is answered it can be said that outliners methods are helpful and allowing an increase of estimation ability and accuracy. When a MAPE (Fig. 4) is used it can be seen that accuracy of models trained on IQR or MAD was increased significantly (45%, 30% vs 95% or 92%).

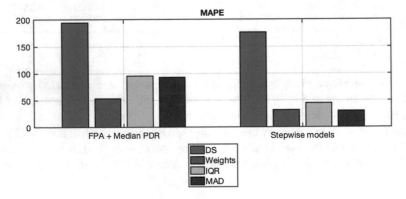

Fig. 4. MAPE comparison for evaluated models

Similar trends can be seen when an MMRE (Fig. 5) criterion was applied. Models trained on simulated dataset benefits from outliner removing.

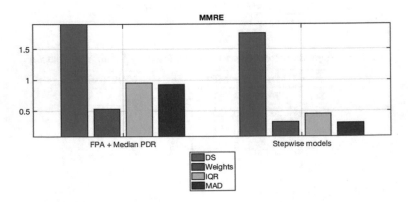

Fig. 5. MMRE comparison for evaluated models

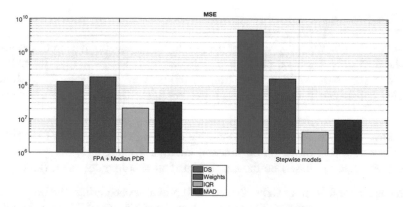

Fig. 6. MSE comparison for evaluated models

When an MSE criterion is used for evaluation, then it can be seen that the FPA model achieved better accuracy than the stepwise model for DS (Fig. 6). It may be justified by a huge error for some observations, which influences an MSE value. In Fig. 6 the logarithmic scale for y axes is used, because the scale is too broad and cannot be compared in basis linear scale.

When a PRED criterion is studied, then again can be confirmed that simulated in-house dataset allows increasing an estimation accuracy (Fig. 7).

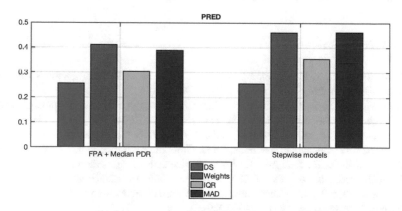

Fig. 7. PRED(0.25) comparison for evaluated models

To answer the RQ2 (Is there any difference in accuracy when each method is compared to each other?) it can be said that there is a difference in the ability to improve estimation. The method based on observation weights has an impact on accuracy, but the effect size is low. When compared to IQR or MAD, then it can be clearly seen MAD based method brings the most significant impact. The most of compared criterion confirms MAD methods outperform all other tested approaches.

6 Conclusion

The purpose of this work is evaluating the possibility of obtaining a simulated in-house dataset, when only a general public available dataset is available (e.g. ISBSG).

In this paper the outliner detection methods were evaluated. This allows for preparing a baseline dataset for further experiments. Those datasets were prepared using:

- Outliner detection based on observation's relative percentage error (Weights),
- Outliner detection based on the adapted interquartile range (IQR),
- Outliner detection based on the adapted median absolute deviation (MAD).

The results of this study show, that MAD based approach brings the best estimation accuracy when a stepwise model is used. It outperforms the FPA method when all criterions are evaluated. MAD based approach achieved MAPE 30.5% on SR model, which is significantly better than FPA (MAPE = 92.8). PRED(0.25) shows that 7% more project were estimated with an error lower then 25% (0.462 vs 0.388).

The achieved results allow concluding that in-house datasets have a positive impact on estimation model accuracy (more accurate estimation than public datasets) and those in-house datasets may be simulated, when needed. MAD based is the best option for the simulation of the in-house dataset – from tested approaches.

In future research the relationship of improving accuracy and removing outliners will be studied. It may be expected that if more outliners were removed then accuracy will be increased again. The limits, stopping condition and other methods for outliner detection will be investigated in our future research.

References

1. Silhavy, R., Silhavy, P., Prokopova, Z.: Evaluating subset selection methods for use case points estimation. Inf. Softw. Technol. **97**, 1–9 (2018)
2. Azzeh, M., Nassif, A.B., Banitaan, S.: Comparative analysis of soft computing techniques for predicting software effort based use case points. Iet Softw. **12**, 19–29 (2018)
3. Borandag, E., Yucalar, F., Erdogan, S.Z.: A case study for the software size estimation through MK II FPA and FP methods. Int. J. Comput. Appl. Technol. **53**, 309–314 (2016)
4. Celar, S., Mudnic, E., Kalajdzic, E.: Software size estimating method based on MK II FPA 1.3 unadjusted. In: Annals of DAAAM for 2009 and 20th International DAAAM Symposium "Intelligent Manufacturing and Automation: Focus on Theory, Practice and Education", Vienna, pp. 1939–1940 (2009)
5. Prokopova, Z., Silhavy, R., Silhavy, P.: The effects of clustering to software size estimation for the use case points methods. Adv. Intell. Syst. Comput. **575**, 479–490 (2017)
6. Azzeh, M., Nassif, A.B., Banitaan, S.: Comparative analysis of soft computing techniques for predicting software effort based use case points. IET Softw. **12**, 19–29 (2017)
7. Azzeh, M., Nassif, A.B.: A hybrid model for estimating software project effort from Use Case Points. Appl. Soft Comput. **49**, 981–989 (2016)
8. Nassif, A., et al.: Neural network models for software development effort estimation: a comparative study. Neural Comput. Appl. **27**(8), 2369–2381 (2016)

9. Azzeh, M., Nassif, A.B.: Project productivity evaluation in early software effort estimation. J. Softw.: Evol. Process **30**, e2110 (2018)
10. Urbanek, T., Kolcavova, A., Kuncar, A.: Inferring productivity factor for use case point method. In: 28th DAAAM International Symposium on Intelligent Manufacturing and Automation, DAAAM 2017, pp. 597–601 (2017)
11. Agouris, P., Mountrakis, G., Stefanidis, A.: Automated spatiotemporal change detection in digital aerial imagery. In: Automated Geo-Spatial Image and Data Exploitation, vol. 4054, pp. 2–13. International Society for Optics and Photonics (2000)
12. Alameddine, I., Kenney, M.A., Gosnell, R.J., Reckhow, K.H.: Robust multivariate outlier detection methods for environmental data. J. Environ. Eng. **136**, 1299–1304 (2010)
13. Seo, Y.S., Bae, D.H.: On the value of outlier elimination on software effort estimation research. Empir. Softw. Eng. **18**, 659–698 (2013)
14. Malinowski, E.R.: Determination of rank by median absolute deviation (DRMAD): a simple method for determining the number of principal factors responsible for a data matrix. J. Chemometr. **23**, 1–6 (2009)
15. Chen, C., Liu, L.-M.: Joint estimation of model parameters and outlier effects in time series. J. Am. Stat. Assoc. **88**, 284–297 (1993)
16. Rousseeuw, P.J., Hubert, M.: Robust statistics for outlier detection. Wiley Interdisciplinary Rev.: Data Mining Knowl. Discov. **1**, 73–79 (2011)
17. ISBSG: ISBSG Development & Enhancement Repository – Release 13. International Software Benchmarking Standards Group (ISBSG) (2015)
18. Silhavy, R., Silhavy, P., Prokopova, Z.: Analysis and selection of a regression model for the Use Case Points method using a stepwise approach. J. Syst. Softw. **125**, 1–14 (2017)
19. Silhavy, P., Silhavy, R., Prokopova, Z.: Evaluation of data clustering for stepwise linear regression on use case points estimation. Adv. Intell. Syst. Comput. **575**, 491–496 (2017)
20. de Myttenaere, A., Golden, B., Le Grand, B., Rossi, F.: Mean absolute percentage error for regression models. Neurocomputing **192**, 38–48 (2016)

Analysis of the Software Project Estimation Process: A Case Study

Zdenka Prokopova(✉) ⓘ, Petr Silhavy ⓘ, and Radek Silhavy ⓘ

Faculty of Applied Informatics, Tomas Bata University in Zlin,
nam T. G. Masaryka 5555, Zlin, Czech Republic
{prokopova, psilhavy, rsilhavy}@utb.cz

Abstract. In this paper is presented an analysis of the software project estimation process for the case of conditions in Czech companies. Key findings of the survey were that (a) the most widely used estimation method is Expert judgments with 57% incidence, (b) the most of surveyed companies (39%) operate in Web applications, and (c) the 54% companies fail to complete a quarter of projects in the estimated timeframe.

Keywords: Software development effort · Estimation methods ·
Software project estimation · Survey

1 Introduction

The estimation of a software project development effort is a critical part of software project management. Software developers require an effective effort-estimation model in order to facilitate project planning. Even though much existing software size and effort measurement methods are not widely adopted in practice. Many studies prove this claim and bring other important facts regarding the software effort estimation process.

Moløkken and Jørgensen in the paper [1] summarise estimation knowledge through a review of surveys on software effort estimation. Their main findings were that: most projects (60–80%) encounter effort and/or schedule overruns, the estimation methods in most frequent use are expert judgment-based, and that there is a lack of surveys including extensive analyses of the reasons for effort and schedule overruns.

Boehm et al. [2] summarise several classes of software cost estimation approaches, i.e., parametric models, expertise-based techniques, learning-oriented techniques, dynamics-based models, regression-based models, and composite-Bayesian techniques. Their survey results are suggestive that no single technique is best for all situations and that a comparison of several approaches is most likely to produce realistic estimates.

Trendowicz et al. in the paper [3] present the results of a study of software effort estimation from an industrial perspective. Their study included surveys of industrial objectives, the abilities of software organizations to apply certain estimation methods, and applied practices of software effort estimation.

© Springer Nature Switzerland AG 2019
R. Silhavy (Ed.): CSOC 2019, AISC 984, pp. 456–467, 2019.
https://doi.org/10.1007/978-3-030-19807-7_44

In the paper [4], Rastogi et al. present a review of general metrics and models regarding effort estimation. The authors found that no single metric can be globally accepted. They recommend hybridization of several approaches as an alternative to produce realistic estimates.

Bardsiri et al. [5] describe a study that is related to the extensive descriptive discovery of the models; also, they described many of the well-known accessible and utilised parametric models and a number of non-parametric methods.

In order to identify the most prominent software estimation models and their characteristics, authors Vera and Ochoa in [6] conducted a systematic literature review on taxonomies of effort estimation methods, reported between 2000 and 2017. The goals were mainly to determine the start-of-the-art in this area and to identify the main variables used to classify these methods. Based on this review, the authors elaborate on how software companies can use those taxonomies and classification criteria to identify suitable methods in support of their estimations.

Idri et al. [7] performed a systematic review of ensemble effort estimation (EEE) studies published between 2000 and 2016. They found that EEE techniques could be categorized into two types: homogeneous and heterogeneous. They found machine learning single models to be the most frequently employed in constructing EEE techniques.

Idri et al. [8] identified 65 studies published in the period 1990–2012. The study revealed that most researchers focus on addressing problems related to the first step of Analogy-based Software development Effort Estimation (ASEE) process. The results of the analysis show that acceptable estimates accuracy can be obtained when combining ASEE techniques with Fuzzy Logic or Genetic Algorithms.

The main aim of Munialo and Muketha in the paper [9] is to review existing agile software development estimation methods exhaustively by exploring estimation methods suitable for new software development methods.

Jorgensen in [10] provides an extensive review of studies related to Expert Estimation (EE) of software development effort. The review shows that EE is the most frequently applied estimation strategy for software projects. The author presents twelve expert estimations of "best practice" principles and provides suggestions on how to implement them in software organizations.

Based on our practical experience, it can be stated that estimating the development effort of software projects with the use of relevant methods is not very widespread among companies in the Czech Republic. In addition, there are neither guidelines nor methodologies and a unified approach that could be used as examples in competitions. For this reason, companies operating in the Czech Republic carried out a survey on the project estimation process and the use of estimation methods. The first results of the survey were partly used in [11]. The survey aimed to find out the status quo in Czech conditions with the prospect of creating ancillary tools for companies. These tools would simplify and refine the planning of software project development.

2 Problem Statement

The main aim of our investigation is an analysis of the process of estimating software projects development effort in Czech companies. The survey involved 91 companies from the Czech Republic. Enlisted companies had a different size, worked in a different business area or using different estimation methods.

2.1 Questionnaire

Research data were gathered using a survey method. The survey design was based on open questions, which allows to receive as much information as possible. The survey was processed by data analysis approaches including aggregation or data coding, where needed. The questionnaire sent to the companies comprises the following questions:

1. What size (how many employees) does your company have?
2. What types of projects (business areas) does your company take?
3. How does your company determine the workload (effort) of the project?
4. Does your company use software tools to calculate the cost of a project?
5. What do you prefer (describe) when calculating the cost of a project?
6. Based on what parameters does your company estimate the project?
7. What does your company include when calculating project?
8. What inputs are used to estimate costs?
9. How many people are involved in cost estimation?
10. How many percents of projects are not delivered in time?
11. How many percents of projects are underestimated?
12. What is the most common reason for the underestimation of the project?

3 Exploration Strategy

The responses gained from companies on the questions of the survey were treated as follows.

3.1 Distribution of Companies by Size

For the purpose of the survey evaluation, the responding companies were divided into four groups according to their size:

- Size1 – 1 to 10 workers
- Size2 – 11 to 25 workers
- Size3 – 26 to 50 workers
- Size4 – 51 or more workers.

As shown on the graph Fig. 1, 75% of companies (total 68) belong to the group Size1, 21% (total 19) to the group Size2, and the remaining 4% (total 4) to the groups Size3 and Size4.

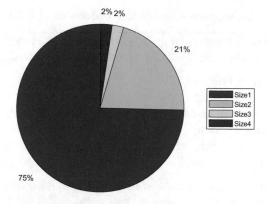

Fig. 1. Distribution of companies by size

3.2 Distribution of Companies by the Business Area

The surveyed companies operate in various business areas. Based on the questionnaire, six main business areas in which the companies operate were specified:

- Area1 – Database applications
- Area2 – Desktop applications
- Area3 – Mobile applications
- Area4 – Business Information Systems
- Area5 – Industrial systems and automation
- Area6 – Web applications

In Fig. 2 is shown the distribution of companies based on the business area. The most represented is Area6 (web applications), with a ratio 39% of companies. The second most represented is Area2 (desktop applications) with 30% and third, in rank is Area3 (mobile applications) with 17% representation. Most of the companies that operate in Area6 were also dedicated to Area2 or Area3.

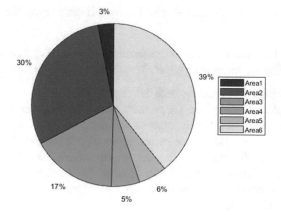

Fig. 2. Distribution of companies by the business area

3.3   Distribution of Companies by the Used Methods of Estimation

In the survey, companies were mainly asked about the used methods of estimation. According to the questionnaire, methods were classified into four groups:

- Method1 – Function Points
- Method2 – Use Case Points
- Method3 – Expert judgments (including a spectrum of methods like Wideband Delphi to simple estimation)
- Method4 – Analogy (includes estimates on the principle of comparison with existing systems or the base of experience from previous projects).

In Fig. 3, is apparent that the most widely used method is Method3 (expert judgments) with 57% incidence. The second most commonly used method is Method4 (analogy) with a 29% incidence. The results also show that the Method1 (Function Points) and Method2 (Use Case Points) are used only at 7% or 8% of companies, respectively.

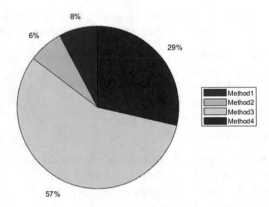

Fig. 3. Distribution of companies by the used methods of estimation

3.4   Distribution of Companies by the Factors Determining the Final Effort

Another aim of the research was to determine what other factors play a role in determining the final development effort. These factors can be divided into groups:

- Factor1 – Analogy (comparison with other projects)
- Factor2 – Implementation time (expected time of development)
- Factor3 – Capacity (decrease or increase estimate)
- Factor4 – Competition (lower estimates at the higher competition)
- Factor5 – Quality of documents (higher quality affects the estimation)
- Factor6 – Client's options (taking into account the client's creditworthiness)
- Factor7 – Subcontracts (whether the development needs subcontractors)
- Factor8 – Not reported.

According to the results presented in Fig. 4, almost half of the companies (48%) have as their determining factor the Factor2 (implementation time), 14% Factor5 (quality of the documents), 10% Factor1 (analogies), and 7% Factor4 (expected competition). Another interesting finding is that 8% of companies consider the client's creditworthiness (Factor6). Creditworthiness has an impact on the final ratio between the estimated work effort and the price per person-hour.

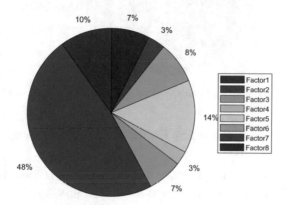

Fig. 4. Distribution of companies by the factors used to determine the estimate

3.5 Distribution of Companies by the Ratio of Projects not Delivered in Time

The next investigated aim was the inaccuracy of the companies' estimate in terms of what percents of the projects were not delivered on-time. For this category were created, four groups:

- Ratio1 – 0 to 25%
- Ratio2 – 26 to 50%
- Ratio3 – 51 to 75%
- Ratio4 – 76 to 100%

Based on the information presented in Fig. 5, more than half of the companies (54%) fail to complete a quarter of their projects (Ratio1) in the estimated timeframe. For 36% of companies, the failure rate is as much as are two-quarters of projects (Ratio2). Finally, 10% of companies fail to complete more than half of the projects (Ratio3 and Ratio4) in the estimated time.

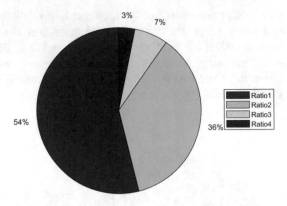

Fig. 5. Distribution of companies by the projects not delivered in time

3.6 Distribution of Companies by the Main Reasons for Delaying Delivery of Projects

Finally, the main reasons for delaying delivery of projects were investigated. The different types of reasons were distributed into the groups:

- Reason1 – Insufficient analysis
- Reason2 – Changing requirements
- Reason3 – Scope of the project
- Reason4 – Bad estimate
- Reason5 – Not reported

As shown in Fig. 6, the most common reason for the delay in delivery of the project is the Reason2 (changing requirements: 39%). Other common reasons are Reason4 (bad estimates: 23%) and Reason1 (insufficient analysis: 18%).

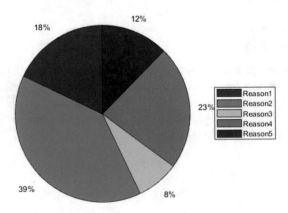

Fig. 6. Distribution of companies by the main reasons for delaying delivery of projects

4   Discussion

Obtained results show interesting facts about software development process with the Czech Republic. As was described in the Exploratory Strategy the survey was analysed using various attributes. It allows to resolve a further finding, which arising from question analysis.

- When examining the dependency between business area and company size, it is clear (see Fig. 7) that the Area6 (web application) is the most common business area, irrespective of the size of the company. In the first and second size groups, web applications, desktop applications, and mobile applications are among the three most important domains. Other business areas, i.e., enterprise information systems and industrial systems and automation, are then represented only in the first two size groups (Size1 and Size2).

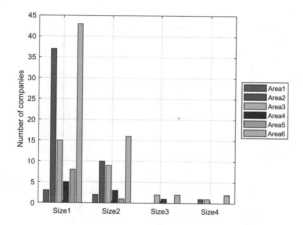

Fig. 7. The relationship between the business areas in each size group

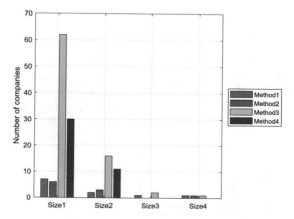

Fig. 8. The relationship between the used estimation methods in each size group

- When analysing the use of estimation methods, taking into account size groups of companies, it is apparent (Fig. 8) that Method3 (expert judgment) is the most widely used, irrespective of the size of the company. Companies (regardless of size) also use more than one estimation method in most cases.
- Another finding concerns the relationship between the business area and the used estimation methods. In Fig. 9 is shown what methods are used in companies that have responded that they are working in a given business area. If a company operates in multiple business areas, it is counted in all areas. The Method3 (expert judgments) and Method4 (analogies) predominant in the majority of areas. A). A more prominent representation of the use of Method1 (Function Points) or Method2 (Use Case Points) can be tracked by Area6 (web applications).
- Looking at the factors determining the final work effort from a company size perspective (Fig. 10), it can be said that just the Factor2 (implementation time) are affected by estimation independently of size group.

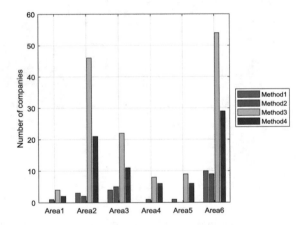

Fig. 9. The relationship between the used estimation methods in each business area

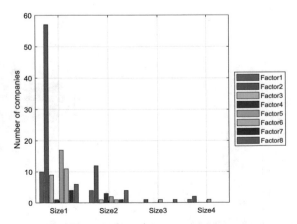

Fig. 10. The relationship between the factors determining the final work effort in each size group

- The relationship between the size of the company and the ratio of projects that were not completed in time is shown in the graph (Fig. 11). All of the size groups (except Size2) are represented in the population of companies that do not complete a quarter of the projects on time (Ratio1).
- Finally, we focus on the relationship between the main reasons for delaying delivery of projects and the size of the company. Looking at individual size groups (Fig. 12), it is found that the Reason2 (change of requirements) is the main cause of failure to provide correct company estimates among all groups.

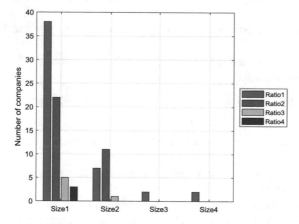

Fig. 11. The relationship between the ratios of projects that were not completed in time in each size group

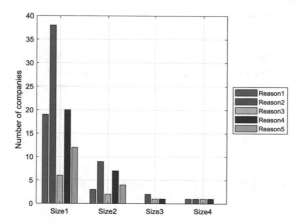

Fig. 12. The relationship between the reasons for delaying delivery of projects in each size group

5 Conclusion

This paper presents a case study of the software project estimation process in Czech companies based on a questionnaire survey.

The presented case study shows that:

- Most of the companies (75%) are small size companies (1 to 10 workers).
- The majority of companies (39%) operate in Web applications.
- The most widely used estimation method is Expert judgments, with 57% incidence.
- Overall 48% of companies report implementation time as a key factor for the estimation.
- Overall 54% of companies fail to complete a quarter of projects in the estimated timeframe.
- The most common reason (39%) for the delay in delivery of the project is the changing requirements.

When we examined the dependency between company size and other attributes, we found out that:

- The Web Application is the most common business area, irrespective of the size of the company.
- Expert judgments are the most widely used estimation method, irrespective of the size of the company.
- Different business areas dominate Expert judgments.
- The implementation time affects the estimation irrespective of the size of the company.
- Companies belonging to all size groups (except Size2) report that one-quarter of the projects are not completed on time.
- The change of requirements is considered the main cause of failure to provide company estimates of all groups.

Revealed results of this survey are also used in other authors' research work. The research aims to create an auxiliary tool for simple and accurate planning of the project development effort in the companies. All calculations used in the paper were performed in the Matlab environment.

References

1. Molokken, K., Jorgensen, M.: A review of surveys on software effort estimation. In: 2003 International Symposium on Empirical Software Engineering, Proceedings, pp. 223–230 (2003)
2. Boehm, B., Abts, C., Chulani, S.: Software development cost estimation approaches – a survey. Ann. Softw. Eng. **10**, 177–205 (2000)
3. Trendowicz, A., Münch, J., Jeffery, R.: State of the practice in software effort estimation: a survey and literature review. In: IFIP Central and East European Conference on Software Engineering Techniques, pp. 232–245. Springer (2008)

4. Rastogi, H., Dhankar, S., Kakkar, M.: A survey on software effort estimation techniques. In: 2014 5th International Conference Confluence the Next Generation Information Technology Summit (Confluence), pp. 826–830 (2014)
5. Bardsiri, A.K., Hashemi, S.M.: Software effort estimation: a survey of well-known approaches. Int. J. Comput. Sci. Eng. (IJCSE) **3**, 46–50 (2014)
6. Vera, T., Ochoa, S.F., Perovich, D.: Survey of software development effort estimation taxonomies. Technical report. Pending ID. Computer Science Department, University of Chile (2017)
7. Idri, A., Hosni, M., Abran, A.: Systematic literature review of ensemble effort estimation. J. Syst. Softw. **118**, 151–175 (2016)
8. Idri, A., Amazal, F.A., Abran, A.: Analogy-based software development effort estimation: a systematic mapping and review. Inf. Softw. Technol. **58**, 206–230 (2015)
9. Munialo, S.W., Muketha, G.M.: A review of agile software effort estimation methods. Int. J. Comput. Appl. Technol. Res. **5**, 612–618 (2016)
10. Jorgensen, M.: A review of studies on expert estimation of software development effort. J. Syst. Softw. **70**, 37–60 (2004)
11. Spalkova, T.: Methods for the Estimation of Software Projects. UPKS, Tomas Bata University in Zlin, Zlin (2015)

Author Index

© Springer Nature Switzerland AG 2019
R. Silhavy (Ed.): CSOC 2019, AISC 984, pp. 469–470, 2019.
https://doi.org/10.1007/978-3-030-19807-7

Printed in the United States
By Bookmasters